INTRODUCTION TO
M⊙DERN
PHYSICS
Solutions to Problems

INTRODUCTION TO
MODERN
PHYSICS
Solutions to Problems

Paolo Amore
Universidad de Colima, Mexico

John Dirk Walecka
College of William and Mary, USA

 **World Scientific**

NEW JERSEY · LONDON · SINGAPORE · BEIJING · SHANGHAI · HONG KONG · TAIPEI · CHENNAI

Published by

World Scientific Publishing Co. Pte. Ltd.

5 Toh Tuck Link, Singapore 596224

USA office: 27 Warren Street, Suite 401-402, Hackensack, NJ 07601

UK office: 57 Shelton Street, Covent Garden, London WC2H 9HE

British Library Cataloguing-in-Publication Data
A catalogue record for this book is available from the British Library.

INTRODUCTION TO MODERN PHYSICS
Solutions to Problems

ISBN 978-981-4520-31-7 (pbk)

Printed in Singapore

Preface

A text entitled *Introduction to Modern Physics: Theoretical Foundations*, authored by John Dirk Walecka, was published by World Scientific Publishing Company (WSPC) in 2008. This book, aimed at the very best students, provides an introduction to the physics of the twentieth century. There are numerous problems included in the text, and although the steps required in solving the problems are clearly laid out, it is evidently advantageous to have prepared solutions available. This is useful for both students and instructors, in order to calibrate their own solutions and to see how someone else would go about solving the problems.

Paolo Amore served as a reader of the manuscript (as well as of the two subsequent books in the modern physics series), and he is very familiar with the text, having actually taught from it. When Dr. K. K. Phua, Executive Chairman of WSPC, suggested the possibility of having a solutions manual prepared, the two of us discussed it and decided we would like to take on the task together. Although draft solutions to the problems were available to the authors upon publication of the text, it does take a significant amount of time and effort to get the solutions into publishable form.

The modern physics book assumes a good first-year calculus-based freshman physics course, along with a good first-year course in calculus. In preparing this solutions manual, we have seen that this is a consistent assumption, although some knowledge of multi-variable calculus is also required, especially familiarity with multiple integrals. The use of readily available mathematics programs for personal computers provides a powerful tool in problem solving and pursuing the implications of the solutions, particularly those programs available to integrate both linear and non-linear differential equations.

Solutions are presented for all of the problems in the modern physics book. The problems are first restated, and then the solutions given. We try, as far as possible, to stick to the notation in the text. We believe these solutions are clear and concise. In a few cases, some useful additional material, not required in the solutions, has been included; it is so indicated. A couple of mistakes in the original problems have been corrected here.

We would like to thank Dr. K. K. Phua, and our editor Ms. Lakshmi Narayanan, for their help and support on this project. It is our hope that the present book will contribute significantly to both the utilization and enjoyment of the modern physics text.

March 4, 2013

Paolo Amore
Facultad de Ciencias
Universidad de Colima
Colima, Mexico

John Dirk Walecka
College of William and Mary
Williamsburg, Virginia

Contents

Chapter 1

Introduction

The book *Introduction to Modern Physics: Theoretical Foundations* starts with the following two paragraphs [Walecka (2008)]:

"At the end of the 19th century, one could take pride in the fact that the laws of physics were now understood. With newtonian mechanics, the statistical analysis of Boltzmann, and Maxwell's equations for electromagnetism, one had an excellent description of the world we see around us. At that point, a series of experiments were performed whose results simply could not be understood within this classical framework. These experiments probed a microscopic, or high-velocity, world that went far beyond our everyday perception. The resolution of these paradoxes led to the theories of quantum mechanics and special relativity, which provide the foundation of modern physics. Atomic, nuclear, particle, and condensed-matter physics are all built on this foundation. Einstein's extension of his theory of special relativity to general relativity provides the current basis for our understanding of gravitation and cosmology—physics to the farthest reaches of the universe. The theories of quantum mechanics, special relativity, and general relativity have been remarkably successful. All current experiments can be understood within this framework. The goal of this book is to introduce a reader, with an assumed knowledge of classical physics, to modern physics.[1]

It is assumed that the reader has had a good one-year, calculus-based freshman physics course, along with a good one-year course in calculus. A sufficient number of appendices are included to bring the reader up to speed on any additional mathematics required at the outset. Over 175 problems are included in the book, some for each chapter. While there

---

[1]There is a nice irony in the parallel statement, "At the end of the 20th century, one could take pride in the fact that the laws of physics were now understood."

are many problems that directly amplify the material in the text, there are also a great number of them that will take dedicated readers just as far as they want to go in modern physics. Although the book is designed so that one can, in principle, read and follow the text without doing any of the problems, the reader is strongly urged to attempt as many of them as possible in order to obtain some confidence in his or her understanding of the basics of modern physics. With very few exceptions, the reader should find that the text, appendices, and problems form a self-contained volume."[2]

As stated in the preface, the publisher has requested that we prepare a solutions manual for the problems in this book. Although the steps required in solving the problems are clearly laid out in the text, it is advantageous to have prepared solutions available. This is useful for instructors as well as students, both to calibrate their own solutions and to see how someone else would go about solving the problems. We write the present book in response to the publisher's request. Solutions are presented here for all of the problems in the modern physics book. The problems are first restated, and then the solutions given. We try, as far as possible, to follow the notation in the text.

Chapter 2 is on *classical physics*. Here, motivated by Newton's laws, there are problems on numerical methods and numerical integration, much facilitated by the mathematics programs now available for personal computers. The motion of a string provides the paradigm for the discussion of continuum mechanics, wave motion, and quantum field theory in the book, and there are several problems on string motion. This also leads directly to the study of Fourier series and Fourier integrals, for which appendices provide a foundation.[3] There are problems in classical statistics, and on the basic concepts in classical electricity and magnetism, both again supported in the appendices.

Chapter 3 is concerned with *some contradictions* with classical physics. Here problems expand on the classical paradoxes encountered in the specific heat of solids, black-body radiation, the Compton effect, and the discrete lines in atomic spectra, paradoxes that led to the Planck hypothesis for the spectral distribution and the Bohr theory of the atom.

Chapter 4 on *quantum mechanics* contains a large number of supporting

---

[2]Two subsequent volumes in this modern physics series have since been published: *Advanced Modern Physics: Theoretical Foundations* [Walecka (2010)], and *Topics in Modern Physics: Theoretical Foundations* [Walecka (2013)].

[3]There are also problems in all but two appendices, whose solutions are included here.

problems. The study of wave motion and Fourier integrals provides a base for the development of the Schrödinger equation. The continuity equation, and appropriate boundary conditions, are examined. Exercises are given on partial integration and hermitian operators, and on commutation relations. Ehrenfest's theorem, which exhibits the proper classical limit, is derived. The rigorous statement of the uncertainty principle is derived and investigated in the motion of the minimal packet. Barrier penetration is examined, and the Schrödinger equation is also solved for both the one- and three-dimensional boxes, and the wave functions studied, as are those for the one-dimensional oscillator and the $1/r$ potential. The elements of angular momentum are developed. The creation and destruction operators for the oscillator are introduced. Finally there are problems on the low-temperature behavior of non-interacting Bose and Fermi gases.

Chapter 5 in on *atomic physics*. The spin-orbit interaction is examined, and the Landé g-factor is derived. There are several problems on the Thomas-Fermi model of the atom, where readily available mathematics programs allow one to numerically integrate the non-linear Thomas-Fermi equation. This model provides great insight into the many-body structure of the atom. There are also problems on determining the electronic configuration of atoms and on deducing the implications of that configuration.

Chapter 6 is concerned with *nuclear physics*. Here the two-body problem is examined, and solved for simple potentials. There are problems on nuclear stability, and the use and applications of the atomic mass tables. There are exercises on the use of the semi-empirical mass formula, and simple derivations of both the Coulomb and symmetry-energy terms appearing in it. The nuclear form factor is calculated. Predictions of the shell model are investigated, and the Schmidt lines for nuclear magnetic moments derived and applied.

Chapter 7 is on *particle physics*. The Yukawa interaction is re-derived. Properties of the S-matrix are investigated, while the theory of transition rates is developed in an appendix. Conservation laws are applied, and quark configurations are examined. The effective weak point-coupling is derived. The basic elements of $SU(2)$ symmetry are examined in a problem. Another appendix describes neutrino mixing.

Chapter 8 is on *special relativity*, with many problems supporting and enhancing the discussion. Several striking applications of Lorentz contraction and time dilation are given. Lorentz scalars are examined. The covariant form of Maxwell's equations is derived, as well as that of Newton's law for a charged particle in an E-M field. Several applications of relativistic kinematics are studied, including particle production, elastic scattering, and Compton scattering. A series of problems complements the discussion in the text, and develops the theory of white dwarf stars in the non-relativistic limit (NRL). Available programs again allow the reader to integrate the resulting non-linear differential equations.

Chapter 9 is concerned with *relativistic quantum mechanics*. Here several exercises give the reader some facility with the Dirac equation and Dirac spinors, supported by an appendix. Problems show how the non-relativistic reduction of the Dirac equation leads to the correct magnetic moment and spin-orbit interaction of the electron. A similar reduction exhibits the main features of the scalar-vector theory of nuclear interactions. The essential elements of the vector–axial-vector (V-A) theory of the weak interactions are illustrated. One problem develops invariant phase space. Another illustrates the kinematics of the quark-parton model and Bjorken scaling.

Chapter 10 is on *general relativity*. The Doppler shift is first derived in special relativity. Problems then develop the calculus of variations, Hamilton's equation, and the Euler-Lagrange equation for the dynamical path, all important background for both mechanics and general relativity.[4] Exercises are presented on time dilation in the Schwarzschild metric and properties of the horizon in the Robertson-Walker metric (with $k = 0$).

Chapter 11 is on *quantum fluids*. One problem examines both the sound and critical velocities in a condensed Bose gas. Another involves the numerical integration of the non-linear Gross-Pitaevskii equation for the density distribution in a vortex with unit circulation in that fluid. An exercise concerns London's phenomenological equation for a superconductor and the corresponding derivation of the penetration depth. Properties of the BCS gap equation in a superconductor are examined in another problem. The Schrödinger equation for a particle in a magnetic field is derived, which provides the basis for the discussion of flux quantization in the text.

The final chapter 12 is on *quantum fields*. The quantization of the motion of the string serves as our elementary model of a quantum field, and the problems guide the reader through the details of just how this is accomplished. Our previous analysis of the creation and destruction operators stands us in good stead here. Inclusion of a higher bending correction in the string provides an example of a non-linear interaction in quantum field theory. Additional problems work through the quantization of the E-M field

---

[4]See [Walecka (2007)].

and of the Dirac field, where anticommutation relations must be imposed on the creation and destruction operators. There are exercises that obtain the transition rate for an arbitrary photon transition. The $\cos^2 \theta/2$ factor in the Mott scattering of an electron is derived from the Dirac spinors. The quantum field formulation of the non-relativistic many-body problem is investigated, and an appendix supplies many important underlying results for non-relativistic many-body systems. Finally, an exercise develops the lagrangian density, action, and Lagrange's equation for the string, which exhibits the basics of continuum mechanics and field theory.

After this overview, we proceed to the solutions of the problems.

Chapter 2

Classical Physics

Problem 2.1 Write, or obtain, a computer program that iterates the finite difference Eqs. (2.6) for a given set of initial conditions. It is always good practice to first convert the problem to dimensionless variables. Examine the following cases:

(a) A one-dimensional simple harmonic oscillator moving in the x-direction with $F = -\kappa x$ and initial conditions $x_0 = a$ and $[dx/dt]_0 = 0$. Use variables $\zeta = x/a$ and $\tau = \sqrt{\kappa/m}\,t$;

(b) A particle performing vertical motion in the upward z-direction in a gravitational field with $F = -mg$ and initial conditions $z_0 = 0$ and $[dz/dt]_0 = v_0$. Use variables $\zeta = zg/v_0^2$ and $\tau = gt/v_0$;

(c) Investigate the limit as the interval $h \to 0$ in each case.

Solution to Problem 2.1

(a) Newton's second law for the harmonic oscillator is

$$m\frac{d^2x}{dt^2} = -\kappa x$$

Use the variables ζ and τ to obtain the dimensionless differential equation

$$\frac{d^2\zeta}{d\tau^2} + \zeta = 0$$

subject to the initial conditions

$$\zeta(0) = 1 \qquad ; \qquad \left[\frac{d\zeta}{d\tau}\right]_0 = 0$$

The exact solution satisfying these conditions is

$$\zeta(\tau) = \cos\tau$$

The differential equation is converted to a difference equation with the use of Eqs. (2.6), which can now be solved iteratively on a computer, starting at the time $\tau = 0$. One has

$$\zeta(h) = 1 \qquad\qquad ; \quad \left[\frac{d\zeta}{d\tau}\right]_h = -h$$

$$\zeta(2h) = 1 - h^2 \qquad ; \quad \left[\frac{d\zeta}{d\tau}\right]_{2h} = -2h$$

$$\zeta(3h) = 1 - 3h^2 \qquad ; \quad \left[\frac{d\zeta}{d\tau}\right]_{3h} = h\left(h^2 - 3\right) \qquad etc.$$

The precision of the approximation depends on h, since Eqs. (2.6) correspond to performing a Taylor expansion of $x(t)$ and $dx(t)/dt$ at an initial time. One can appreciate this point by looking at Fig. 2.1, where the numerical solutions obtained with $h = \pi/50$ and $h = \pi/200$ are compared to the exact solution $\zeta(\tau) = \cos\tau$.

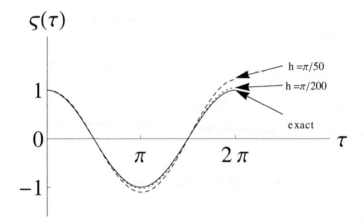

Fig. 2.1   Comparison between the exact solution $\zeta(\tau) = \cos\tau$ and the numerical solutions obtained with $h = \pi/50$ and $h = \pi/200$.

(b) Newton's second law in this case is

$$m\frac{d^2z}{dt^2} = -mg$$

After introducing the dimensionless variables ζ and τ, the differential equation reduces to

$$\frac{d^2\zeta}{d\tau^2} + 1 = 0$$

with the initial conditions

$$\zeta(0) = 0 \qquad ; \qquad \left[\frac{d\zeta}{d\tau}\right]_0 = 1$$

The exact solution of the differential equation is then

$$\zeta(\tau) = -\frac{1}{2}\tau^2 + \tau$$

The use of Eqs. (2.6) converts the original differential equation into a system of algebraic equations which can be solved iteratively. Actually, for this simple problem it is possible to obtain the general form of the solution to these finite difference equations in the form

$$\zeta(nh) = nh\left[1 - \frac{(n-1)h}{2}\right] \qquad ; \qquad \left[\frac{d\zeta}{d\tau}\right]_{nh} = 1 - nh$$

Figure 2.2 compares the numerical solutions corresponding to $h = 1/10$ and $h = 1/20$ with the exact solution.

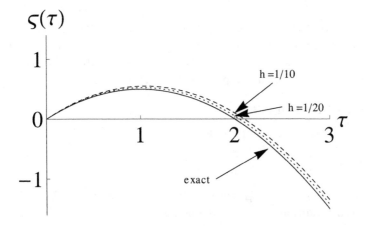

Fig. 2.2   Comparison between the exact solution $\zeta(\tau) = -\tau^2/2 + \tau$ and the numerical solutions obtained with $h = 1/10$ and $h = 1/20$.

(c) It is a useful numerical exercise to investigate convergence by increasing the range in τ, and decreasing h, in the previous examples. Convergence can also be investigated in the following interesting manner.

In the case of the harmonic oscillator, it is evident from Fig. 2.1 that the amplitude of the oscillations of the numerical solutions *increases*, and therefore the total energy of the numerical solution is not constant. We may convince ourselves of this fact by looking at the expression for the energy obtained with the discretized solution

$$E_n^{(\text{num})} = \frac{1}{2}\left\{ \left[\frac{d\zeta}{d\tau}\right]_{nh}^2 + \zeta^2(nh) \right\} = \frac{1}{2}\left(1 + h^2\right)^n$$

If we assume that $\tau(\bar{n}) = \bar{n}h$ is the exact period of the oscillation, that is $h = 2\pi/\bar{n}$, the numerical value for the energy at this time will be

$$E_{\bar{n}}^{(\text{num})} = \frac{1}{2}\left(1 + \frac{4\pi^2}{\bar{n}^2}\right)^{\bar{n}}$$

which behaves for large $\bar{n}$ as

$$E_{\bar{n}}^{(\text{num})} = \frac{1}{2} + \frac{2\pi^2}{\bar{n}} + O\left[\frac{1}{\bar{n}^2}\right]$$

Therefore one needs $\bar{n} \approx 4000$ to be able to obtain the energy with a 1% error after one period. The conservation of the energy is only recovered in the limit $\bar{n} \to \infty$ (or equivalently $h \to 0$), where the finite difference $\{\zeta(nh) - \zeta[(n-1)h]\}/h$ reduces to the derivative $d\zeta/d\tau$.

Problem 2.2 A particle of mass m and charge e moves along the x-axis subject to an electric field $\mathcal{E}$ which points along that axis. Define $x \equiv q$ and $mdx/dt \equiv p$. Assume that at $t = 0$ the particle starts from the origin $(p, q)_0 = 0$. Calculate and plot the phase orbit $p(q)$ in phase space.

Solution to Problem 2.2

Newton's second law for the particle is

$$m\frac{d^2x}{dt^2} = e\mathcal{E}$$

and therefore the particle moves with a constant acceleration. With the use of the initial conditions, $(p, q)_0 = 0$, one has

$$q(t) = \frac{e\mathcal{E}}{2m}t^2 \qquad ; \; p(t) = m\frac{dq}{dt} = e\mathcal{E}t$$

After eliminating the time from these equations, we obtain the phase-space orbit

$$p = \alpha\sqrt{q} \qquad ; \; \alpha \equiv \sqrt{2me\mathcal{E}}$$

which is readily plotted for various values of the constant α.

Problem 2.3 The following is a trigonometric identity

$$\cos a + \cos b = 2\cos\left(\frac{a+b}{2}\right)\cos\left(\frac{a-b}{2}\right)$$

Consider the superposition of two traveling waves of slightly different wavelengths $\lambda_1 \equiv \lambda$ and $\lambda_2 \equiv \lambda + \Delta\lambda$.

$$q(x,t) = \cos\frac{2\pi}{\lambda_1}(x - ct) + \cos\frac{2\pi}{\lambda_2}(x - ct)$$

Use the above identity to show that the resulting disturbance oscillates with a wavelength

$$\frac{1}{\lambda_{\text{osc}}} = \frac{1}{2}\left(\frac{1}{\lambda_1} + \frac{1}{\lambda_2}\right) \approx \frac{1}{\lambda}$$

Show there is an *amplitude modulation* of the disturbance with wavelength

$$\frac{1}{\lambda_{\text{amp}}} = \frac{1}{2}\left(\frac{1}{\lambda_1} - \frac{1}{\lambda_2}\right) \approx \frac{\Delta\lambda}{2\lambda^2}$$

Solution to Problem 2.3

With the use of the trigonometric identity, we have

$$q(x,t) = \cos\frac{2\pi}{\lambda_1}(x - ct) + \cos\frac{2\pi}{\lambda_2}(x - ct)$$
$$= 2\cos\left[\pi\left(\frac{1}{\lambda_1} + \frac{1}{\lambda_2}\right)(x - ct)\right]\cos\left[\pi\left(\frac{1}{\lambda_1} - \frac{1}{\lambda_2}\right)(x - ct)\right]$$
$$\equiv 2\cos\left[\frac{2\pi}{\lambda_{\text{osc}}}(x - ct)\right]\cos\left[\frac{2\pi}{\lambda_{\text{amp}}}(x - ct)\right]$$

Now notice that $1/\lambda_{\text{amp}} \approx \Delta\lambda/2\lambda^2 \ll 1/\lambda_{\text{osc}}$, and therefore the disturbance oscillates with a wavelength λ_{osc}, while its amplitude is modulated with a wavelength λ_{amp}.

Problem 2.4 A string of length L is stretched around a cylinder and the ends joined. It is free to oscillate without friction in a direction parallel to the axis of the cylinder (Fig. 2.3).

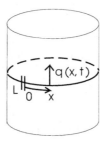

Fig. 2.3 String of length L stretched around a cylinder free to oscillate without friction in a direction parallel to the axis of the cylinder. Illustration of periodic boundary conditions.

(a) Show that the physical situation is described by *periodic boundary conditions*

$$q(x + L) = q(x) \qquad ; \text{ periodic boundary conditions}$$

(b) Show that the eigenfunctions and eigenvalues in this case can be taken as

$$\phi_n(x) = \frac{1}{\sqrt{L}} \exp\left(ik_n x\right)$$

$$k_n = \frac{2\pi}{L} n \qquad ; n = 0, \pm1, \pm2, \cdots$$

(c) Show that these eigenfunctions satisfy the orthonormality relation

$$\int_0^L \phi_n^\star(x)\phi_m(x)\, dx = \delta_{mn}$$

Here $\phi^\star$ is the complex conjugate, and δ_{mn} is the Kronecker delta.

(d) Show that if one had chosen to specify the coordinate interval by $-L/2 \le x \le L/2$ rather than by $0 \le x \le L$, these eigenfunctions continue to satisfy the orthonormality relation

$$\int_{-L/2}^{L/2} \phi_n^\star(x)\phi_m(x)\, dx = \delta_{mn}$$

Solution to Problem 2.4

(a) Call $q(x, t)$ the displacement of the string in a direction parallel to the axis of the cylinder and x the coordinate specifying a point on the surface of the cylinder, belonging to a plane which is orthogonal to the axis of the cylinder. Because of the symmetry of the problem, there is no preferential direction, and the origin of x can be set arbitrarily at a given point. As one moves on the circle, after traveling a distance L, one returns to the original point. If this operation is carried out at a fixed time (that is, the string is frozen), one needs to recover the initial displacement of the string as the initial point is reached. Given the arbitrary nature of the initial point, this argument holds for all points. Thus

$$q(x + L) = q(x) \qquad ; \text{periodic boundary conditions}$$

(b) To start with, one needs to verify that the $\phi_n(x)$ obey *periodic boundary conditions*. We observe that

$$\phi_n(x + L) = \frac{1}{\sqrt{L}} e^{ik_n(x+L)} = e^{ik_n L} \phi_n(x)$$

The periodic boundary conditions require that the phase $e^{ik_n L}$ be unity, and therefore $k_n = 2\pi n/L$, with $n = 0, \pm 1, \pm 2, \cdots$.

We next need to verify that $\phi_n(x)$ and k_n are respectively eigenfunctions and eigenvalues of the scalar Helmholtz Eq. (2.33). This is done by simply calculating the second derivative of $\phi_n(x)$

$$\frac{d^2 \phi_n(x)}{dx^2} = -k_n^2 \phi_n(x)$$

which proves the point.

(c) The orthogonality relation requires the evaluation of the integral

$$\int_0^L \phi_n^*(x) \phi_m(x) dx = \frac{1}{L} \int_0^L e^{2\pi i (m-n)x/L} dx$$

There are two cases: if $n = m$ then $e^{2\pi i (m-n)x/L} = 1$, and one has

$$\int_0^L \phi_n^*(x) \phi_m(x) dx = \frac{1}{L} \int_0^L dx = 1 \qquad ; n = m$$

In the second case $n \neq m$, and the exponential becomes $e^{2\pi i (m-n)x/L} = \cos 2\pi (m - n)x/L + i \sin 2\pi (m - n)x/L$. It is clear that the integral of both terms vanishes, and thus

$$\int_0^L \phi_n^*(x)\phi_m(x)dx = 0 \qquad ; n \neq m$$

The two results can be combined using the Kronecker delta

$$\int_0^L \phi_n^*(x)\phi_m(x)dx = \delta_{mn}$$

(d) We start by observing that

$$\phi_n\left(x - \frac{L}{2}\right) = \frac{1}{\sqrt{L}}e^{ik_n(x-L/2)} = \phi_n(x)e^{-ik_n L/2}$$

where the last factor is a pure phase. Therefore, with a change of variable $x' = x + L/2$,

$$\int_{-L/2}^{L/2} \phi_n^*(x)\phi_m(x)dx = \int_0^L \phi_n^*\left(x' - \frac{L}{2}\right)\phi_m\left(x' - \frac{L}{2}\right)dx' \quad (2.1)$$
$$= e^{i(k_n-k_m)L/2}\delta_{mn}$$
$$= \delta_{mn}$$

Problem 2.5 A uniform string under tension τ with fixed endpoints is plucked in the middle giving rise to an initial displacement that can be approximated by (here $h \ll L$)

$$q(x,0) = \frac{2hx}{L} \qquad ; 0 \leq x \leq \frac{1}{2}L$$
$$= \frac{2h(L-x)}{L} \qquad ; \frac{1}{2}L \leq x \leq L$$

It is released from rest. Determine the Fourier coefficients in Eqs. (2.49).

Solution to Problem 2.5

Since the string is released from rest, one has $[\partial q(x,t)/\partial t]_{t=0} = 0$, and the second of Eqs. (2.49) then implies that the phase $\delta_m = 0$. Therefore, the first of Eqs. (2.49) reduces to

$$A_m = \int_0^L q(x,0)\sqrt{\frac{2}{L}}\sin\left(\frac{m\pi x}{L}\right)dx$$

It is convenient to study the behavior of the integrand with respect to the change of variables $x \rightarrow L - x$. Under this change

$$q(x,0) \rightarrow q(x,0) \qquad ; \; x \rightarrow L - x$$
$$\sin\left(\frac{m\pi x}{L}\right) \rightarrow (-1)^{m+1} \sin\left(\frac{m\pi x}{L}\right)$$

Thus the integrand in the expression for A_m will be even only if m is odd; for m even, the integrand will be odd with respect to the change of variables, and therefore the integral will vanish.

To select the odd values of m we introduce a new integer k ($k = 0, 1, 2, \cdots$), with $m = 2k + 1$, and evaluate the integral over half the region (multiplying by a factor 2 to take into account the second half)

$$A_{2k+1} = \sqrt{\frac{2}{L}} \frac{4h}{L} \int_0^{L/2} x \sin\left[\frac{(2k+1)\pi x}{L}\right] dx$$
$$= \sqrt{\frac{2}{L}} \frac{4h}{L} \frac{(-1)^k L^2}{(2\pi k + \pi)^2}$$

The initial profile, and resulting profile at $t = 0.25$, are shown in Fig. 2.4.

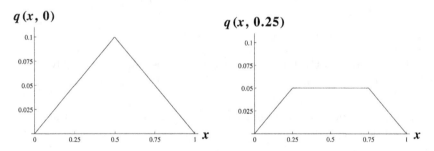

$q(x, 0)$ $\qquad\qquad$ $q(x, 0.25)$

Fig. 2.4 (a) Initial profile of the string at $t = 0$; (b) Profile of the string at $t = 0.25$. Here $h = 1/10$, $L = 1$, and $\omega_n = k_n c$ with $c = 1$ in Eq. (2.43) [see appendix A].

Problem 2.6 Consider other boundary conditions for the string in Fig. 2.9 in the text. Suppose the end of the string at $x = L$ is attached to a mass m_0 and also to a massless restoring spring with force constant κ. Assume that it is constrained to move without friction along a vertical wire while under a tension τ.

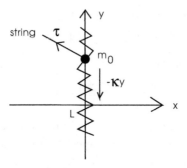

Fig. 2.5   End of string at $x = L$ attached to a mass m_0 and massless spring with force constant κ, constrained to move without friction along a vertical wire while under a tension τ.

(a) Write Newton's second law for the mass m_0. Assume small displacements, and show

$$m_0 \frac{d^2y}{dt^2} = -\kappa y - \tau \frac{dy}{dx}$$

(b) Look for a normal mode, with angular frequency ω, of the whole system of mass m_0 and string. Show

$$\left[\frac{1}{y}\frac{dy}{dx}\right]_{x=L} = \frac{m_0\omega^2 - \kappa}{\tau} \qquad ; \text{ homogeneous B.C.}$$

$$\rightarrow \frac{-\kappa}{\tau} \qquad ; m_0 \rightarrow 0$$

The second line follows in the limit that the mass m_0 becomes vanishingly small. This is known as a *homogeneous boundary condition* on the end of the string.

(c) Suppose one goes to the limit $(m_0, \kappa) \rightarrow 0$ at finite τ. Show the boundary condition becomes

$$y'|_{x=L} = 0 \qquad ; \text{ free end}$$

Here $y' = dy/dx$. This is known as a *free end*.

Solution to Problem 2.6

(a) There are two forces acting on the mass m_0: one is the force due to the spring with spring constant κ, the other is the force due to the tension of the string. While the force of the spring acts in the y direction, the tension τ acts on the direction which is tangent to the string at $x = L$.

Since the mass m_0 moves along the y direction, we have to consider the component of the tension τ along this direction. Let us call θ the angle that τ forms with the x-axis. As we see from the plot, for small displacements we may approximate $\sin \theta \approx dy/dx|_{x=L}$ and express the component of the tension in the positive y-direction as $-\tau dy/dx|_{x=L}$.

Therefore Newton's second law for the mass m_0 reads

$$m_0 \frac{d^2y}{dt^2} = -\kappa y - \tau \frac{dy}{dx} \qquad ; x = L$$

where this relation holds at $x = L$, and the r.h.s. is evaluated at a given t.

(b) Let us look for a normal-mode solution of the form

$$y(x,t) = y(x)e^{i\omega t}$$

After substituting this solution in the above equation, canceling the factor $e^{i\omega t}$, and setting $x = L$, we obtain

$$(m_0\omega^2 - \kappa)y(L) = \tau \left.\frac{dy}{dx}\right|_{x=L}$$

which is the desired *homogeneous boundary condition*

$$\left[\frac{1}{y}\frac{dy}{dx}\right]_{x=L} = \frac{(m_0\omega^2 - \kappa)}{\tau}$$

(c) If we consider the limit $(m_0, \kappa) \to 0$, and keep τ finite, this equation reduces to

$$\left[\frac{1}{y}\frac{dy}{dx}\right]_{x=L} = 0$$

which is equivalent to

$$\left.\frac{dy}{dx}\right|_{x=L} = 0$$

since $y(L)$ is finite.

Problem 2.7 Suppose one has a string with a fixed end at $x = 0$, while the end at $x = L$ is free [Prob. 2.6(c)].

(a) Show the eigenfunctions and eigenvalues are

$$q_n(x) = \left(\frac{2}{L}\right)^{1/2} \sin \frac{(2n+1)\pi x}{2L} \qquad ; n = 0, 1, 2, \cdots, \infty$$

$$\omega_n = \frac{(2n+1)\pi c}{2L}$$

Sketch the first few eigenfunctions.

(b) Show the eigenfunctions are again orthonormal.

Solution to Problem 2.7

(a) The $q_n(x)$ given in part (a) are eigenfunctions of the scalar Helmholtz equation since

$$\frac{d^2 q_n(x)}{dx^2} = -\left[\frac{(2n+1)\pi}{2L}\right]^2 q_n(x)$$

In addition, we see that

$$q_n(0) = 0$$

$$\left[\frac{dq_n}{dx}\right]_{x=L} = \left(\frac{2}{L}\right)^{1/2} \frac{(2n+1)\pi}{2L} \cos \frac{(2n+1)\pi}{2}$$

$$= 0 \qquad\qquad\qquad ;\ n = 0, 1, 2, \ldots, \infty$$

In Fig. 2.6 we display the first four eigenfunctions for a string of unit length.

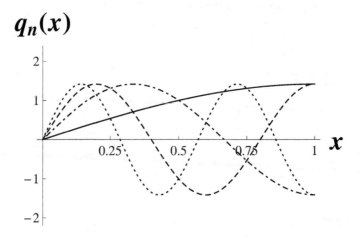

Fig. 2.6   First four eigenfunctions with mixed boundary conditions. Here $L = 1$.

(b) We need to verify that the real $q_n(x)$ are normalized to unity and that each one is orthogonal to all the others. If we first consider the normalization integral, we may exploit trigonometric relations to write

$$q_n^2(x) = \left(\frac{2}{L}\right) \sin^2 \frac{(2n+1)\pi x}{2L} = \frac{1}{L}\left[1 - \cos \frac{(2n+1)\pi x}{L}\right]$$

The cosine term in this expression clearly does not contribute to the integral, and therefore

$$\int_0^L q_n^2(x)dx = \frac{1}{L}\int_0^L dx = 1 \qquad ; n = m$$

Now consider the case of two distinct eigenfunctions $(n \neq m)$

$$q_n(x)q_m(x) = \left(\frac{2}{L}\right)\sin\frac{(2n+1)\pi x}{2L}\sin\frac{(2m+1)\pi x}{2L}$$

$$= \frac{1}{L}\left[\cos\frac{(m-n)\pi x}{L} - \cos\frac{(m+n+1)\pi x}{L}\right]$$

Once again, the trigonometric functions do not contribute to the integral, and we therefore have

$$\int_0^L q_n(x)q_m(x)dx = 0 \qquad ; n \neq m$$

The two results can be combined in the orthonormality relation

$$\int_0^L q_n(x)q_m(x)dx = \delta_{mn}$$

Problem 2.8 The hamiltonian for a classical, one-dimensional simple harmonic oscillator is given by Eq. (2.12). According to classical statistics, the differential probability in phase space at a temperature T is

$$dP(p,q) = \frac{e^{-H(p,q)/k_{\mathrm{B}}T}\,dpdq}{\int_{-\infty}^{\infty}dq\int_{-\infty}^{\infty}dp\,e^{-H(p,q)/k_{\mathrm{B}}T}}$$

(a) Show by explicit evaluation of the integrals that the mean value of the energy is

$$\langle E \rangle = \langle H \rangle = \frac{1}{2}k_{\mathrm{B}}T + \frac{1}{2}k_{\mathrm{B}}T = k_{\mathrm{B}}T$$

(b) Relate this result to the classical equipartition theorem.

Solution to Problem 2.8

The hamiltonian of the simple harmonic oscillator is given in Eq. (2.12)

$$H = \frac{p^2}{2m} + \frac{\kappa q^2}{2}$$

(a) The mean value of the energy is obtained as

$$\langle E \rangle = \langle H \rangle = \int dP(p,q) H(p,q)$$

$$= \frac{\int_{-\infty}^{\infty} dq \int_{-\infty}^{\infty} dp\ H(p,q)\ e^{-H(p,q)/k_{\mathrm{B}}T}}{\int_{-\infty}^{\infty} dq \int_{-\infty}^{\infty} dp\ e^{-H(p,q)/k_{\mathrm{B}}T}}\, ,$$

Concentrate first on the two-dimensional integral appearing in the denominator, and observe that because of the particular form of the hamiltonian, it can be expressed as the product of two one-dimensional integrals

$$\int_{-\infty}^{\infty} dq \int_{-\infty}^{\infty} dp\ e^{-H(p,q)/k_{\mathrm{B}}T} = \int_{-\infty}^{\infty} dq \int_{-\infty}^{\infty} dp\ e^{-(p^2/2m + \kappa q^2/2)/k_{\mathrm{B}}T}$$

$$= \int_{-\infty}^{\infty} dq\ e^{-\kappa q^2/2k_{\mathrm{B}}T} \int_{-\infty}^{\infty} dp\ e^{-p^2/2mk_{\mathrm{B}}T}$$

With two simple changes of variables, one obtains

$$\int_{-\infty}^{\infty} dq \int_{-\infty}^{\infty} dp\ e^{-H(p,q)/k_{\mathrm{B}}T} = 2k_{\mathrm{B}}T\sqrt{\frac{m}{\kappa}} \left[\int_{-\infty}^{\infty} dx\ e^{-x^2} \right]^2$$

$$= 2\pi k_{\mathrm{B}}T\sqrt{\frac{m}{\kappa}}$$

where the gaussian integral has already been given in Eq. (2.65).

Although we could proceed in a similar fashion with the calculation of the integral appearing in the numerator, we may avoid its explicit calculation with a simple trick. Let us define $\beta \equiv 1/k_{\mathrm{B}}T$, and then

$$\int_{-\infty}^{\infty} dq \int_{-\infty}^{\infty} dp\ H(p,q)\ e^{-\beta H(p,q)} = -\frac{\partial}{\partial \beta} \int_{-\infty}^{\infty} dq \int_{-\infty}^{\infty} dp\ e^{-\beta H(p,q)}$$

$$= 2\pi (k_{\mathrm{B}}T)^2 \sqrt{\frac{m}{\kappa}}$$

We thus obtain $\langle E \rangle = k_{\mathrm{B}}T$, as anticipated.

It is instructive to obtain the same result exploiting the particular form of the hamiltonian, which is quadratic both in the momentum and in the coordinate. With a simplification of the terms appearing both in the numerator and in the denominator, and the use of the integrals of Eq. (2.65), we obtain

$$\langle E \rangle = \frac{\int_{-\infty}^{\infty} dp \, (p^2/2m) \, e^{-p^2/2mk_BT}}{\int_{-\infty}^{\infty} dp \, e^{-p^2/2mk_BT}} + \frac{\int_{-\infty}^{\infty} dq \, (\kappa q^2/2) \, e^{-\kappa q^2/2k_BT}}{\int_{-\infty}^{\infty} dq \, e^{-\kappa q^2/2k_BT}}$$

$$= \frac{1}{2}k_BT + \frac{1}{2}k_BT = k_BT$$

(b) According to the equipartition theorem, there is a contribution of $k_BT/2$ to the mean energy for each quadratic degree of freedom in the hamiltonian. For the simple harmonic oscillator in one dimension, there are just two quadratic degrees of freedom, and therefore $\langle E \rangle = k_BT$, as we have explicitly shown in part (a).

Problem 2.9 (a) Use Gauss' theorem in Eq. (F.7) to derive Gauss' law from the first of Maxwell's Eqs. (2.90)

$$\oint_S d\mathbf{S} \cdot \mathbf{E} = \frac{1}{\varepsilon_0}Q \qquad ; \text{ Gauss' law}$$

Here Q is the total charge contained in the volume V.

(b) What is the corresponding physical content of the second of Maxwell's equations?

Solution to Problem 2.9

(a) Let V be the volume enclosed by the surface S. Then with the use of Gauss' theorem and the first of Maxwell's Eqs. (2.90), one obtains

$$\oint_S d\mathbf{S} \cdot \mathbf{E} = \int_V d^3x \, \boldsymbol{\nabla} \cdot \mathbf{E}$$

$$= \int_V d^3x \, \frac{\rho}{\varepsilon_0} = \frac{1}{\varepsilon_0}Q$$

where Q is the total charge contained in the volume V.

(b) The r.h.s. of the second of Maxwell's equations vanishes, implying that there is *no magnetic charge*.

Problem 2.10 Use Stokes' theorem in Eq. (F.10) to derive Faraday's law of induction from the third of Maxwell's Eqs. (2.90)

$$\Delta \mathcal{V} = \oint_C \mathbf{E} \cdot d\mathbf{l} = -\frac{d}{dt}\Phi_{\text{mag}} \qquad ; \text{ Faraday's law}$$

Here $\Delta \mathcal{V}$ is the change in voltage around the closed loop C, and Φ_{mag} is the total magnetic flux through that loop.

Solution to Problem 2.10

We use Stokes' theorem, together with the third of Maxwell's equations (2.90), to obtain

$$\Delta \mathcal{V} = \oint_C \mathbf{E} \cdot d\mathbf{l} = \int_A d\mathbf{A} \cdot (\boldsymbol{\nabla} \times \mathbf{E})$$

$$= -\int_A d\mathbf{A} \cdot \frac{\partial \mathbf{B}}{\partial t} = -\frac{d}{dt}\Phi_{\text{mag}}$$

Here $\Delta \mathcal{V}$ and Φ_{mag} are the change in voltage around the closed loop C and magnetic flux through that loop defined respectively by

$$\Delta \mathcal{V} \equiv \oint_C \mathbf{E} \cdot d\mathbf{l} \qquad ; \; \Phi_{\text{mag}} \equiv \int_A d\mathbf{A} \cdot \mathbf{B}$$

Problem 2.11 (a) Neglect the last term in the last of Maxwell's Eqs. (2.90) (his "displacement current"), and use Stokes' theorem in Eq. (F.10) to derive Ampere's law

$$\oint_C \mathbf{B} \cdot d\mathbf{l} = \mu_0 i \qquad ; \; \text{Ampere's law}$$

Here i is the total current flowing through the area bounded by the curve C.

(b) Under what conditions can that second term be neglected?

Solution to Problem 2.11

(a) We use Stokes' theorem, together with the fourth of Maxwell's Eqs. (2.90) with the last term neglected, to obtain

$$\oint_C \mathbf{B} \cdot d\mathbf{l} = \int_A d\mathbf{A} \cdot (\boldsymbol{\nabla} \times \mathbf{B})$$

$$= \mu_0 \int_A d\mathbf{A} \cdot \mathbf{j} = \mu_0 i$$

where from its definition, i is the total current flowing through the area bounded by the curve C

$$i \equiv \int_A d\mathbf{A} \cdot \mathbf{j}$$

(b) The divergence of the last of Maxwell's Eqs. (2.90) leads to the continuity equation

$$\nabla \cdot \mathbf{j} + \frac{\partial \rho}{\partial t} = 0$$

which expresses the conservation of electric charge. Ampere's law therefore corresponds to the case where $\nabla \cdot \mathbf{j} = 0$, and the electric charge density ρ does not change with time.

Problem 2.12 Show that the plane-wave radiation field in Eqs. (2.94) provides a solution to Maxwell's Eqs. (2.90) in free space.

Solution to Problem 2.12

We have

$$E_y = A \cos k(x - ct)$$
$$cB_z = A \cos k(x - ct)$$

Let us start with the first of Maxwell's Eqs. (2.90). Since

$$\nabla \cdot \mathbf{E} = \frac{\partial E_y}{\partial y} = 0$$

the plane-wave radiation field provides a solution to the first of Maxwell's equations in free space where $\rho = 0$.

In a similar fashion, one concludes that the second of Maxwell's Eqs. (2.90) is also fulfilled

$$\nabla \cdot \mathbf{B} = \frac{\partial B_z}{\partial z} = 0$$

We next observe that

$$\nabla \times \mathbf{E} = \frac{\partial E_y}{\partial x} \mathbf{e}_z = -Ak \sin k(x - ct) \, \mathbf{e}_z$$
$$-\frac{\partial \mathbf{B}}{\partial t} = -Ak \sin k(x - ct) \, \mathbf{e}_z$$

where $\mathbf{e}_z$ is a unit vector in the z-direction. Therefore the plane-wave radiation field is also a solution to the third of Maxwell's Eqs. (2.90).

Finally, we have

$$\nabla \times \mathbf{B} = -\frac{\partial B_z}{\partial x} \mathbf{e}_y = \frac{Ak}{c} \sin k(x - ct) \, \mathbf{e}_y$$
$$\frac{\partial \mathbf{E}}{\partial t} = Akc \sin k(x - ct) \, \mathbf{e}_y$$

where $\mathbf{e}_y$ is similarly a unit vector in the y-direction. Thus the plane-wave radiation field is also a solution to the fourth of Maxwell's Eqs. (2.90) in free space where $\mathbf{j} = 0$

$$\boldsymbol{\nabla} \times \mathbf{B} = \frac{1}{c^2} \frac{\partial \mathbf{E}}{\partial t}$$

Problem 2.13 Show that the cavity standing wave in Eqs. (2.96) provides a solution to Maxwell's Eqs. (2.90) in free space.

Solution to Problem 2.13

We are given

$$E_y = A \, \sin \frac{n\pi x}{L} \, \cos \frac{n\pi ct}{L}$$
$$-cB_z = A \, \cos \frac{n\pi x}{L} \, \sin \frac{n\pi ct}{L}$$

As in Problem 2.12, we then have

$$\boldsymbol{\nabla} \cdot \mathbf{E} = \frac{\partial E_y}{\partial y} = 0$$
$$\boldsymbol{\nabla} \cdot \mathbf{B} = \frac{\partial B_z}{\partial z} = 0$$

Therefore the cavity standing wave in Eqs. (2.96) provides a solution to the first two of Maxwell's equations in free space.

With respect to the third of Maxwell's equations, we observe

$$\boldsymbol{\nabla} \times \mathbf{E} = \frac{\partial E_y}{\partial x}\mathbf{e}_z = A\frac{n\pi}{L} \cos \frac{n\pi x}{L} \, \cos \frac{n\pi ct}{L}\mathbf{e}_z$$
$$-\frac{\partial \mathbf{B}}{\partial t} = A\frac{n\pi}{L} \cos \frac{n\pi x}{L} \, \cos \frac{n\pi ct}{L}\mathbf{e}_z$$

and therefore the cavity standing wave in Eqs. (2.96) is also a solution to the third of Maxwell's equations (2.90).

Finally, we have

$$\boldsymbol{\nabla} \times \mathbf{B} = -\frac{\partial B_z}{\partial x}\mathbf{e}_y = -A\frac{n\pi}{cL} \, \sin \frac{n\pi x}{L} \, \sin \frac{n\pi ct}{L}\mathbf{e}_y$$
$$\frac{\partial \mathbf{E}}{\partial t} = -A\frac{n\pi c}{L} \, \sin \frac{n\pi x}{L} \, \sin \frac{n\pi ct}{L}\mathbf{e}_y$$

Therefore the cavity standing wave is also a solution to the fourth of Maxwell's Eqs. (2.90) in free space

$$\nabla \times \mathbf{B} = \frac{1}{c^2} \frac{\partial \mathbf{E}}{\partial t}$$

Problem 2.14 A collection of particles with masses m_j located at $\mathbf{r}_j$ interact through forces $\mathbf{F}_{kj}(\mathbf{r}_j - \mathbf{r}_k)$ and obey Newton's laws in an inertial frame. Consider the transformation to a new frame moving with constant velocity $\mathbf{V}$ relative to the first so that $\mathbf{r}'_j = \mathbf{r}_j - \mathbf{V}t$. Show that Newton's laws continue to hold in the new frame, and hence conclude that it is again inertial.

Solution to Problem 2.14

Let us start by writing Newton's second law for the jth particle

$$m_j \frac{d^2 \mathbf{r}_j}{dt^2} = \sum_{k \neq j} \mathbf{F}_{kj}(\mathbf{r}_j - \mathbf{r}_k)$$

Now consider the transformation to a new frame moving with constant velocity $\mathbf{V}$ relative to the first so that $\mathbf{r}'_j = \mathbf{r}_j - \mathbf{V}t$. Under this transformation one has

$$\mathbf{r}'_j - \mathbf{r}'_k = \mathbf{r}_j - \mathbf{r}_k$$
$$\frac{d^2 \mathbf{r}'_j}{dt^2} = \frac{d^2 \mathbf{r}_j}{dt^2}$$

Therefore Newton's law continues to hold in the new frame since

$$m_j \frac{d^2 \mathbf{r}'_j}{dt^2} = \sum_{k \neq j} \mathbf{F}_{kj}(\mathbf{r}'_j - \mathbf{r}'_k)$$

Chapter 3

Some Contradictions

Problem 3.1 N molecules of a diatomic gas A-B are in contact with a heat bath at temperature T. In addition to the motion of the molecules themselves in the gas, they are free to perform internal one-dimensional simple harmonic motion of frequency ν_0 along the line joining them.

(a) What is the additional mean vibrational energy of the molecules according to the classical equipartition theorem?

(b) What is the additional vibrational contribution to the specific heat of the gas?

(c) What is the additional mean vibrational energy of the molecules in the Einstein model?

(d) What is the additional vibrational contribution to the specific heat in the Einstein model?

Solution to Problem 3.1

(a) According to the equipartition theorem, each quadratic degree of freedom in the hamiltonian makes a contribution of $k_B T/2$ to the average energy of the molecule. If we call $\mathbf{r}_A$ and $\mathbf{r}_B$ the coordinates of the atoms and $\mathbf{p}_A$ and $\mathbf{p}_B$ their momenta, the hamiltonian for this molecule reads

$$H = \frac{\mathbf{p}_A^2}{2m_A} + \frac{\mathbf{p}_B^2}{2m_B} + \frac{1}{2}\kappa(\mathbf{r}_A - \mathbf{r}_B)^2$$

To count the quadratic degrees of freedom it is convenient to introduce the coordinates of the center-of-mass and the relative separation between the atoms

$$\mathbf{R} = \frac{m_A \mathbf{r}_A + m_B \mathbf{r}_B}{m_A + m_B} \qquad ; \mathbf{r} = \mathbf{r}_A - \mathbf{r}_B$$

which allows us to write the hamiltonian as

$$H = \frac{\mathbf{P}^2}{2M} + \frac{\mathbf{p}^2}{2\mu} + \frac{1}{2}\kappa r^2$$

where $M = m_A + m_B$ is the total mass of the molecule and $1/\mu = 1/m_A + 1/m_B$ is the reduced mass.

It is now simple to see that in addition to translation and rotation of the molecule, there is only one vibrational degree of freedom along its axis, and therefore[1]

$$\langle \epsilon^{(\text{vibr})} \rangle = \frac{1}{2}k_B T + \frac{1}{2}k_B T = k_B T$$

(b) We first need to specify the total vibrational energy of the gas, which is simply obtained from the previous result as

$$E^{(\text{vibr})} = N\langle \epsilon^{(\text{vibr})} \rangle = Nk_B T$$

The vibrational contribution to the specific heat at constant volume is then

$$C_v^{(\text{vibr})} = \frac{dE^{(\text{vibr})}}{dT} = Nk_B$$

(c) In the Einstein model the oscillators (molecules) follow the Planck distribution, and therefore the additional mean vibrational energy of the molecules is

$$E^{(\text{vibr})} = N\frac{h\nu_0}{e^{h\nu_0/k_B T} - 1}$$

(d) The additional vibrational contribution to the molar specific heat in the Einstein model is then

$$C_v^{(\text{vibr})} = \frac{dE}{dT} = N_A k_B \left[\frac{x^2 \, e^x}{(e^x - 1)^2} \right] \qquad ; \, x \equiv \frac{h\nu_0}{k_B T}$$

Problem 3.2 (a) Derive the Debye expression for the molar heat capacity

$$C_v = 9N_A k_B \left(\frac{T}{\theta_D} \right)^3 \int_0^{\theta_D/T} \frac{u^4 \, e^u \, du}{(e^u - 1)^2} \qquad ; \text{ Debye theory}$$

[*Hint*: start from Eq. (3.39).]

(b) What are the high-temperature and low-temperature limits of this result?

---

[1] See Prob. 2.8.

Solution to Problem 3.2

(a) In line with the suggestion, we start from Eq. (3.39) in the text

$$\frac{E}{3N} = \frac{h\nu_{max}}{1/3} \int_0^1 x^3 dx \frac{1}{e^{h\nu_{max}x/k_BT} - 1} \qquad ; x \equiv \frac{\nu}{\nu_{max}}$$

After introducing the Debye temperature $\theta_D \equiv h\nu_{max}/k_B$, this equation becomes

$$\frac{E}{3N} = \frac{k_B\theta_D}{1/3} \int_0^1 x^3 dx \frac{1}{e^{\theta_D x/T} - 1}$$

The molar heat capacity at constant volume can then be obtained as

$$C_v = \frac{dE}{dT} = 9N_A k_B \left(\frac{\theta_D}{T}\right)^2 \int_0^1 \frac{x^4 e^{\theta_D x/T} dx}{(e^{\theta_D x/T} - 1)^2}$$

$$= 9N_A k_B \left(\frac{T}{\theta_D}\right)^3 \int_0^{\theta_D/T} \frac{u^4 e^u du}{(e^u - 1)^2} \qquad ; u \equiv \frac{\theta_D x}{T}$$

The integrand of the expression for C_v is plotted in Fig. 3.1. This plot will be useful in the discussion of the low- and high-temperature limits.

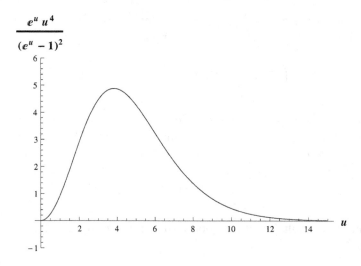

Fig. 3.1   Integrand of the expression for C_v.

(b) Let us now study the high-temperature and low-temperature limits of this formula for C_v.

In the high-temperature limit we have $T \gg \theta_D$, and therefore the integral is performed on a narrow region $0 \le u \le \theta_D/T \ll 1$. We can obtain a good approximation to the integral by expanding the integrand around $u = 0$ (see Fig. 3.1)

$$\frac{u^4 e^u}{(e^u - 1)^2} = u^2 + O\left(u^3\right)$$

In this limit, we recover the result predicted by the equipartition theorem for the molar heat capacity

$$C_v = 3N_A k_B \qquad ; T \gg \theta_D$$

In the low-temperature limit we have $T \ll \theta_D$, and the upper limit of integration becomes very large. For large values of u, however, the integrand decays exponentially fast, allowing one to simply replace infinity as the upper limit in the integration. It is useful to make a change of variable $v = e^{-u}$ and transform the integral into

$$C_v \approx 9N_A k_B \left(\frac{T}{\theta_D}\right)^3 \int_0^1 \frac{(\log 1/v)^4\, dv}{(1-v)^2}$$

For $0 \le v < 1$ the series $S(v) = \sum_{j=0}^\infty (j+1)\, v^j$ converges uniformly to the the function $1/(1-v)^2$, and therefore the expression for C_v becomes

$$C_v = 9N_A k_B \left(\frac{T}{\theta_D}\right)^3 \sum_{j=0}^\infty (j+1) \int_0^1 (\log 1/v)^4\, v^j\, dv$$

where the uniform convergence of the series allows us to move the sum outside the integral.[2] It is convenient to calculate the integral by switching back to the variable u and then performing repeated integrations by parts

$$\int_0^1 (\log 1/v)^4\, v^j\, dv = \int_0^\infty u^4\, e^{-(j+1)u}\, du = \frac{24}{(j+1)^5}$$

Notice that this result can also be obtained directly from the definition of the Γ-function

$$\int_0^\infty e^{-\alpha x} x^r dx = \frac{\Gamma(r+1)}{\alpha^{r+1}} \qquad ; \alpha > 0 \qquad ; r > -1$$

---

[2]See, for instance, the discussion in [Arfken, Weber, and Harris (2012)]. The integral is well-protected as $v \to 1$.

We thus obtain

$$C_v = 216 N_A k_B \left(\frac{T}{\theta_D}\right)^3 \sum_{j=0}^{\infty} \frac{1}{(j+1)^4}$$

$$= 216 N_A k_B \left(\frac{T}{\theta_D}\right)^3 \zeta(4)$$

$$= \frac{12}{5}\pi^4 N_A k_B \left(\frac{T}{\theta_D}\right)^3 \qquad ; T \ll \theta_D$$

where we have substituted the value of the Riemann zeta function, $\zeta(4) = \pi^4/90$.

Problem 3.3 A Debye temperature of $\theta_D = 1890\,^\circ\text{K}$ is found to give an excellent fit to the molar heat capacity of diamond (Fig. 3.8 in the text). Diamond is pure carbon with an atomic weight of 12 (so that 12 gm = 1 mole), and the measured mass density of diamond is approximately $3.25\,\text{gm/cm}^3$. Use Debye's theory to deduce the speed of sound c_s in diamond in m/s.

Solution to Problem 3.3

The Debye frequency for diamond is obtained from Eq. (3.40)

$$\nu_{\text{max}} = \frac{k_B \theta_D}{h} = 3.94 \times 10^{13}\ \text{s}^{-1}$$

The second of Eqs. (3.38) relates c_s to the Debye frequency [see Eq. (3.15)]

$$c_s = \left(\frac{4\pi V}{3N}\right)^{1/3} \nu_{\text{max}} \qquad ; \text{sound velocity}$$

Since there are $(3.25/12)N_A \approx 1.63 \times 10^{23}$ atoms in 3.25 grams of diamond, the number density for diamond is obtained from the mass density as

$$\frac{N}{V} = 1.63 \times 10^{29}\ \text{m}^{-3}$$

Thus the sound velocity in diamond is determined from Debye theory to be

$$c_s = 1.16 \times 10^4\ \frac{\text{m}}{\text{s}}$$

Problem 3.4 The velocities of the transverse and longitudinal waves in a solid are actually distinct. In the counting of normal modes, one should therefore replace $3/c_s^3 \to 2/c_t^3 + 1/c_l^3$. A single ν_{max} is still employed.

(a) What is the corresponding modification of Eq. (3.41) and Fig. 3.8 in the text?

(b) What is now the content of the second of Eqs. (3.38)?

Solution to Problem 3.4

(a) Under the replacement $3/c_s^3 \rightarrow 2/c_t^3 + 1/c_l^3$, Eqs. (3.38) become

$$\frac{E}{V} = \left(\frac{2}{c_t^3} + \frac{1}{c_l^3} \right) \int_0^{\nu_{max}} 4\pi\nu^2 d\nu \left[\frac{h\nu}{e^{h\nu/k_B T} - 1} \right]$$

$$\frac{3N}{V} = \left(\frac{2}{c_t^3} + \frac{1}{c_l^3} \right) \int_0^{\nu_{max}} 4\pi\nu^2 d\nu$$

Since Eq. (3.41) was obtained by taking the ratio of the two Eqs. (3.38), we see that Eq. (3.41) is *unchanged* when the velocities of the transverse and longitudinal waves in the solid are distinct. Therefore Fig. 3.8 in the text is also *unchanged*.

(b) The second of Eqs. (3.38) now relates the number density N/V to the Debye frequency ν_{max} and to the transverse and longitudinal speeds of sound c_t and c_l.

Problem 3.5 An energy spectrum $u(\nu, T)$, known to be that of a black body, is plotted against ν and found to have its maximum at ν_B. Determine the corresponding temperature of the distribution. [*Hint*: see Fig. 3.6 in the text.]

Solution to Problem 3.5

The black-body energy density is given in Eq. (3.31) in terms of the dimensionless function

$$u(x) \equiv \frac{8\pi x^3}{e^x - 1}$$

where $x \equiv h\nu/k_B T$. This function, which is plotted in Fig. 3.6 in the text, has a maximum at $x_B \approx 2.82$. Therefore the temperature of the distribution is

$$T = \frac{h\nu_B}{k_B x_B} \approx 0.35 \frac{h\nu_B}{k_B}$$

Problem 3.6 Classically, when light is scattered from a free electron, the electron is first accelerated according to [compare Eq. (3.43)]

$$\frac{d^2 y}{dt^2} = \frac{e\mathcal{E}_0}{m} \cos(\omega t)$$

The accelerated electron then radiates. The result is the Thomson cross section for the scattering of light by an electron

$$\frac{d\sigma}{d\Omega} = r_0^2 \frac{1}{2} \left(1 + \cos^2\theta\right) \qquad \text{; Thomson cross section}$$

$$r_0 = \frac{e^2}{4\pi\varepsilon_0} \frac{1}{m_e c^2} = 2.819 \times 10^{-15}\,\text{m} \qquad \text{; classical electron radius}$$

Discuss, in qualitative terms, the dependence of the frequency shift of the scattered light on the incident light *intensity*, which one might expect in this picture. Here, there is a radiation pressure on the electron and it is recoiling as it re-radiates. Compare with the corresponding paradox in the classical treatment of the photoelectric effect.

Solution to Problem 3.6

In this picture, the velocity of the recoiling electron would be proportional to the radiation pressure, which, in turn, is proportional to the incident light intensity $\mathcal{E}_0^2$. The frequency of the re-radiation would then be Doppler shifted, with a shift proportional to $\mathcal{E}_0^2$. As with the photoelectric effect, this is in disagreement with experiment.

Problem 3.7 Consider the Compton effect for back-scattered light. What is the fractional shift in wavelength for light of incident wavelength $10^{-10}\,\text{m}$? Of 4000Å?

Solution to Problem 3.7

The Compton formula for the shift in wavelength of the light scattered by an electron is given in Eq. (3.54)

$$\lambda_1 - \lambda_0 = \frac{h}{m_e c} \left(1 - \cos\theta\right)$$

where $\lambda_C = h/m_e c = 2.427 \times 10^{-12}\,\text{m}$ is the Compton wavelength of the electron. Thus the fractional shift in wavelength for back-scattered light $(\theta = \pi)$ is

$$\frac{\Delta\lambda}{\lambda_0} = \frac{2\lambda_C}{\lambda_0}$$

In the case of an incident wavelength of $10^{-10}\,\text{m}$, one then has

$$\frac{\Delta\lambda}{\lambda_0} = 4.85 \times 10^{-2}$$

With an incident wavelength of $4000 \text{ Å} = 4 \times 10^{-7}$ m, the result is

$$\frac{\Delta\lambda}{\lambda_0} = 1.21 \times 10^{-5}$$

Problem 3.8 The Balmer formula (1885) is an empirical relation for the frequencies of light observed in atomic transitions. For the Lyman-α series in hydrogen, it can be re-written in terms of a photon energy as

$$\varepsilon_n = -13.61 \text{ eV} \left(1 - \frac{1}{n^2}\right) \qquad \text{; Balmer formula}$$

$$n = 2, 3, \cdots$$

(a) Calculate the corresponding frequencies (in s^{-1}), and wavelengths (in m and Å) of the radiation from the $n = 2$ and $n = 3$ states;

(b) Explain how the Bohr model and Bohr hypotheses yield this result.

Solution to Problem 3.8

(a) The frequency of the radiation is obtained as[3]

$$\nu_n = -\frac{\varepsilon_n}{h} \equiv \nu_\infty \left(1 - \frac{1}{n^2}\right)$$

where we have defined

$$\nu_\infty \equiv \frac{13.61 \text{ eV}}{h} = 3.291 \times 10^{15} \text{ s}^{-1}$$

Thus, for the $n = 2$ state we obtain

$$\nu_2 = 2.468 \times 10^{15} \text{ s}^{-1}$$

and for the $n = 3$ state

$$\nu_3 = 2.925 \times 10^{15} \text{ s}^{-1}$$

Similarly, the wavelengths of radiation from these states are

$$\lambda_2 = \frac{c}{\nu_2} = 1.215 \times 10^{-7} \text{ m}$$

for the $n = 2$ state and

$$\lambda_3 = \frac{c}{\nu_3} = 1.025 \times 10^{-7} \text{ m}$$

---

[3]Unfortunately, with the given expression for ε_n, the photon frequency is determined from $\varepsilon_n = -h\nu_n$.

for the $n = 3$ state. This is readily converted to Å with 10^{-10} m $= 1$Å.

(b) In the Bohr model, the energy of the electron in the hydrogen atom is given by

$$E_n = -\frac{13.61 \text{ eV}}{n^2} \qquad ; n = 1, 2, \cdots, \infty$$

With the use of this formula, we see that an electron in an excited state (corresponding to $n > 1$) can reach the lowest orbit (with $n = 1$) by emitting a photon with frequency

$$h\nu_n = E_n - E_1 = 13.61 \text{ eV} \left(1 - \frac{1}{n^2}\right)$$

Therefore the empirical Balmer formula can be derived directly from the Bohr model.

Problem 3.9 A completely isolated hydrogen atom in space can have arbitrarily large bound orbits. Assume the Bohr model with $l = n\hbar$.

(a) How large must n be to get an orbit of radius $r = 0.529$ m?

(b) What is the binding energy of the orbit in part (a)?[4]

Solution to Problem 3.9

(a) We use Eq. (3.62), which expresses the radius of the orbit in terms of the quantum number n and of the Bohr radius a_0

$$r = \frac{n^2}{Z} a_0$$

where $a_0 = 0.5292 \times 10^{-10}$ m, and $Z = 1$ for hydrogen. Thus

$$n = \sqrt{\frac{r}{a_0}} = 10^5$$

(b) The binding energy of the orbit in part (a) is obtained from Eq. (3.66)

$$E_n = -13.61 \text{ eV} \frac{Z^2}{n^2} = -1.361 \times 10^{-9} \text{ eV}$$

Problem 3.10 The muon is a particle just like the electron except that it has a mass $m_\mu = 206.8 \, m_e$. What is the ratio of the radius of the first Bohr orbit of a muon relative to that of an electron for a nucleus with charge Ze_p?

---

[4]Such huge atoms are indeed created and studied with modern laser spectroscopy.

Solution to Problem 3.10

The Bohr radius for a muon is

$$a_0^{(\mu)} = \frac{\hbar^2}{m_\mu(e^2/4\pi\varepsilon_0)}$$

Therefore

$$\frac{a_0^{(\mu)}}{a_0^{(e)}} = \frac{m_e}{m_\mu} = \frac{1}{206.8}$$

Problem 3.11 An alternate method of counting the number of normal modes with fixed boundaries is as follows: The wave number vector is defined as $\mathbf{k} = (\pi/L)\mathbf{n}$ with $\mathbf{n} = (n_x, n_y, n_z)$. One can uniquely assign a *unit cube in n-space* to each triplet of integers (n_x, n_y, n_z) by using, say, the point in the most positive direction in each cube. If $L \to \infty$, then for a given wavenumber volume d^3k around $\mathbf{k}$, one is counting a very large number of normal modes, or equivalently, one is computing a very large volume in n-space. Work in this limit.

(a) Show that the total volume of the first octant in n-space out to a given $n = |\mathbf{n}|$ is[5]

$$\mathcal{N} = g\frac{1}{8}\left(\frac{4\pi n^3}{3}\right)$$

(b) Rewrite this expression in terms of $k = |\mathbf{k}|$, and then differentiate it with respect to k to obtain

$$\frac{d\mathcal{N}}{V} = \frac{g}{(2\pi)^3}4\pi k^2\, dk$$

Compare with Eq. (3.16).

Solution to Problem 3.11

(a) In Fig. 3.2 we provide a visualization of the alternate method of counting the normal modes with fixed boundaries. We consider modes with $1 \leq (n_x, n_y, n_z) \leq n_{max}$, which are contained in the portion of a sphere of radius n_{max}, falling in the first octant. We call N_{sphere} the number of these modes and N_{cube} the number of modes falling in a cube of side n_{max}. In the latter case, we know how to perform an exact counting of the modes. As $n_{max} \to \infty$, the ratio N_{sphere}/N_{cube} tends to the ratio between the volume

---

[5]Here we include the degeneracy factor g.

of an octant of a sphere of radius $n_{\max}$ and the volume of a cube of side $n_{\max}$. Thus

$$\operatorname{Lim}_{n_{\max}\to\infty} \frac{N_{\text{sphere}}}{N_{\text{cube}}} = \frac{\pi}{6}$$

For instance, for a small cube of side $n_{\max} = 10$, corresponding to the case presented in Fig. 3.2, $N_{\text{sphere}} = 410$ and $N_{\text{cube}} = 1000$. Their ratio falls about 22% below the asymptotic value.

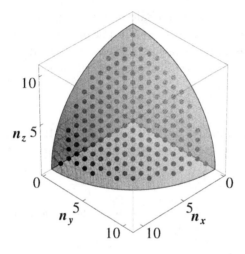

Fig. 3.2 Visualization of the alternate method of counting the number of normal modes with $n_{\max} = 10$.

For a larger cube of side $n_{\max} = 100$, $N_{\text{sphere}} = 511776$ and $N_{\text{cube}} = 10^6$, and their ratio falls just 2% below the asymptotic value. Therefore, if we consider a large number of modes, the error that we make using the asymptotic formula is certainly small.

With a degeneracy factor of g, each triplet (n_x, n_y, n_z) can accommodate g modes, leading to a total number of modes

$$\mathcal{N} = g\frac{1}{8}\left(\frac{4\pi n^3}{3}\right)$$

as claimed.

(b) We rewrite the total number of modes in terms of the magnitude of the wavenumber $|\mathbf{k}|$ through $n = Lk/\pi$

$$\mathcal{N} = g\frac{1}{8}\left(\frac{4\pi}{3}\right)\left(\frac{Lk}{\pi}\right)^3 = \frac{g}{6\pi^2}Vk^3$$

If we differentiate with respect to k, and divide by the volume V, we have

$$\frac{d\mathcal{N}}{V} = \frac{g}{(2\pi)^3}4\pi k^2 dk$$

which is equivalent to Eq. (3.16).

Problem 3.12 There exists a cosmic microwave background (CMB) that fills all of space. It remains from the early hot era of the universe after the radiation decoupled and adiabatically expanded with time. The CMB exhibits an almost perfect *black-body spectrum* with an energy density

$$d\varepsilon_\gamma = \frac{8\pi\nu^2\,d\nu}{c^3}\frac{h\nu}{e^{h\nu/k_BT}-1}$$

and a current temperature $T(t_p) = 2.73\,°K$ [Ohanian (1995)]. Show that when integrated over frequencies, this gives rise to a radiation density of

$$\varepsilon_\gamma = \frac{4\sigma}{c}T^4 \qquad ;\ \sigma = \frac{\pi^2 k_B^4}{60\hbar^3 c^2}$$

Here σ is the Stefan-Boltzmann constant, with the value in cgs units of $\sigma = 5.670 \times 10^{-5}\,\text{erg/sec-cm}^2\text{-°K}^4$.

Solution to Problem 3.12

We need to evaluate the integral

$$\epsilon_\gamma = \frac{8\pi h}{c^3}\int_0^\infty \frac{\nu^3 d\nu}{e^{h\nu/k_BT}-1}$$

It is useful to re-write this expression as

$$\epsilon_\gamma = \frac{8\pi h}{c^3}\int_0^\infty \frac{e^{-h\nu/k_BT}\nu^3 d\nu}{1-e^{-h\nu/k_BT}}$$

Since $0 < e^{-h\nu/k_BT} \le 1$, we can expand the denominator and obtain[6]

$$\epsilon_\gamma = \frac{8\pi h}{c^3}\sum_{j=0}^\infty\int_0^\infty \nu^3\,e^{-(j+1)h\nu/k_BT}\,d\nu$$

---

[6] As for Prob. 3.2, the series obtained converges uniformly in the region of integration, and therefore it is possible to move the summation outside the integral.

Notice that the integrals appearing in this expression are of the same kind discussed in Prob. 3.2, and they can be expressed directly in terms of a Γ-function, which gives

$$\int_0^\infty e^{-\alpha x} x^3 dx = \frac{3!}{\alpha^4} \qquad ; \alpha > 0$$

With the use of this relation, we obtain

$$\epsilon_\gamma = \frac{48\pi h}{c^3} \left(\frac{k_B T}{h}\right)^4 \sum_{j=0}^\infty \frac{1}{(j+1)^4}$$

$$= \frac{48\pi h}{c^3} \left(\frac{k_B T}{h}\right)^4 \zeta(4)$$

where $\zeta(4) = \pi^4/90$. Therefore

$$\epsilon_\gamma = \frac{8\pi^5 T^4 k_B^4}{15 c^3 h^3} = \frac{4\sigma}{c} T^4$$

where σ is the Stefan-Boltzmann constant given above.[7]

---

[7] Recall $h = 2\pi\hbar$.

Chapter 4

Quantum Mechanics

Problem 4.1 A particle moves in a real potential $V(x)$. Show that the continuity equation and probability flux are unchanged from those of a free particle in Eq. (4.46).

Solution to Problem 4.1

If $V(x)$ is real, the Schrödinger equation and its complex conjugate read

$$i\hbar\frac{\partial}{\partial t}\Psi(x,t) = \left[-\frac{\hbar^2}{2m}\frac{\partial^2}{\partial x^2} + V(x)\right]\Psi(x,t)$$

$$-i\hbar\frac{\partial}{\partial t}\Psi^\star(x,t) = \left[-\frac{\hbar^2}{2m}\frac{\partial^2}{\partial x^2} + V(x)\right]\Psi^\star(x,t)$$

The additional terms in passing from Eq. (4.43) to Eq. (4.45) then *cancel*

$$\Psi^\star(x,t)V(x)\Psi(x,t) - [V(x)\Psi^\star(x,t)]\,\Psi(x,t) = 0$$

and the continuity equation and probability flux are unchanged from those of a free particle.

Problem 4.2 This problem demonstrates that there is enough flexibility in the Fourier transform to match an arbitrary set of initial conditions in the classical string problem. Consider a superposition of the solutions in Eq. (4.14)

$$q(x,t) = \text{Re}\int_{-\infty}^{\infty} dk\,a(k)\exp\left\{i(kx - \omega_k t)\right\}$$

$$= \frac{1}{2}\int_{-\infty}^{\infty} dk\,[a(k)\exp\left\{i(kx - \omega_k t)\right\} + a^\star(k)\exp\left\{-i(kx - \omega_k t)\right\}]$$

Here we take the frequency $\omega_k = |k|c$ to be positive, and this real expression evidently satisfies the wave equation.[1]

(a) Show that this function and its time derivative at the initial time $t = 0$ are given by

$$q(x,0) = \frac{1}{2} \int_{-\infty}^{\infty} dk \, [a(k) + a^*(-k)] \, e^{ikx}$$

$$\dot{q}(x,0) \equiv \left. \frac{\partial q(x,t)}{\partial t} \right|_{t=0} = \frac{1}{2} \int_{-\infty}^{\infty} dk \, (-i\omega_k) \, [a(k) - a^*(-k)] \, e^{ikx}$$

(b) Invert these Fourier transforms, and solve for $a(k)$ to obtain

$$a(k) = \frac{1}{2\pi} \int_{-\infty}^{\infty} dx \left[q(x,0) + \frac{i}{\omega_k} \dot{q}(x,0) \right] e^{-ikx}$$

This result uniquely determines the complex Fourier amplitude $a(k)$ in terms of the initial conditions of the string.

Solution to Problem 4.2

(a) At $t = 0$ we have

$$q(x,0) = \frac{1}{2} \int_{-\infty}^{\infty} dk \, [a(k) \exp\{ikx\} + a^*(k) \exp\{-ikx\}]$$

$$= \frac{1}{2} \int_{-\infty}^{\infty} dk \, [a(k) + a^*(-k)] \exp\{ikx\}$$

Here the last line has been obtained by performing the change of variable $k' = -k$ in the second term of the expression.

Similarly, we may differentiate $q(x,t)$ with respect to the time and then set $t = 0$ to obtain

$$\dot{q}(x,0) = \frac{1}{2} \int_{-\infty}^{\infty} dk \, (-i\omega_k) \, [a(k) \exp\{ikx\} - a^*(k) \exp\{-ikx\}]$$

$$= \frac{1}{2} \int_{-\infty}^{\infty} dk \, (-i\omega_k) \, [a(k) - a^*(-k)] \exp\{ikx\}$$

Again, the last line has been obtained by performing the change of variable $k' = -k$ in the second term of the expression (recall $\omega_k = |k|c$).

---

[1] Note that the solution now contains waves propagating in both directions.

(b) The inversion relation for Fourier transform pairs in Eqs. (4.41) then gives

$$\frac{1}{2\pi} \int_{-\infty}^{\infty} dx\, q(x,0) e^{-ikx} = \frac{1}{2} [a(k) + a^\star(-k)]$$

$$\frac{1}{2\pi} \int_{-\infty}^{\infty} dx\, \dot{q}(x,0) e^{-ikx} = \frac{1}{2}(-i\omega_k)\, [a(k) - a^\star(-k)] \qquad (4.1)$$

A linear combination of these relations then yields

$$\frac{1}{2\pi} \int_{-\infty}^{\infty} dx \left[q(x,0) + \frac{i}{\omega_k}\dot{q}(x,0) \right] e^{-ikx} = a(k)$$

Problem 4.3 (a) Substitute the second of Eqs. (4.41) into the first to obtain the following relation[2]

$$q(x) = \int_{-\infty}^{\infty} dy \left[\frac{1}{2\pi} \int_{-\infty}^{\infty} dk\, e^{ik(x-y)} \right] q(y)$$

(b) The r.h.s. must reproduce $q(x)$ at the point x no matter what $q(y)$ does at every other point $y \neq x$ in the integral. Hence, conclude that the "kernel" in this relation must have the form

$$\frac{1}{2\pi} \int_{-\infty}^{\infty} dk\, e^{ik(x-y)} = \delta(x - y) \qquad \text{; Dirac delta function}$$

where the *Dirac delta function* is defined to have the following properties:

$$\delta(x - y) = 0 \qquad ; x \neq y$$
$$\neq 0 \qquad x = y$$
$$\int_{-\infty}^{\infty} dy\, q(y)\delta(x - y) = q(x)$$

This is an expression of the *completeness* of the Fourier transform.[3]

(c) Use an analogous argument to show

$$\frac{1}{2\pi} \int_{-\infty}^{\infty} dx\, e^{i(k-k')x} = \delta(k - k')$$

---

[2] It is assumed here that $q(x)$ is a well-behaved function.
[3] Two additional useful properties of the Dirac delta function are

$$\delta(-x) = \delta(x)$$
$$\delta(ax) = \frac{1}{|a|}\delta(x) \qquad ; a \text{ real}$$

(d) Use Eqs. (4.41) and the result in part (c) to prove *Parseval's theorem*

$$\int_{-\infty}^{\infty} dx\, |q(x)|^2 = \int_{-\infty}^{\infty} dk\, |a(k)|^2$$

Solution to Problem 4.3

(a) We write the second of Eqs. (4.41) as

$$a(k) = \frac{1}{\sqrt{2\pi}} \int_{-\infty}^{\infty} dy\, q(y) e^{-iky}$$

and then substitute it into the first of Eqs. (4.41) to obtain

$$q(x) = \int_{-\infty}^{\infty} dy \left[\frac{1}{2\pi} \int_{-\infty}^{\infty} dk\, e^{ik(x-y)} \right] q(y)$$

(b) After identification of the expression in brackets with the Dirac delta function, the identity in part (a) can be re-cast as

$$q(x) = \int_{-\infty}^{\infty} dy\, \delta(x - y) q(y)$$

Since this expression must hold for a fixed x, the only possibility is that $\delta(x - y)$ must vanish unless $x = y$, proving our point.

(c) The same argument applies for $(2\pi)^{-1} \int_{-\infty}^{\infty} dx\, e^{i(k-k')x}$, since the only difference is that the integration is now carried out in x rather then in k.

(d) Substitution of the Fourier transform on the l.h.s. of the following expression, and manipulation of the resulting multiple integrals, leads to

$$
\begin{aligned}
\int_{-\infty}^{\infty} dx\, |q(x)|^2 &= \int_{-\infty}^{\infty} dx \left| \frac{1}{\sqrt{2\pi}} \int_{-\infty}^{\infty} dk\, a(k) e^{ikx} \right|^2 \\
&= \int_{-\infty}^{\infty} dk \int_{-\infty}^{\infty} dk'\, a(k) a^\star(k') \left[\frac{1}{2\pi} \int_{-\infty}^{\infty} dx\, e^{i(k-k')x} \right] \\
&= \int_{-\infty}^{\infty} dk \int_{-\infty}^{\infty} dk'\, a(k) a^\star(k') \delta(k - k') \\
&= \int_{-\infty}^{\infty} dk\, |a(k)|^2
\end{aligned}
$$

This is Parseval's theorem for Fourier transforms.

Problem 4.4 This problem concerns *partial integration*. Start from the identity

$$\frac{d(uv)}{dx} = u\frac{dv}{dx} + \frac{du}{dx}v$$

Now integrate this expression on x between the boundaries $[a, b]$ appropriate to the problem at hand.

(a) Explain how the boundary contribution $[uv]_a^b$ *vanishes* under the following boundary conditions: localized disturbance, fixed endpoints, periodic boundary conditions;

(b) Show that under the conditions in (a), the derivative can be transferred to the other term in the integrand, with a minus sign

$$\int_a^b dx \left(u\frac{dv}{dx}\right) = -\int_a^b dx \left(\frac{du}{dx}v\right)$$

Solution to Problem 4.4

(a) Consider the three cases:

(1) For localized disturbances in the interval $[a, b] = [-\infty, \infty]$, the functions vanish as one goes off to infinity in either direction, so that $\lim_{|x|\to\infty} u(x) = \lim_{|x|\to\infty} v(x) = 0$. Therefore $[uv]_a^b = 0$;

(2) For fixed endpoints where x lies in the interval $[a, b]$, the functions vanish at the boundaries $u(a) = u(b) = v(a) = v(b) = 0$. Therefore $[uv]_a^b = 0$;

(3) For periodic boundary conditions with x in the interval $[a, b]$, the functions are equal on the boundaries $u(a) = u(b)$ and $v(a) = v(b)$. Hence, again in this case, one has $[uv]_a^b = u(b)v(b) - u(a)v(a) = 0$.

(b) Under the conditions in (a) we have

$$\int_a^b dx \left(u\frac{dv}{dx}\right) = [uv]_a^b - \int_a^b dx \left(\frac{du}{dx}v\right) = -\int_a^b dx \left(\frac{du}{dx}v\right)$$

Problem 4.5 In quantum mechanics, classical quantities become *operators*. For example,

$$p \to \frac{\hbar}{i} \frac{\partial}{\partial x}$$

$$T(p) = \frac{p^2}{2m} \to -\frac{\hbar^2}{2m} \frac{\partial^2}{\partial x^2}$$

$$H(p, x) = T(p) + V(x) \to -\frac{\hbar^2}{2m} \frac{\partial^2}{\partial x^2} + V(x) \qquad \text{; etc.}$$

Given a wave function $\Psi(x, t)$, the *expectation value* of an operator $O(p, x)$ is defined as

$$\langle O \rangle \equiv \frac{\int_a^b dx \, \Psi^\star(x, t) \, O(p, x) \, \Psi(x, t)}{\int_a^b dx \, \Psi^\star(x, t) \Psi(x, t)} \qquad \text{; expectation value}$$

An operator is said to be *hermitian* if it has the following property[4]

$$\int_a^b dx \, \Psi^\star(x, t) O(p, x) \Psi(x, t) = \int_a^b dx \, [O(p, x) \Psi(x, t)]^\star \Psi(x, t)$$

$$\text{; hermitian}$$

The boundaries (a, b) on x are, again, appropriate to the problem at hand.

Show that under the conditions in Prob. 4.4, the momentum p, kinetic energy $T(p)$, and hamiltonian $H(p, x) = T(p) + V(x)$ with a real $V(x)$, are *hermitian operators*.

Solution to Problem 4.5

Let us start with the momentum operator

$$\int_a^b dx \, \Psi^\star(x, t) p \Psi(x, t) = \int_a^b dx \, \Psi^\star(x, t) \frac{\hbar}{i} \frac{\partial}{\partial x} \Psi(x, t)$$

$$= \frac{\hbar}{i} [\Psi^\star(x, t) \Psi(x, t)]_a^b - \frac{\hbar}{i} \int_a^b dx \, \left[\frac{\partial}{\partial x} \Psi^\star(x, t) \right] \Psi(x, t)$$

$$= \int_a^b dx \, \left[-\frac{\hbar}{i} \frac{\partial}{\partial x} \Psi^\star(x, t) \right] \Psi(x, t)$$

$$= \int_a^b dx \, [p \Psi(x, t)]^\star \Psi(x, t)$$

Here the surface term $[\Psi^\star(x, t) \Psi(x, t)]_a^b$ in the second line vanishes under the boundary conditions discussed in Prob. 4.4. We thus see that the momentum operator $p = (\hbar/i) \partial / \partial x$ is hermitian.

---

[4]More generally, this relation must hold for any acceptable initial and final wave functions in the integrand.

Consider next the kinetic-energy operator

$$\int_a^b dx \ \Psi^\star(x,t) T(p) \Psi(x,t) = \int_a^b dx \ \Psi^\star(x,t) \left[-\frac{\hbar^2}{2m} \frac{\partial^2}{\partial x^2} \right] \Psi(x,t)$$

$$= -\frac{\hbar^2}{2m} \left[\Psi^\star(x,t) \frac{\partial}{\partial x} \Psi(x,t) \right]_a^b + \frac{\hbar^2}{2m} \int_a^b dx \ \frac{\partial}{\partial x} \Psi^\star(x,t) \ \frac{\partial}{\partial x} \Psi(x,t)$$

$$= \frac{\hbar^2}{2m} \left[\Psi(x,t) \frac{\partial}{\partial x} \Psi^\star(x,t) \right]_a^b - \frac{\hbar^2}{2m} \int_a^b dx \ \left[\frac{\partial^2}{\partial x^2} \Psi^\star(x,t) \right] \Psi(x,t)$$

$$= \int_a^b dx \ [T(p)\Psi(x,t)]^\star \ \Psi(x,t)$$

Here all the surface terms vanish under the boundary conditions of Prob. 4.4.[5]

Finally we come to the hamiltonian operator. In this case, having already proven that $T(p)$ is hermitian, it is sufficient to prove that $V(x)$ is hermitian. This proof is straightforward, since it does not involve any partial integration and holds as long as $V(x)$ is real.

Problem 4.6 (a) A stationary state is an *eigenstate* of the hamiltonian with eigenvalue E [see Eq. (4.69)]

$$H(p,x)\,\psi(x) = E\,\psi(x)$$

Show that if $H(p,x)$ is hermitian (Prob. 4.5), then the eigenvalue E is *real*;

(b) Show that the expectation value of *any* hermitian operator is a real quantity.

Solution to Problem 4.6

(a) Let us multiply the eigenvalue equation for H on the left by $\psi^\star(x)$ and integrate over x

$$\int_a^b dx \ \psi^\star(x) H(p,x)\psi(x) = E \int_a^b dx \ \psi^\star(x)\psi(x)$$

We may similarly consider the complex conjugate of the eigenvalue equation, multiply it by $\psi(x)$, and integrate over x

$$\int_a^b dx \ [H(p,x)\psi(x)]^\star \ \psi(x) = E^\star \int_a^b dx \ \psi^\star(x)\psi(x)$$

---

[5] With periodic boundary conditions, the spatial derivatives are also equal on the boundaries.

If we subtract the two equations, taking into account the hermiticity of $H(p, x)$, we have

$$(E - E^\star) \int_a^b dx \; \psi^\star(x)\psi(x) = 0$$

Since $\int_a^b dx \; \psi^\star(x)\psi(x)$ is strictly positive, this equation implies that $E^\star = E$, and hence E is real.

(b) This proof can be generalized to *any* hermitian operator O, starting from its eigenvalue equation

$$O\psi(x) = \mathcal{O}\psi(x)$$

and repeating the same steps as in part (a). The fact that $\langle O \rangle = \langle O \rangle^\star$ follows directly from the hermiticity of O.

Problem 4.7 Operators can be characterized by their *commutation relations*. The commutator of two operators A and B is defined by

$$[A, B] \equiv AB - BA \qquad ; \text{ commutator}$$

where this expression has meaning when applied to a wave function

$$[A, B]\Psi = (AB - BA)\Psi$$

One does not necessarily get the same result when these operators are applied in opposite order.

Show that the canonical commutation relation between the coordinate x and the momentum $p = (\hbar/i)\partial/\partial x$ is

$$[p, x] = \frac{\hbar}{i} \qquad ; \text{ canonical commutation relation}$$

Solution to Problem 4.7

We apply the commutator of p and x to a wave function $\Psi(x)$

$$[p, x]\Psi(x) = \frac{\hbar}{i} \left[\frac{\partial}{\partial x}, x \right] \Psi(x) = \frac{\hbar}{i} \frac{\partial}{\partial x} [x\Psi(x)] - \frac{\hbar}{i} x \frac{\partial}{\partial x} \Psi(x)$$
$$= \frac{\hbar}{i} \Psi(x)$$

Therefore the canonical commutation relation between p and x is

$$[p, x] = \frac{\hbar}{i}$$

Problem 4.8 This problem generalizes the discussion of *momentum space*. Introduce the following Fourier transform relations at a given instant in time

$$\Psi(x,t) = \frac{1}{\sqrt{2\pi}} \int_{-\infty}^{\infty} dk \, A(k,t) \, e^{ikx} \qquad ; \text{ given } t$$

$$A(k,t) = \frac{1}{\sqrt{2\pi}} \int_{-\infty}^{\infty} dx \, \Psi(x,t) \, e^{-ikx}$$

Furthermore, extend the assumption that appears above Eq. (4.55) to read

The quantity $|A(k,t)|^2 \, dk$ is the probability of finding the particle with momentum between $\hbar k$ and $\hbar k + d(\hbar k)$.

(a) Show that for a free particle

$$A(k,t) = a(k) \, e^{-i\omega_k t} \qquad ; \text{ free particle}$$

Note that in the presence of a potential this relation is no longer valid.

(b) Use the completeness relations in Prob. 4.3 to prove the following:

$$\int_{-\infty}^{\infty} |A(k,t)|^2 \, dk = \int_{-\infty}^{\infty} |\Psi(x,t)|^2 \, dx$$

$$\int_{-\infty}^{\infty} A^\star(k,t) \, \hbar k \, A(k,t) \, dk = \int_{-\infty}^{\infty} \Psi^\star(x,t) \frac{\hbar}{i} \frac{\partial \Psi(x,t)}{\partial x} \, dx$$

$$\int_{-\infty}^{\infty} A^\star(k,t) \, i \frac{\partial A(k,t)}{\partial k} \, dk = \int_{-\infty}^{\infty} \Psi^\star(x,t) \, x \, \Psi(x,t) \, dx$$

The first relation is the generalization of Parseval's theorem.

(c) Use the results in (b) to make the following identification of the operators (p, q) in coordinate space and momentum space

$$(p, q) \to \left(\frac{\hbar}{i} \frac{\partial}{\partial x}, \, x \right) \qquad ; \text{ coordinate space}$$

$$\to \left(\hbar k, \, i \frac{\partial}{\partial k} \right) \qquad ; \text{ momentum space}$$

(e) Show that the canonical commutation relation of Prob. 4.7 holds in either case

$$[p, q] = \frac{\hbar}{i} \qquad ; \text{ canonical commutation relation}$$

Thus, coordinate space and momentum space simply provide different *representations* of these commutation relations.

Solution to Problem 4.8

(a) If we substitute the expression for $\Psi(x,t)$ in the Schrödinger equation for a free particle

$$i\hbar\frac{\partial}{\partial t}\Psi(x,t) = -\frac{\hbar^2}{2m}\frac{\partial^2}{\partial x^2}\Psi(x,t)$$

and then take the inverse Fourier transform, we obtain the following differential equation

$$i\hbar\frac{\partial A(k,t)}{\partial t} = \frac{\hbar^2 k^2}{2m}A(k,t) = \hbar\omega_k A(k,t)$$

The solution to this equation is

$$A(k,t) = a(k)e^{-i\omega_k t}$$

with $a(k)$ arbitrary. In the presence of a potential this relation is no longer valid, since the inverse Fourier transform of the product $V(x)\Psi(x,t)$ complicates the analysis.

(b) We may prove these relations by direct substitution of the Fourier transform of $A(k,t)$; for example

$$\int_{-\infty}^{\infty}|A(k,t)|^2 dk = \int_{-\infty}^{\infty}dx\int_{-\infty}^{\infty}dy\,\Psi^\star(x,t)\Psi(y,t)\left[\frac{1}{2\pi}\int_{-\infty}^{\infty}dk\,e^{ik(x-y)}\right]$$

$$= \int_{-\infty}^{\infty}dx\int_{-\infty}^{\infty}dy\,\Psi^\star(x,t)\Psi(y,t)\delta(x-y)$$

$$= \int_{-\infty}^{\infty}dx\,|\Psi(x,t)|^2$$

Similarly, we may consider the second relation

$$\int_{-\infty}^{\infty}\hbar k|A(k,t)|^2 dk$$

$$= \hbar\int_{-\infty}^{\infty}dx\int_{-\infty}^{\infty}dy\,\Psi^\star(x,t)\Psi(y,t)\left[\frac{1}{2\pi}\int_{-\infty}^{\infty}dk\,k\,e^{ik(x-y)}\right]$$

$$= \hbar\int_{-\infty}^{\infty}dx\int_{-\infty}^{\infty}dy\,\Psi^\star(x,t)\Psi(y,t)i\frac{\partial}{\partial y}\left[\frac{1}{2\pi}\int_{-\infty}^{\infty}dk\,e^{ik(x-y)}\right]$$

$$= \frac{\hbar}{i}\int_{-\infty}^{\infty}dx\int_{-\infty}^{\infty}dy\,\Psi^\star(x,t)\frac{\partial\Psi(y,t)}{\partial y}\delta(x-y)$$

$$= \int_{-\infty}^{\infty}dx\,\Psi^\star(x,t)\frac{\hbar}{i}\frac{\partial\Psi(x,t)}{\partial x}$$

where the third equality is obtained with a partial integration on y.

Finally, we will now prove the last relation of point (b)

$$\int_{-\infty}^{\infty} A^\star(k,t) i \frac{\partial A(k,t)}{\partial k} dk$$

$$= \int_{-\infty}^{\infty} dx \int_{-\infty}^{\infty} dy \, \Psi^\star(x,t) y \Psi(y,t) \left[\frac{1}{2\pi} \int_{-\infty}^{\infty} dk \, e^{ik(x-y)} \right]$$

$$= \int_{-\infty}^{\infty} dx \, \Psi^\star(x,t) x \Psi(x,t)$$

(c) By looking at the second and third relations proven in part (b), we may easily obtain the representation of the momentum and of the position operators both in coordinate and momentum space

$$(p,q) \rightarrow \left(\frac{\hbar}{i} \frac{\partial}{\partial x}, x \right) \qquad \text{; coordinate space}$$

$$\rightarrow \left(\hbar k, i \frac{\partial}{\partial k} \right) \qquad \text{; momentum space}$$

(e) In Prob. 4.7 we calculated the commutator $[p, q]$ working in coordinate space. We may repeat this calculation in momentum space by picking a wave function $a(k)$ and applying the commutator to it

$$[p,q]a(k) = i\hbar \left[k, \frac{\partial}{\partial k} \right] a(k) = i\hbar k \frac{\partial}{\partial k} a(k) - i\hbar \frac{\partial}{\partial k} [ka(k)]$$

$$= \frac{\hbar}{i} a(k)$$

Therefore the canonical commutation relation between p and q is

$$[p,q] = \frac{\hbar}{i}$$

both in coordinate and momentum space.

Problem 4.9 Start from the definition of the expectation value in Prob. 4.5.

(a) Take the time derivative, and use the Schrödinger equation to establish *Ehrenfest's theorem* [recall Eq. (4.48)]

$$\frac{d}{dt} \langle O \rangle = \left\langle \frac{i}{\hbar} [H, O] \right\rangle \qquad \text{; Ehrenfest's theorem}$$

(b) Use the canonical commutation relation in Prob. 4.7 to show

$$\frac{i}{\hbar}[H, x] = \frac{p}{m} \qquad\qquad ; \ \frac{i}{\hbar}[H, p] = -\frac{\partial}{\partial x}V(x)$$

where $H = T(p) + V(x)$.

(c) Conclude from the above results that a well-localized wave packet obeys Newton's second law, thus satisfying the *correspondence principle*.

Solution to Problem 4.9

(a) In taking the time derivative of $\langle O \rangle$ we observe that the normalization of the wave function of a particle moving in a real potential $V(x)$ does not change with time (this is a consequence of the equation of continuity), and we use the Schrödinger equation to express $\partial\Psi(x, t)/\partial t$ and $\partial\Psi^*(x, t)/\partial t$ [6]

$$
\begin{aligned}
\frac{d}{dt}\langle O \rangle &= \frac{\int_a^b dx\,\{[\partial\Psi^\star(x, t)/\partial t]O\Psi(x, t) + \Psi^\star(x, t)O[\partial\Psi(x, t)/\partial t]\}}{\int_a^b dx\,\Psi^\star(x, t)\Psi(x, t)} \\
&= \frac{1}{i\hbar}\frac{\int_a^b dx\,[-(H\Psi)^\star O\Psi + \Psi^\star OH\Psi]}{\int_a^b dx\,\Psi^\star\Psi} \\
&= \frac{i}{\hbar}\frac{\int_a^b dx\,\Psi^\star[H, O]\,\Psi}{\int_a^b dx\,\Psi^\star\Psi} \\
&= \left\langle \frac{i}{\hbar}[H, O] \right\rangle
\end{aligned}
$$

(b) We consider a general hamiltonian $H = p^2/2m + V(x)$. As we have seen in Prob. 4.8, one may work either in momentum or coordinate space, the physics being independent of the chosen representation.

To evaluate the first commutator, we work in the momentum representation and write

$$\frac{i}{\hbar}[H, x]a(p) = -\frac{1}{2m}\left[p^2, \frac{\partial}{\partial p}\right]a(p) = \frac{p}{m}a(p)$$

The same result may also be obtained without making any explicit reference to representation by using the general property of the commutator

$$[A^2, B] = A[A, B] + [A, B]A$$

---

[6] It is assumed here that the operator O has no explicit dependence on the time t, and that H is hermitian.

which yields

$$[p^2, x] = \frac{2\hbar}{i} p$$

Similar arguments may be applied to calculate the second commutator. An evaluation of the commutator in coordinate space gives

$$\frac{i}{\hbar}[H, p]\psi(x) = [V(x), \frac{\partial}{\partial x}]\psi(x) = -\frac{\partial V}{\partial x}\psi(x)$$

This same result can be obtained using the following general property of the commutator[7]

$$[A^n, B] = \sum_{k=1}^{n} A^{n-k} [A, B] A^{k-1}$$

With $A = x$ and $B = p$, together with the canonical commutation relation, this implies

$$[x^n, p] = in\hbar x^{n-1}$$

For a general potential $V(x) = \sum_{n=0}^{\infty} v_n x^n / n!$ one then has

$$\frac{i}{\hbar}[V(x), p] = -\sum_{n=1}^{\infty} \frac{v_n}{(n-1)!} x^{n-1} = -\frac{\partial V}{\partial x}$$

which confirms our previous result.

(c) We apply Ehrenfest's theorem, using first $O = p$ and then $O = x$, to obtain the following equations

$$\frac{d}{dt}\langle p \rangle = \left\langle -\frac{\partial V}{\partial x} \right\rangle$$

$$\frac{d}{dt}\langle x \rangle = \frac{1}{m}\langle p \rangle$$

If the time derivative of the second equation is substituted into the first, one obtains

$$m\frac{d^2}{dt^2}\langle x \rangle = \left\langle -\frac{\partial V}{\partial x} \right\rangle$$

For a well-localized wave packet we may approximate

$$\left\langle -\frac{\partial V}{\partial x} \right\rangle \approx -\frac{\partial V(\langle x \rangle)}{\partial \langle x \rangle}$$

---

[7] Just write out the sum on the r.h.s.

which, when substituted back in the previous equation, yields Newton's second law. Here $\langle x \rangle$ plays the role of the classical coordinate of a particle of mass m, moving under the effect of a potential V.

Problem 4.10[8] (a) Prove Schwarz's inequality, which for two complex functions (f, g) states that[9]

$$\int |g(x)|^2 dx \cdot \int |f(x)|^2 dx \geq \left| \int g^\star(x) f(x)\, dx \right|^2$$

(b) Define the mean-square deviations of (x, p) for a localized wave packet as (see Prob. 4.5)

$$(\Delta x)^2 \equiv \langle (x - \langle x \rangle)^2 \rangle$$
$$(\Delta p)^2 \equiv \langle (p - \langle p \rangle)^2 \rangle$$

Use Schwarz's inequality, the hermiticity of (x, p), and the reality of $(\langle x \rangle, \langle p \rangle)$, to prove

$$(\Delta x)^2 (\Delta p)^2 \geq \frac{1}{4} \left| \langle [p, x] \rangle \right|^2$$

(c) Now use the canonical commutation relation in Prob. 4.7 to establish the rigorous form of the uncertainty principle for a localized wave packet

$$\Delta x \cdot \Delta p \geq \frac{\hbar}{2} \qquad ; \text{ uncertainty principle}$$

Solution to Problem 4.10

(a) To simplify the discussion, we introduce the definitions

$$\alpha \equiv \int |f(x)|^2 dx \quad ; \beta \equiv \int |g(x)|^2 dx \quad ; \gamma \equiv \int g^\star(x) f(x) dx$$

where α and β are real, positive numbers, while γ is complex.

---

[8]Problems 4.10 and 4.11 are more difficult, but they are also very instructive.
[9]*Hints*: Start from

$$\int \left| f(x) - g(x) \frac{\int g^\star(y) f(y)\, dy}{\int |g(y)|^2 dy} \right|^2 dx \geq 0$$

Also, in (c) use

$$|a|^2 + |b|^2 \equiv \frac{1}{2}\left[|a - b|^2 + |a + b|^2 \right] \geq \frac{1}{2} |a - b|^2$$

We use this notation to write

$$\int \left| f(x) - g(x)\frac{\int g^*(y)f(y)dy}{\int |g(y)|^2\,dy} \right|^2 dx = \int \left| f(x) - \frac{\gamma}{\beta}\,g(x) \right|^2 dx$$

$$= \int |f(x)|^2\,dx - \frac{\gamma}{\beta}\int f^*(x)g(x)dx$$

$$- \frac{\gamma^*}{\beta}\int g^*(x)f(x)dx + \frac{|\gamma|^2}{\beta^2}\int |g(x)|^2\,dx$$

$$= \alpha - \frac{|\gamma|^2}{\beta} \geq 0$$

Since $\beta > 0$, both sides of this inequality can be multiplied by β to obtain

$$\alpha\beta \geq |\gamma|^2$$

which is precisely Schwarz's inequality.

(b) As we have seen in Prob. 4.5, $\langle O \rangle$ indicates the expectation value of the operator O in a state described by a wave function $\Psi(x,t)$

$$\langle O \rangle = \frac{\int_a^b dx\, \Psi^*(x,t)O(p,x)\Psi(x,t)}{\int_a^b dx\, \Psi^*(x,t)\Psi(x,t)}$$

The mean-square-deviation of O in a state described by $\Psi(x,t)$ is then

$$(\Delta O)^2 \equiv \left\langle (O - \langle O \rangle)^2 \right\rangle = \frac{\int_a^b dx\, \Psi^*(x,t)\,(O - \langle O \rangle)^2\,\Psi(x,t)}{\int_a^b dx\, \Psi^*(x,t)\Psi(x,t)}$$

which can be re-cast in the form

$$(\Delta O)^2 = \langle O^2 \rangle - \langle O \rangle^2$$

Now assume the wave function is normalized so that $\int_a^b dx\, \Psi^*\Psi = 1$. Then with the following identification

$$f(x) \equiv O\Psi(x) \qquad ;\ g(x) \equiv \Psi(x)$$

we can use Schwarz's inequality to obtain the relation[10]

$$\langle O^2 \rangle \geq \langle O \rangle^2$$

This implies

$$(\Delta O)^2 \geq 0$$

---

[10]Recall that the operator O is hermitian. The equality holds when $\Psi(x)$ is an eigenfunction of O.

We will now use Schwarz's inequality a second time, identifying

$$f(x) \equiv (p - \langle p \rangle)\Psi(x) \qquad ; \; g(x) \equiv (x - \langle x \rangle)\Psi(x)$$

With the use of the hermiticity of p and x, we see that

$$\int |f(x)|^2 dx = (\Delta p)^2 \qquad ; \; \int |g(x)|^2 dx = (\Delta x)^2$$

which confirm our previous finding that the mean-square-deviation of a given operator O must be either positive or null. Schwarz's inequality then reads

$$(\Delta x)^2 (\Delta p)^2 \geq \left| \int \Psi^\star(x) \; (x - \langle x \rangle) \; (p - \langle p \rangle)\Psi(x) dx \right|^2$$
$$= |\langle xp \rangle - \langle x \rangle \langle p \rangle|^2$$

Observe that since x and p are hermitian operators, their expectation values must be real numbers.

On the other hand we may write

$$\langle xp \rangle = \frac{1}{2}\langle xp + px \rangle + \frac{1}{2}\langle [x, p] \rangle$$

We may easily convince ourselves that $\langle xp + px \rangle$ is real, since $\langle px \rangle = \langle xp \rangle^\star$. The commutator $[p, x]$, on the other hand, is an imaginary quantity. Therefore, the Schwarz inequality can be written as

$$(\Delta x)^2 (\Delta p)^2 \geq \left(\frac{\langle xp \rangle + \langle xp \rangle^\star}{2} - \langle x \rangle \langle p \rangle \right)^2 + \left| \frac{1}{2}\langle [x, p] \rangle \right|^2$$
$$\geq \frac{1}{4} |\langle [x, p] \rangle|^2$$

where the final inequality is clearly satisfied by the preceding expression.

(c) If the canonical commutation relation is substituted in this inequality, one obtains

$$(\Delta x)^2 (\Delta p)^2 \geq \frac{\hbar^2}{4}$$

or, equivalently

$$\Delta x \, \Delta p \geq \frac{\hbar}{2}$$

This is the rigorous form of the Heisenberg uncertainty principle for a localized packet.

Problem 4.11 In contrast to the classical string problem where a small-amplitude pulse propagates without change in shape, a quantum wave packet satisfying the Schrödinger equation *spreads* with time. To illustrate this, consider a specific form of the Fourier amplitude in Eq. (4.20)[11]

$$a(k) = \exp\left\{-\alpha^2(k - k_0)^2\right\} \qquad ; \text{ minimal packet}$$

$$\Psi(x,t) = \int_{-\infty}^{\infty} dk\, a(k) \exp\left\{i\left(kx - \omega_k t\right)\right\}$$

(a) Complete the square in the exponent in the Fourier transform, and show that[12]

$$\Psi(x,0) = \frac{\sqrt{\pi}}{\alpha} \exp\left\{ik_0 x - \frac{x^2}{4\alpha^2}\right\}$$

$$|\Psi(x,0)|^2 = \frac{\pi}{\alpha^2} \exp\left\{-\frac{x^2}{2\alpha^2}\right\}$$

(b) Now repeat the calculation for finite time in $\Psi(x,t)$, and show

$$|\Psi(x,t)|^2 = \frac{\pi}{\alpha\alpha_t} \exp\left\{-\frac{(x - v_{\mathrm{gp}} t)^2}{2\alpha_t^2}\right\}$$

$$v_{\mathrm{gp}} = \frac{\hbar k_0}{m} = \frac{p_0}{m} \qquad ; \ \alpha_t^2 = \alpha^2 + \frac{\hbar^2 t^2}{4m^2\alpha^2}$$

Solution to Problem 4.11

(a) The exponent in the Fourier transform at $t = 0$ may be simplified by completing the square

$$-\alpha^2(k - k_0)^2 + ikx = -\alpha^2(k - k_0)^2 + i(k - k_0)x + ik_0 x$$

$$= -\left[\alpha(k - k_0) - i\frac{x}{2\alpha}\right]^2 - \frac{x^2}{4\alpha^2} + ik_0 x$$

---

[11] Although we leave the demonstration of the fact to the dedicated reader, at $t = 0$ this *minimal packet* actually achieves the lower bound in the uncertainty relation in Prob. 4.10(c); here α is real.

[12] Make use of the following definite integral

$$\int_{-\infty+iy}^{\infty+iy} dz\, e^{-\alpha^2 z^2} = \int_{-\infty}^{\infty} dx\, e^{-\alpha^2 x^2} = \frac{\sqrt{\pi}}{\alpha}$$

You will learn later in your career how to evaluate such an integral, as well as that for complex α required in part (b), using complex variable techniques (see [Fetter and Walecka (2003)]). For now, just accept the definite integrals, or go on to Prob. 4.12.

Therefore

$$\Psi(x,0) = \int_{-\infty}^{\infty} dk \, \exp\left\{-\left[\alpha(k-k_0) - i\frac{x}{2\alpha}\right]^2 - \frac{x^2}{4\alpha^2} + ik_0 x\right\}$$

With the change of variable $z \equiv k - k_0 - ix/2\alpha^2$ the integral reduces into the form specified in the footnote

$$\Psi(x,0) = \int_{-\infty-ix/2\alpha^2}^{\infty-ix/2\alpha^2} dz \, \exp\left\{-\alpha^2 z^2 - \frac{x^2}{4\alpha^2} + ik_0 x\right\}$$

$$= \frac{\sqrt{\pi}}{\alpha} \exp\left\{ik_0 x - \frac{x^2}{4\alpha^2}\right\}$$

The corresponding initial probability density is

$$|\Psi(x,0)|^2 = \frac{\pi}{\alpha^2} \exp\left\{-\frac{x^2}{2\alpha^2}\right\}$$

****

(*Aside*) Although not required in the problem, let us verify that this wave packet achieves the lower bound of the uncertainty relation at $t = 0$. First we observe that $\langle x \rangle = 0$, since its integrand is an odd function of x integrated over a symmetric interval. We evaluate the remaining integrals using the expressions for the gaussian integrals given in Eqs. (2.65) in the text

$$\int_{-\infty}^{\infty} dx |\Psi(x,0)|^2 = \frac{\pi}{\alpha^2} \int_{-\infty}^{\infty} e^{-x^2/2\alpha^2} dx = \frac{\sqrt{2\pi^3}}{\alpha}$$

$$\int_{-\infty}^{\infty} dx \Psi^\star(x,0) x^2 \Psi(x,0) = \frac{\pi}{\alpha^2} \int_{-\infty}^{\infty} x^2 e^{-x^2/2\alpha^2} dx = \sqrt{2\pi^3}\,\alpha$$

The uncertainty in position therefore reads

$$(\Delta x)^2 = \frac{\int_{-\infty}^{\infty} dx \, \Psi^\star(x,0) x^2 \Psi(x,0)}{\int_{-\infty}^{\infty} dx \, |\Psi(x,0)|^2} = \alpha^2$$

Similarly, we need to calculate both

$$\int_{-\infty}^{\infty} dx \Psi^\star(x,0) \frac{\hbar}{i} \frac{\partial}{\partial x} \Psi(x,0) = \int_{-\infty}^{\infty} \left(\frac{\hbar k_0 \pi}{\alpha^2} + i\frac{\hbar \pi x}{2\alpha^4}\right) e^{-x^2/2\alpha^2} dx$$

$$= \frac{\sqrt{2\pi^3}}{\alpha} \hbar k_0$$

and

$$\int_{-\infty}^{\infty} dx\, \Psi^\star(x,0) \left(-\hbar^2 \frac{\partial^2}{\partial x^2}\right) \Psi(x,0)$$

$$= -\frac{\hbar^2 \pi}{\alpha^2} \int_{-\infty}^{\infty} \left(\frac{x^2}{4\alpha^4} + \frac{k_0 x}{i\alpha^2} - k_0^2 - \frac{1}{2\alpha^2}\right) e^{-x^2/2\alpha^2} dx$$

$$= \frac{\sqrt{2\pi^3}}{\alpha} \left(\frac{\hbar^2}{4\alpha^2} + \hbar^2 k_0^2\right)$$

Notice that the terms which are linear in x in the above integrands do not contribute to the integrals by symmetry. Therefore the uncertainty in momentum reads

$$(\Delta p)^2 = \frac{\int_{-\infty}^{\infty} dx\, \Psi^\star(x,0)(-\hbar^2 \partial^2/\partial x^2)\Psi(x,0)}{\int_{-\infty}^{\infty} dx\, |\Psi(x,0)|^2}$$

$$- \left[\frac{\int_{-\infty}^{\infty} dx\, \Psi^\star(x,0)(\hbar\partial/i\partial x)\Psi(x,0)}{\int_{-\infty}^{\infty} dx\, |\Psi(x,0)|^2}\right]^2$$

$$= \left(\frac{\hbar}{2\alpha}\right)^2$$

Thus we see that the lower bound of the uncertainty relation is indeed achieved

$$\Delta x\, \Delta p = \frac{\hbar}{2}$$

****

(b) The exponent in the Fourier transform at finite t reads

$$-\left(\alpha^2 + i\frac{\hbar t}{2m}\right) k^2 + 2\left(\alpha^2 k_0 + i\frac{x}{2}\right) k - \alpha^2 k_0^2 \equiv -ak^2 + 2bk + c$$

where the coefficients a, b and c may easily be read off. We may then complete the square as

$$-ak^2 + 2bk + c = -a\left(k - \frac{b}{a}\right)^2 + \frac{b^2}{a} + c$$

The Fourier transform is then evaluated as[13]

$$\Psi(x,t) = \sqrt{\frac{\pi}{a}} \exp\left\{\frac{b^2}{a} + c\right\}$$

---

[13]See the previous footnote.

Substitution of the expressions for a, b and c reduces the exponent to

$$\frac{b^2}{a} + c = -\alpha^2 k_0^2 + \frac{(2\alpha^2 k_0 + ix)^2}{4(\alpha^2 + i\hbar t/2m)}$$

$$= -\alpha^2 k_0^2 + \frac{(2\alpha^2 k_0 + ix)^2 (\alpha^2 - i\hbar t/2m)}{4\alpha^4 + \hbar^2 t^2/m^2}$$

Notice that the imaginary term in this expression will not appear in the probability density, since it gives rise to a pure phase in the amplitude. The real part of this expression is given by

$$\mathrm{Re}\left(\frac{b^2}{a} + c\right) = -\frac{\alpha^2(\hbar k_0 t - mx)^2}{4\alpha^4 m^2 + \hbar^2 t^2}$$

$$= -\frac{(x - v_{\mathrm{gp}} t)^2}{4\alpha_t^2}$$

To complete the calculation we need to simplify the factor

$$\frac{\pi}{a} = \frac{2\pi m}{2\alpha^2 m + i\hbar t} = \frac{4\pi m^2 (\alpha^2 - i\hbar t/2m)}{4\alpha^4 m^2 + \hbar^2 t^2}$$

Since we are interested in the probability density, it is sufficient to calculate the norm of this expression

$$\left|\sqrt{\frac{\pi}{a}}\right|^2 = \left|\frac{\pi}{a}\right| = \frac{\pi}{\alpha \alpha_t}$$

With the use of the above results, we finally arrive at

$$|\Psi(x, t)|^2 = \frac{\pi}{\alpha \alpha_t} \exp\left\{-\frac{(x - v_{\mathrm{gp}} t)^2}{2\alpha_t^2}\right\}$$

Problem 4.12 (a) Sketch the behavior of the probability density for the packet in Prob. 4.11(b) as a function of time. Notice that while it indeed moves with the group velocity, the additional terms in the second of Eqs. (4.24) lead to a broadening of the packet as it moves along.

(b) Take α_t^2 as a measure of the width of the packet in coordinate space

$$(\Delta x)^2 \sim \alpha_t^2 \sim \alpha^2 + \frac{\hbar^2 t^2}{4m^2 \alpha^2}$$

Minimize this expression with respect to α^2 and show that the best one can do to localize the particle at the time t, if it starts off as a free particle at

$t = 0$, is

$$(\Delta x)^2_{\min} = \frac{\hbar t}{m}$$

Evaluate the quantity $\hbar/m$ for the following masses: $1\,\mathrm{gm}, m_p, m_e$. Discuss.

Solution to Problem 4.12

(a) The probability density for the wave packet of Prob. 4.11(b) is a gaussian, whose width increases with time because of the dependence of α_t on the time; moreover, the factor $\pi/\alpha\alpha_t$ in the probability density makes the peak decrease with time.

It is useful to define a time scale t_0 as the time required for the wave packet to have its width doubled

$$t_0 = \frac{2\sqrt{3}m\alpha^2}{\hbar}$$

Furthermore, to facilitate our discussion, we give arbitrary values to the physical constants and set $\hbar = m = \alpha = k_0 = 1$.

At the time t_0 we have $\alpha_{t_0} = 2\alpha$, and the maximum value of the probability density is $\pi/2\alpha^2$ (half the value at $t = 0$). With the arbitrary values chosen, we have $t_0 = 2\sqrt{3} \approx 3.46$. The probability density $\rho(x,t) \equiv |\Psi(x,t)|^2$ at the times $t = 0$, $t = t_0$, $t = 2t_0$ and $t = 3t_0$ is shown in Fig. 4.1.

(b) The minimum of $\alpha_t^2(\alpha^2)$ corresponds to the value of α^2 for which the derivative vanishes

$$\frac{\partial \alpha_t^2}{\partial \alpha^2} = 1 - \frac{\hbar^2 t^2}{4m^2\alpha^4} = 0$$

This gives

$$\alpha^2 = \frac{\hbar t}{2m} \quad ; \text{ for minimum}$$

$$(\alpha_t^2)_{\min} = \frac{\hbar t}{m}$$

Thus

$$(\Delta x)^2_{\min} = \frac{\hbar t}{m}$$

Let us evaluate the quantity $\hbar/m$ for $m = 1\,\text{gm}$, m_p, and m_e

$$\frac{\hbar}{m} = 1.06 \times 10^{-31}\,\frac{\text{m}^2}{\text{s}} \qquad ; m = 1\,\text{gm}$$

$$= 6.31 \times 10^{-8}\,\frac{\text{m}^2}{\text{s}} \qquad ; m = m_p$$

$$= 1.16 \times 10^{-4}\,\frac{\text{m}^2}{\text{s}} \qquad ; m = m_e$$

Thus while the spread in the wave packet is completely negligible for a rifle bullet, it is of the order of a centimeter after one second for an electron.

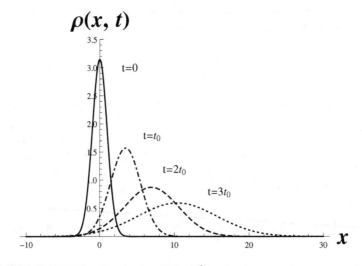

Fig. 4.1   Probability density $\rho(x,t) \equiv |\Psi(x,t)|^2$ at the times $t = 0$, $t = t_0$, $t = 2t_0$, and $t = 3t_0$ for the minimal packet.

Problem 4.13 Make a good plot, to scale, of the energy levels, wave functions, and probability densities (absolute squares of the wave functions) for the first four energy levels of a particle in a one-dimensional box of length L.

Solution to Problem 4.13

The energy levels of a particle in a one-dimensional box are

$$E_n = \frac{\hbar^2 \pi^2 n^2}{2mL^2} \qquad ; n = 1, 2, \cdots$$

and its wave functions are

$$\psi_n(x) = \sqrt{\frac{2}{L}} \sin \frac{n\pi x}{L} \qquad ; n = 1, 2, \cdots$$

where $0 \leq x \leq L$ and $\psi_n(0) = \psi_n(L) = 0$.

In Fig. 4.2 we plot the ratios $E_n/E_1 = n^2$ for the first four states of a particle in a one-dimensional box of length L, and in Fig. 4.3 we plot the re-scaled wave functions $\sqrt{L/2}\,\psi_n(x) = \sin(n\pi x/L)$ for these states.

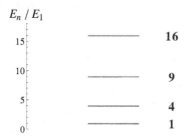

Fig. 4.2   Ratios $E_n/E_1 = n^2$ for the first four states of a particle in a one-dimensional box of length L.

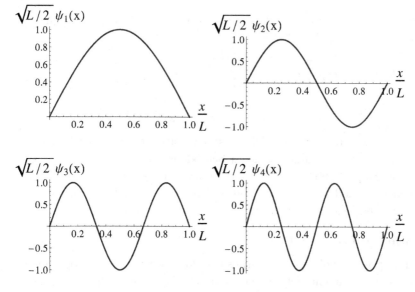

Fig. 4.3   First four wave functions of a particle in a one-dimensional box of length L.

In Fig. 4.4 we plot the corresponding re-scaled probability densities $(L/2)\rho_n(x) \equiv (L/2)|\psi_n(x)|^2$.

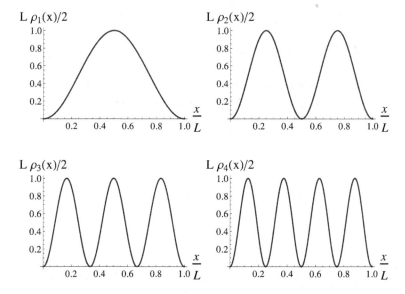

Fig. 4.4 Probability densities corresponding to the first four wave functions of a particle in a one-dimensional box of length L.

Problem 4.14 Consider the following wave function for a particle in a one-dimensional box of length L

$$\Psi(x,t) = a_1\sqrt{\frac{2}{L}}\,e^{-i\omega_1 t}\sin\frac{\pi x}{L} + a_2\sqrt{\frac{2}{L}}\,e^{-i\omega_2 t}\sin\frac{2\pi x}{L}$$

$$a_1 = a_2 = \frac{1}{\sqrt{2}} \qquad ; \quad \hbar\omega_n = \frac{\hbar^2\pi^2}{2mL^2}\,n^2$$

(a) Use the general principle of linear superposition to demonstrate that this is a solution to the time-dependent Schrödinger equation;

(b) Plot the probability density at the initial time $t = 0$;

(c) Calculate and plot the probability density at the subsequent times $(\omega_2 - \omega_1)t = m\pi/2$ for $m = 1, 2, 3, 4$.

Solution to Problem 4.14

(a) The principle of linear superposition follows from the observation that the time-dependent Schrödinger equation is a linear equation and therefore a linear combination of independent solutions is still a solution. For instance, let $\Psi_a(x,t)$ and $\Psi_b(x,t)$ be distinct solutions of the Schrödinger equation

$$i\hbar\frac{\partial\Psi_a(x,t)}{\partial t} = -\frac{\hbar^2}{2m}\frac{\partial^2\Psi_a(x,t)}{\partial x^2} + V(x)\Psi_a(x,t)$$

$$i\hbar\frac{\partial\Psi_b(x,t)}{\partial t} = -\frac{\hbar^2}{2m}\frac{\partial^2\Psi_b(x,t)}{\partial x^2} + V(x)\Psi_b(x,t)$$

Then the wave function obtained from the linear combination $\Psi(x,t) = c_a\Psi_a(x,t) + c_b\Psi_b(x,t)$ is still a solution of the Schrödinger equation

$$i\hbar\frac{\partial\Psi(x,t)}{\partial t} = i\hbar c_a\frac{\partial\Psi_a(x,t)}{\partial t} + i\hbar c_b\frac{\partial\Psi_b(x,t)}{\partial t}$$
$$= c_a H\Psi_a(x,t) + c_b H\Psi_b(x,t) = H\Psi(x,t)$$

Here $H \equiv -(\hbar^2/2m)\partial^2/\partial x^2 + V(x)$ is the hamiltonian, and $c_{a,b}$ are complex numbers with $|c_a|^2 + |c_b|^2 = 1$.

Now notice that the wave function given in the problem is indeed the linear combination of two independent solutions to the Schrödinger equation for a particle in a one-dimensional box, with $V(x) = 0$ for $0 < x < L$ and $V(x) = \infty$ for $x \leq 0$ or $x \geq L$

$$\Psi_a(x,t) \equiv \sqrt{\frac{2}{L}}\,e^{-i\omega_1 t}\sin\frac{\pi x}{L}$$

$$\Psi_b(x,t) \equiv \sqrt{\frac{2}{L}}\,e^{-i\omega_2 t}\sin\frac{2\pi x}{L}$$

where

$$i\hbar\frac{\partial\Psi_{a,b}(x,t)}{\partial t} = -\frac{\hbar^2}{2m}\frac{\partial^2\Psi_{a,b}(x,t)}{\partial x^2}$$

and $c_a = c_b = 1/\sqrt{2}$.

(b) In Fig. 4.5 we plot the probability density at the initial time, multiplied by L.

(c) The probability density at the time t is given by

$$\rho(x,t) = |\Psi(x,t)|^2$$
$$= \frac{1}{L}\left|\sin\left(\frac{\pi x}{L}\right) + e^{i(\omega_1 - \omega_2)t}\sin\left(\frac{2\pi x}{L}\right)\right|^2$$
$$= \frac{1}{L}\left[\sin^2\left(\frac{\pi x}{L}\right) + \sin^2\left(\frac{2\pi x}{L}\right) + 2\cos\left[(\omega_2 - \omega_1)t\right]\sin\left(\frac{\pi x}{L}\right)\sin\left(\frac{2\pi x}{L}\right)\right]$$

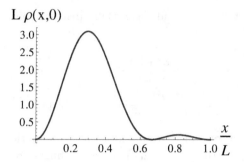

Fig. 4.5   Probability density multiplied by L at the initial time in Prob. 4.14.

At the times $t_m = m\pi/2(\omega_2 - \omega_1)$ this reduces to

$$\rho(x, t_m) = |\Psi(x, t_m)|^2$$
$$= \frac{1}{L}\sin^2\left(\frac{\pi x}{L}\right)\left[4\cos\left(\frac{\pi m}{2}\right)\cos\left(\frac{\pi x}{L}\right) + 2\cos\left(\frac{2\pi x}{L}\right) + 3\right]$$

In Fig. 4.6 we plot the probability density multiplied by L, evaluated at the different times t_m with $m = 1, 2, 3, 4$.

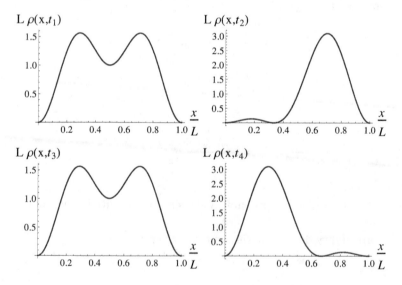

Fig. 4.6   Probability density at the times $t = t_1, t_2, t_3, t_4$.

Notice how the probability density sloshes back and forth in the box with period $\tau = 2\pi/(\omega_2 - \omega_1)$.

Problem 4.15 Consider a potential which vanishes for $x < 0$ and is $V = -V_0$ for $x \geq 0$, as illustrated in Fig. 4.7.

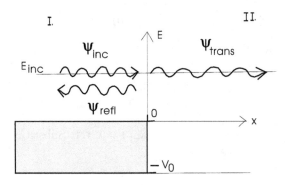

Fig. 4.7   Scattering state in a potential which vanishes for $x < 0$ and is $V = -V_0$ for $x \geq 0$.

A particle of mass m is prepared in an incident plane wave state, with momentum $\hbar k$ and energy $E_{\text{inc}} \equiv E = \hbar^2 k^2/2m$, far to the left of the potential drop. In addition to the incident wave there will be both reflected and transmitted waves

$$\psi_{\text{inc}} = e^{ikx} \qquad\qquad ; \; k^2 = \frac{2mE}{\hbar^2}$$

$$\psi_{\text{refl}} = r\,e^{-ikx}$$

$$\psi_{\text{trans}} = t\,e^{i\kappa x} \qquad ; \; \kappa^2 = \frac{2m}{\hbar^2}(E + V_0)$$

(a) Write the boundary conditions at $x = 0$ in terms of (r, t, k, κ);

(b) Determine r and t;

(c) Compute the transmission and reflection coefficients of Eqs. (4.84);

(d) Show $\mathcal{T} + \mathcal{R} = 1$, and interpret this result;

(e) Show that as $V_0 \to \infty$ at fixed E, then $\mathcal{T} \to 0$. Hence conclude that whereas a particle will always fall into a potential well in classical mechanics, it will never fall into the well in quantum mechanics if only the well is deep enough![14]

---

[14]This effect is actually observed in the scattering of low-energy neutrons from the deep, attractive nuclear potential.

Solution to Problem 4.15

(a) The wave function and its derivative must be continuous at $x = 0$

$$\psi_{\text{inc}}(0) + \psi_{\text{refl}}(0) - \psi_{\text{trans}}(0) = 0$$
$$\psi'_{\text{inc}}(0) + \psi'_{\text{refl}}(0) - \psi'_{\text{trans}}(0) = 0$$

These equations imply

$$1 + r - t = 0$$
$$ik - ikr - i\kappa t = 0$$

(b) The first equation is solved for t to obtain $t = r + 1$. Substitution into the second equation then gives

$$r = \frac{k - \kappa}{k + \kappa} \qquad ; t = \frac{2k}{k + \kappa}$$

(c) With the use of Eqs. (4.84), we have

$$\mathcal{T} = \frac{\kappa}{k}|t|^2 = \frac{4k\kappa}{(k + \kappa)^2}$$
$$\mathcal{R} = |r|^2 = \left(\frac{k - \kappa}{k + \kappa}\right)^2$$

(d) We easily see that

$$\mathcal{T} + \mathcal{R} = \frac{4k\kappa}{(k + \kappa)^2} + \left(\frac{k - \kappa}{k + \kappa}\right)^2 = 1$$

This equation expresses the conservation of probability: the particle can either be reflected or transmitted by the potential.

(e) It is convenient to express $\mathcal{T}$ in terms of the ratio E/V_0:

$$\mathcal{T} = \frac{4\sqrt{E/V_0}\sqrt{1 + E/V_0}}{\left(\sqrt{E/V_0} + \sqrt{1 + E/V_0}\right)^2}$$

$\mathcal{T}$ and $\mathcal{R}$ are plotted as a function of E/V_0 in Fig. 4.8.

If we keep E fixed and let $V_0 \to \infty$, we may approximate

$$\mathcal{T} \approx 4\sqrt{\frac{E}{V_0}}$$

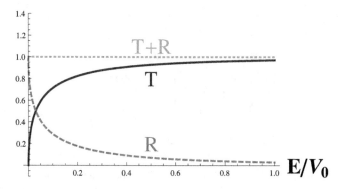

Fig. 4.8 Transmission and reflection coefficients for a particle incident with energy E over a potential $V(x) = 0$ for $x < 0$ and $V(x) = -V_0$ for $x \geq 0$.

Thus we conclude that the transmission coefficient of the potential will always tend to zero if V_0 is large enough (the conservation of probability then requires that the reflection coefficient will tend to one in this limit).

It is interesting to notice that for

$$\frac{E}{V_0} = \frac{3}{4\sqrt{2}} - \frac{1}{2} \approx 0.03$$

the probability of reflection or transmission by the particle are equal: $\mathcal{T} = \mathcal{R} = 1/2$. Thus, for $E \ll 0.03V_0$, there is a larger probability of reflecting the particle at the potential drop.

Problem 4.16 A particle of mass m is prepared in a scattering state and sent against a potential barrier of height $V_0 > E_{\text{inc}}$ as discussed in the text and illustrated in Fig. 4.18 in the text, only now the potential is of a finite thickness d (see Fig. 4.9). Now instead of just two distinct regions, there are three, and the scattering wave function has the form

$$\psi_{\text{I}} = e^{ikx} + r\,e^{-ikx} \qquad ; x < 0 \qquad\qquad ; k = \left(2mE/\hbar^2\right)^{1/2}$$

$$\psi_{\text{II}} = a\,e^{-\kappa x} + b\,e^{\kappa x} \qquad ; 0 < x < d \qquad ; \kappa = \left[2m(V_0 - E)/\hbar^2\right]^{1/2}$$

$$\psi_{\text{III}} = t\,e^{ikx} \qquad\qquad\quad ; d < x$$

(a) Write the boundary conditions at $x = 0$ and $x = d$ in terms of (r, t, a, b, k, κ). How many equations are there, and how many unknowns?[15]

(b) Solve for t. What is the transmission coefficient? Discuss.

---

[15]Note that it is now necessary to keep solutions with *both* roots for κ in the region of the potential in order to satisfy the boundary conditions. There is no argument that rules out the solution $e^{+\kappa x}$ *within the potential.*

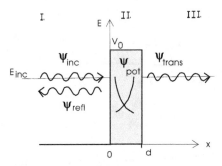

Fig. 4.9   Particle with $E_{\text{inc}} < V_0$ in a scattering state with a potential of finite thickness.

Solution to Problem 4.16

(a) Since we need to impose the continuity of the wave function and of its derivative both at $x = 0$ and at $x = d$, we have four equations

$$\psi_{\text{I}}(0) - \psi_{\text{II}}(0) = 0$$
$$\psi_{\text{I}}'(0) - \psi_{\text{II}}'(0) = 0$$
$$\psi_{\text{II}}(d) - \psi_{\text{III}}(d) = 0$$
$$\psi_{\text{II}}'(d) - \psi_{\text{III}}'(d) = 0$$

This leads to four equations in the four unknowns (r, t, a, b)

$$1 - a - b + r = 0$$
$$a\kappa - b\kappa - ikr + ik = 0$$
$$ae^{-\kappa d} + be^{\kappa d} - te^{ikd} = 0$$
$$-a\kappa e^{-\kappa d} + b\kappa e^{\kappa d} - ikte^{ikd} = 0$$

(b) If we multiply the first equation by κ, and add it to the second equation, we obtain

$$\kappa(-2b + r + 1) - ik(r - 1) = 0$$

which can be solved for b

$$b = \frac{r(k + i\kappa) - (k - i\kappa)}{2i\kappa}$$

Similarly, we multiply the third equation by κ and subtract the fourth equation

$$e^{-\kappa d}\left(2a\kappa + it(k + i\kappa)e^{(\kappa + ik)d}\right) = 0$$

which can be solved for a

$$a = \frac{t(k+i\kappa)e^{(\kappa+ik)d}}{2i\kappa}$$

In this way the parameters a, b of the wave function in the region II are expressed directly in terms of the parameters r and t of the reflected (region I) and transmitted (region II) wave functions. If we substitute these relations in the first equation and solve for r we obtain:

$$r = -\frac{k+i\kappa}{k-i\kappa}\left[te^{(\kappa+ik)d} - 1\right]$$

which also allows us to express b in terms of t:

$$b = \frac{4ik\kappa - t(k+i\kappa)^2 e^{(\kappa+ik)d}}{2i\kappa(k-i\kappa)}$$

With the use of these relations inside the third equation, we finally have an equation involving only t, which can be solved to obtain:

$$t = \frac{4ik\kappa\, e^{(\kappa-ik)d}}{e^{2d\kappa}(k+i\kappa)^2 - (k-i\kappa)^2}$$

$$= \frac{2ik\kappa\, e^{-ikd}}{(k^2-\kappa^2)\sinh(\kappa d) + 2ik\kappa\cosh(\kappa d)}$$

Substitution into the equation for r gives[16]

$$r = \frac{(k^2+\kappa^2)\sinh(\kappa d)}{(k^2-\kappa^2)\sinh(\kappa d) + 2ik\kappa\cosh(\kappa d)}$$

The transmission and reflection coefficients are then obtained as the ratios of the transmitted and reflected fluxes to the incident flux[17]

$$\mathcal{T} = |t|^2 = \frac{4k^2\kappa^2}{(k^2-\kappa^2)^2\sinh^2(\kappa d) + 4k^2\kappa^2\cosh^2(\kappa d)}$$

$$\mathcal{R} = |r|^2 = \frac{(k^2+\kappa^2)^2\sinh^2(\kappa d)}{(k^2-\kappa^2)^2\sinh^2(\kappa d) + 4k^2\kappa^2\cosh^2(\kappa d)}$$

---

[16] Although not required for the subsequent analysis, the explicit solution for (a, b) is

$$a = \frac{2k(k+i\kappa)e^{2\kappa d}}{e^{2\kappa d}(k+i\kappa)^2 - (k-i\kappa)^2}$$

$$b = -\frac{2k(k-i\kappa)}{e^{2\kappa d}(k+i\kappa)^2 - (k-i\kappa)^2}$$

[17] See the discussion for the step barrier on pp. 87-88 of the text.

Notice that $\mathcal{T} + \mathcal{R} = 1$, as required by the conservation of probability.

When the energy of the incident particle gets close to the energy of the barrier, $E \to V_0$, the transmission coefficient tends to

$$\mathrm{Lim}_{E \to V_0} \mathcal{T} = \frac{1}{1 + mV_0 d^2/2\hbar^2}$$

Here the dimensionless parameter $mV_0 d^2/2\hbar^2$ determines the deviation of $\mathcal{T}$ from the classical value of 1.

With the use of $\kappa = [2m(V_0 - E)/\hbar^2]^{1/2}$ and $k = (2mE/\hbar^2)^{1/2}$ the transmission coefficient can be expressed directly in terms of the energy of the incident particle as

$$\mathcal{T} = \frac{8E(E - V_0)}{8E^2 - 8EV_0 + V_0^2 \left[1 - \cosh\left(2d\sqrt{2m(V_0 - E)}/\hbar^2\right)\right]}$$

This formula also holds for $E \geq V_0$: in this case the parameter $\kappa = [2m(V_0 - E)/\hbar^2]^{1/2}$ becomes imaginary, and the hyperbolic function in $\mathcal{T}$ becomes a trigonometric function. As we see in Fig. 4.10, the transmission coefficient has an oscillatory behavior for $E \geq V_0$, and the potential barrier becomes fully transparent at specific energies $E_n = V_0 + \hbar^2\pi^2 n^2/2md^2$.

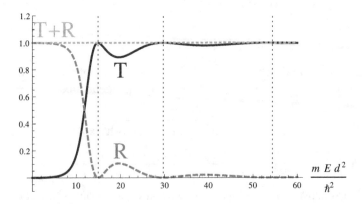

Fig. 4.10 Transmission and reflection coefficients for a particle incident with energy E for a potential barrier of finite thickness (see Fig. 4.9). This figure corresponds to $mV_0 d^2/\hbar^2 = 10$. The potential barrier becomes fully transparent at energies $E_n = V_0 + \hbar^2\pi^2 n^2/2md^2$.

Problem 4.17 The hamiltonian for the one-dimensional simple harmonic oscillator is

$$H = \frac{1}{2m}p^2 + \frac{1}{2}m\omega^2 q^2$$

where both (p, q) are hermitian and satisfy the canonical commutation relations in Prob. 4.8(e). Introduce the new operators[18]

$$a \equiv \frac{p}{(2m\hbar\omega)^{1/2}} - iq\left(\frac{m\omega}{2\hbar}\right)^{1/2}$$

$$a^\dagger \equiv \frac{p}{(2m\hbar\omega)^{1/2}} + iq\left(\frac{m\omega}{2\hbar}\right)^{1/2}$$

$$\mathcal{N} \equiv a^\dagger a$$

(a) Show the hamiltonian takes the form

$$H = \hbar\omega\frac{1}{2}(a^\dagger a + aa^\dagger)$$

(b) Derive the following properties of the new operators

$$[a, a^\dagger] = 1$$
$$[\mathcal{N}, a] = -a$$
$$[\mathcal{N}, a^\dagger] = a^\dagger$$
$$H = \hbar\omega\,(\mathcal{N} + 1/2)$$

(c) Label the eigenstates and eigenvalues of the hermitian operator $\mathcal{N}$ (H is hermitian) by

$$\mathcal{N}\psi_n = n\psi_n$$

Show from the results in part (b) that $a\psi_n$ is an eigenstate of $\mathcal{N}$ with eigenvalue lowered by 1, and $a^\dagger\psi_n$ is an eigenstate of $\mathcal{N}$ with eigenvalue raised by 1;

(d) Prove from the structure of H and the results in Prob. 4.5 that the eigenvalues of H must be non-negative. Hence, conclude that eventually the application of a must lead to a state where the eigenvalue n *vanishes*, otherwise further application of a would lead to an eigenstate and eigenvalue that violates this positivity condition. Hence there must be a *ground state*

---

[18]$(a^\dagger, a)$ are known as the *creation and destruction operators*, while $\mathcal{N}$ is the *number operator*.

where

$$a\psi_0 = 0 \qquad ; \text{ ground state}$$

(e) Since repeated application of $a^\dagger$ on the state ψ_0 raises the eigenvalue of $\mathcal{N}$ by one, conclude that the operators $(\mathcal{N}, H)$ have the following eigenvalue spectrum

$$\mathcal{N}\psi_n = n\psi_n \qquad ; n = 0, 1, 2, \cdots, \infty$$
$$H\psi_n = \hbar\omega\, (n + 1/2)\, \psi_n$$

(f) What is it in the above derivation that leads to the discrete eigenvalue spectrum for these operators?

Solution to Problem 4.17

(a) The expressions defining a and $a^\dagger$ may be easily inverted to give p and q in terms of these operators

$$p = \left(\frac{m\hbar\omega}{2}\right)^{1/2} \left(a + a^\dagger\right)$$

$$q = i \left(\frac{\hbar}{2m\omega}\right)^{1/2} \left(a - a^\dagger\right)$$

After substitution of these expressions in the hamiltonian, taking care with the relative order of the operators a and $a^\dagger$, we obtain

$$H = \frac{\hbar\omega}{4} \left(a + a^\dagger\right)^2 - \frac{\hbar\omega}{4} \left(a - a^\dagger\right)^2$$
$$= \frac{\hbar\omega}{2} \left(a^\dagger a + a a^\dagger\right)$$

(b) The first commutator is easily obtained by substituting the definitions of a and $a^\dagger$ in terms of p and q and then using the canonical commutation relation

$$[a, a^\dagger] = \frac{1}{2\hbar} \left(i\,[p, q] - i\,[q, p]\right) = 1$$

To obtain the remaining two commutators, we use the general property of the commutator

$$[AB, C] = A\,[B, C] + [A, C]\,B$$

which can be proven by using the definition of commutator on the two sides of the equation.

Thus

$$[\mathcal{N}, a] = [a^\dagger, a] \, a = -a$$
$$[\mathcal{N}, a^\dagger] = a^\dagger \, [a, a^\dagger] = a^\dagger$$

The first commutator in (b) tells us that $aa^\dagger = a^\dagger a + 1$, and therefore

$$H = \hbar\omega \left(\mathcal{N} + \frac{1}{2} \right)$$

(c) To prove that the states $a\psi_n$ and $a^\dagger\psi_n$ are eigenstates of $\mathcal{N}$, we use the last two commutators obtained in part (b)

$$\mathcal{N} a\psi_n = [\mathcal{N}, a] \, \psi_n + a\mathcal{N}\psi_n = (n-1)a\psi_n$$
$$\mathcal{N} a^\dagger\psi_n = [\mathcal{N}, a^\dagger] \, \psi_n + a^\dagger\mathcal{N}\psi_n = (n+1)a^\dagger\psi_n$$

(d) Suppose that ψ_0 is the lowest eigenstate of H, and E_0 is the lowest eigenvalue, with $H\psi_0 = E_0\psi_0$. If we multiply both sides of this equation on the left by $\psi_0^*(x)$, and integrate over all space, we obtain

$$E_0 = \frac{\int_{-\infty}^{\infty} \psi_0^*(x) H \psi_0(x) dx}{\int_{-\infty}^{\infty} |\psi_0(x)|^2 dx} = \langle H \rangle = \langle T + V \rangle$$

First we need to convince ourselves that the expectation value of the kinetic-energy operator is always a positive-definite quantity. With the use of the definition of $\langle T \rangle$, and an integration by parts, we have

$$\langle T \rangle = -\frac{\hbar^2}{2m} \frac{\int_{-\infty}^{\infty} \psi_0^*(x)(\partial^2/\partial x^2)\psi_0(x) dx}{\int_{-\infty}^{\infty} |\psi_0(x)|^2 dx}$$
$$= \frac{\hbar^2}{2m} \frac{\int_{-\infty}^{\infty} |\partial\psi_0(x)/\partial x|^2 \, dx}{\int_{-\infty}^{\infty} |\psi_0(x)|^2 dx} > 0$$

Now consider the expectation value of the potential. We assume that at $x = x_0$ the potential has a global minimum, with $V(x) \geq V_0 \equiv V(x_0)$ for all x[19]

$$\langle V \rangle = \frac{\int_{-\infty}^{\infty} \psi_0^*(x) V(x) \psi_0(x) dx}{\int_{-\infty}^{\infty} |\psi_0(x)|^2 dx}$$
$$= V_0 + \frac{\int_{-\infty}^{\infty} [V(x) - V_0] |\psi_0(x)|^2 dx}{\int_{-\infty}^{\infty} |\psi_0(x)|^2 dx} \geq V_0$$

---

[19] Classically, x_0 is a (stable) point of equilibrium.

Here the last inequality follows from $[V(x) - V_0]|\psi_0(x)|^2 \geq 0$ for all x. If the results obtained for $\langle T \rangle$ and $\langle V \rangle$ are combined, we obtain

$$\langle H \rangle > V_0$$

In the case of the simple harmonic oscillator one has $x_0 = 0$ and $V_0 = 0$, thus $E_0 > 0$, as anticipated.

The results obtained so far tell us that if ψ_n is an eigenstate of $\mathcal{N}$ with eigenvalue n, it is also eigenstate of H with eigenvalue $\hbar\omega(n + 1/2)$; moreover, the state $a\psi_n$ is also an eigenstate of $\mathcal{N}$, with eigenvalue $n - 1$, and of H with eigenvalue $\hbar\omega(n-1/2)$. If we repeatedly apply the operator a to the new state, we generate eigenstates of H with lower energy. Since we have just proven that the energy eigenvalues must be positive, we conclude that this process cannot continue indefinitely and there must be a state for which $a\psi_0$ is not a new eigenstate of H, and therefore $a\psi_0 = 0$. We call this state the ground state.

(e) With the use of the result obtained in (d), we have

$$\mathcal{N}\psi_0 = a^\dagger a\psi_0 = 0$$

which tells us that the lowest eigenvalue of $\mathcal{N}$ is zero. As we have seen, the application of $a^\dagger$ to an eigenstate of $\mathcal{N}$ generates a new eigenstate with eigenvalue raised by one. For this reason, the eigenvalues of $\mathcal{N}$ must be $n = 0, 1, 2, \cdots, \infty$. The eigenvalues of H must then be $\hbar\omega(n + 1/2)$.

(f) The discrete eigenvalue spectrum arises from the fact that the acceptable eigenfunctions $\psi_n(x)$ must die off as $x \to \pm\infty$; this enters in the proof through the assumptions that the adjoint operator $a^\dagger$ exists and that the operators H and N are hermitian.

Problem 4.18 One can actually construct the eigenstates of the harmonic oscillator in Prob. 4.17 as follows (with an appropriate choice of phase)

$$\psi_n = \frac{1}{\sqrt{n!}}(a^\dagger)^n \psi_0$$

(a) Use the hermiticity of (p, q) to show

$$\int_{-\infty}^{\infty} dq \, [a^\dagger \psi_m]^\star \, \psi_n = \int_{-\infty}^{\infty} dq \, \psi_m^\star a \, \psi_n$$

(b) Use the commutation relations and definition of the ground state in Prob. 4.17 to show that the wave function ψ_n is correctly normalized;

(c) Prove the following

$$a\psi_n = \sqrt{n}\,\psi_{n-1}$$
$$a^\dagger \psi_n = \sqrt{n+1}\,\psi_{n+1}$$

These relations for the creation and destruction operators acting on the eigenstates of the number operator play an important role in quantum mechanics.

Solution to Problem 4.18

(a) We work on the l.h.s. of the equation and use the hermiticity of p and q to obtain

$$\int_{-\infty}^{\infty} dq\, \left[a^\dagger \psi_m\right]^\star \psi_n = \int_{-\infty}^{\infty} dq\, \left[\frac{p}{(2m\hbar\omega)^{1/2}}\psi_m + iq\left(\frac{m\omega}{2\hbar}\right)^{1/2}\psi_m\right]^\star \psi_n$$

$$= \int_{-\infty}^{\infty} dq\, \psi_m^\star \left[\frac{p}{(2m\hbar\omega)^{1/2}} - iq\left(\frac{m\omega}{2\hbar}\right)^{1/2}\right]\psi_n$$

$$= \int_{-\infty}^{\infty} dq\, \psi_m^\star a\psi_n$$

(b) It is convenient first to derive a useful identity.[20] Let A and B be two operators, and consider the commutator $[A, B^n]$ with $n = 1, 2, \cdots$

$$[A, B^n] = [A, B]B^{n-1} + BAB^{n-1} - B^n A$$
$$= [A, B]B^{n-1} + B[A, B]B^{n-2} + B^2 AB^{n-3} - B^n A$$
$$= \cdots$$
$$= \sum_{k=1}^{n} B^{k-1}[A, B]B^{n-k}$$

In our case $A = a$ and $B = a^\dagger$, and this expression simplifies to

$$[a, (a^\dagger)^n] = n(a^\dagger)^{n-1}$$

Thus in the normalization integral

$$\psi_n^\star \psi_n = \frac{1}{n!}\psi_0^\star a^n (a^\dagger)^n \psi_0 = \frac{1}{n!}\psi_0^\star a^{n-1}\left\{[a, (a^\dagger)^n] + (a^\dagger)^n a\right\}\psi_0$$

$$= \frac{1}{n!}\psi_0^\star a^{n-1}\left\{n(a^\dagger)^{n-1}\right\}\psi_0$$

$$= \frac{1}{(n-1)!}\psi_0^\star a^{n-1}(a^\dagger)^{n-1}\psi_0 \equiv \psi_{n-1}^\star \psi_{n-1}$$

---

[20]See the solution to Prob. 4.9.

Here we have used the property $a\psi_0 = 0$ to eliminate the second term in the curly brackets in the first line.

The identity that we have obtained can now be used on $\psi_{n-1}^\star \psi_{n-1}$, and on all the further terms which will appear at each application, until completely eliminating the creation and destruction operators

$$\psi_n^\star \psi_n = \psi_{n-1}^\star \psi_{n-1} = \psi_{n-2}^\star \psi_{n-2} = \cdots = \psi_0^\star \psi_0$$

We conclude that all the states have the same normalization as the ground state.

(c) We prove the first relation using the commutator obtained in (b)

$$a\psi_n = \frac{1}{\sqrt{n!}} a(a^\dagger)^n \psi_0 = \frac{1}{\sqrt{n!}} \left[a, (a^\dagger)^n \right] \psi_0$$

$$= \sqrt{n} \frac{1}{\sqrt{(n-1)!}} (a^\dagger)^{n-1} \psi_0$$

$$= \sqrt{n}\, \psi_{n-1}$$

To prove the second relation, we only need to express the result in terms of the state ψ_{n+1}

$$a^\dagger \psi_n = \frac{1}{\sqrt{n!}} (a^\dagger)^{n+1} \psi_0 = \sqrt{n+1} \frac{1}{\sqrt{(n+1)!}} (a^\dagger)^{n+1} \psi_0$$

$$= \sqrt{n+1}\, \psi_{n+1}$$

Problem 4.19 Make a good plot, to scale, of the energy levels, wave functions, and probability densities (absolute squares of the wave functions) for the first four energy levels of a particle in a one-dimensional simple harmonic oscillator. The analytic form of the wave functions can be found in any good book on quantum mechanics, for example [Schiff (1968)].

Solution to Problem 4.19

The energy levels of the one-dimensional simple harmonic oscillator are

$$E_n = \hbar\omega \left(n + \frac{1}{2} \right) \qquad\qquad ;\, n = 0, 1, 2, \ldots$$

and its wave functions are

$$\psi_n(x) = \left(\frac{m\omega}{\hbar} \right)^{1/4} \left(\frac{1}{2^n n! \sqrt{\pi}} \right)^{1/2} e^{-\xi^2/2} h_n(\xi) \qquad ;\, \xi = \left(\frac{m\omega}{\hbar} \right)^{1/2} x$$

where $h_n(\xi)$ are the Hermite polynomials of order n.[21]

In Fig. 4.11 we plot the first four energy levels of the simple harmonic oscillator (divided by $\hbar\omega$), and in Fig. 4.12 and 4.13 we plot the wave functions and probability densities corresponding to the first four states.

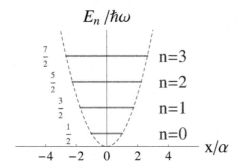

Fig. 4.11 Dimensionless energies $E_n/\hbar\omega = n + 1/2$ for the first four states of the one-dimensional simple harmonic oscillator. The dashed line is $V(x)/\hbar\omega$.

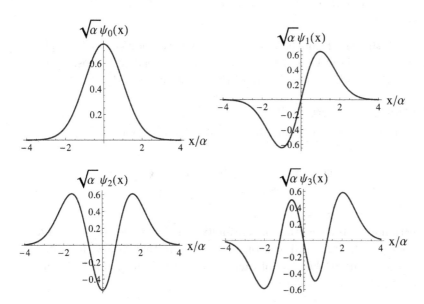

Fig. 4.12 First four wave functions of the one-dimensional simple harmonic oscillator. Here $\alpha \equiv (\hbar/m\omega)^{1/2}$.

---

[21]The first four Hermite polynomials are $h_0(\xi) = 1, h_1(\xi) = 2\xi, h_2(\xi) = 4\xi^2 - 2$, and $h_3(\xi) = 8\xi^3 - 12\xi$. Note that $\xi = (m\omega/\hbar)^{1/2}x$ is dimensionless.

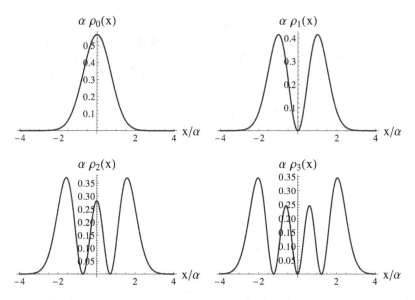

Fig. 4.13   Probability densities corresponding to the first four wave functions of the one-dimensional simple harmonic oscillator. Here $\alpha \equiv (\hbar/m\omega)^{1/2}$.

Problem 4.20 Start from the three-dimensional Schrödinger Eq. (4.109), assume a real potential V, and follow the arguments leading to Eqs. (4.46) to derive the three-dimensional form of the continuity Eq. (4.111).

Solution to Problem 4.20

We consider the three-dimensional Schrödinger equation Eq. (4.109)

$$i\hbar\frac{\partial}{\partial t}\Psi(\mathbf{x}, t) = \left[-\frac{\hbar^2}{2m}\nabla^2 + V(\mathbf{x})\right]\Psi(\mathbf{x}, t)$$

where $V(\mathbf{x}) \equiv V(x, y, z)$ is a real potential.

In section 4.3.7 the equation of continuity was derived for the case of a free particle ($V = 0$) in one dimension. In analogy to that case we interpret $|\Psi(\mathbf{x}, t)|^2 d^3x$ as the probability of finding the particle in a small volume d^3x around $\mathbf{x}$, and we calculate

$$i\hbar\frac{\partial}{\partial t}\Psi^*(\mathbf{x}, t)\Psi(\mathbf{x}, t) = \Psi^*(\mathbf{x}, t)\left[i\hbar\frac{\partial}{\partial t}\Psi(\mathbf{x}, t)\right] + \left[i\hbar\frac{\partial}{\partial t}\Psi^*(\mathbf{x}, t)\right]\Psi(\mathbf{x}, t)$$

With the use of the Schrödinger equation and its complex conjugate, this gives

$$i\hbar\frac{\partial}{\partial t}\Psi^{\star}(\mathbf{x},t)\Psi(\mathbf{x},t) = \Psi^{\star}(\mathbf{x},t)\left[-\frac{\hbar^2}{2m}\nabla^2 + V(\mathbf{x})\right]\Psi(\mathbf{x},t)$$

$$-\left\{\left[-\frac{\hbar^2}{2m}\nabla^2 + V(\mathbf{x})\right]\Psi^{\star}(\mathbf{x},t)\right\}\Psi(\mathbf{x},t)$$

where we have used the fact that V is real. The terms in V now cancel from this equation, and one is left with[22]

$$i\hbar\frac{\partial}{\partial t}\Psi^{\star}(\mathbf{x},t)\Psi(\mathbf{x},t) = -\frac{\hbar^2}{2m}\left\{\Psi^{\star}(\mathbf{x},t)\nabla^2\Psi(\mathbf{x},t) - [\nabla^2\Psi^{\star}(\mathbf{x},t)]\Psi(\mathbf{x},t)\right\}$$

$$= -\frac{\hbar^2}{2m}\nabla\cdot\left\{\Psi^{\star}(\mathbf{x},t)\nabla\Psi(\mathbf{x},t) - [\nabla\Psi^{\star}(\mathbf{x},t)]\,\Psi(\mathbf{x},t)\right\}$$

This corresponds to the three-dimensional continuity equation

$$\frac{\partial\rho}{\partial t} + \nabla\cdot\mathbf{S} = 0$$

with

$$\rho = |\Psi|^2 \qquad\qquad \text{; probability density}$$

$$\mathbf{S} = \frac{\hbar}{2im}\left[\Psi^{\star}\nabla\Psi - (\nabla\Psi^{\star})\,\Psi\right] \qquad \text{; probability current}$$

Problem 4.21 Make good to-scale plots, by any means you find the most informative (density of points, contours, *etc.*), of the probability densities for the first three stationary states of a particle of mass m in a three-dimensional box. Assume $a < b < c$ (see Fig. 4.22 in the text), and examine the states $(n_1, n_2, n_3) = (1,1,1), (1,1,2), (1,2,1)$.

Solution to Problem 4.21

We consider a three-dimensional box of sides $a = 1$, $b = 2$ and $c = 3$. The wave function for a particle in a three-dimensional box is given in Eq. (4.115)

$$\psi_{n_1,n_2,n_3}(x,y,z) = \left(\frac{8}{abc}\right)^{1/2}\sin\frac{n_1\pi x}{a}\sin\frac{n_2\pi y}{b}\sin\frac{n_3\pi z}{c}$$

In Figs. 4.14, 4.15 and 4.16 we plot the regions corresponding to a probability density $|\psi_{n_1 n_2 n_3}(x,y,z)|^2 > 1/4$ for the states with $(n_1, n_2, n_3) = (1,1,1), (1,1,2), (1,2,1)$.

---

[22]Here we use the vector relation $\nabla\cdot(a\nabla b) = \nabla a\cdot\nabla b + a\nabla^2 b$.

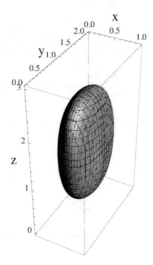

Fig. 4.14   Region corresponding to $|\psi_{111}(x, y, z)|^2 > 1/4$.

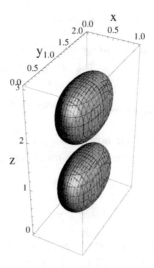

Fig. 4.15   Region corresponding to $|\psi_{112}(x, y, z)|^2 > 1/4$.

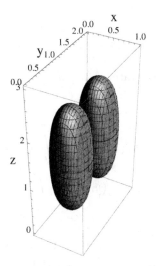

Fig. 4.16   Region corresponding to $|\psi_{121}(x, y, z)|^2 > 1/4$.

Problem 4.22 (a) Show that the eigenfunctions that are given in Eqs. (4.115) for a particle in the box in Fig. 4.22 in the text satisfy the following orthonormality relation

$$\int_0^a dx \int_0^b dy \int_0^c dz \, \psi^\star_{n_1 n_2 n_3}(\mathbf{x}) \psi_{n_1' n_2' n_3'}(\mathbf{x}) = \delta_{n_1 n_1'} \delta_{n_2 n_2'} \delta_{n_3 n_3'}$$

(b) Consider the eigenfunctions obeying periodic boundary conditions for a free particle in a cubical box of side L centered on the origin [Eqs. (4.118)]. Show that they satisfy the orthonormality relation

$$\int_{-L/2}^{L/2} \int_{-L/2}^{L/2} \int_{-L/2}^{L/2} d^3x \, \psi^\star_{\mathbf{k}}(\mathbf{x}) \psi_{\mathbf{k}'}(\mathbf{x}) = \delta_{n_x n_x'} \delta_{n_y n_y'} \delta_{n_z n_z'}$$

Here $d^3x \equiv dxdydz$, and the triple integral goes over all three components.

Solution to Problem 4.22

(a) Since the Schrödinger equation for a particle in a box is separable, its eigenfunctions, given in Eq. (4.115), are the direct product of the eigenfunctions of the one-dimensional problems along each orthogonal direction

$$\psi_{n_1 n_2 n_3}(x, y, z) = \psi_{n_1}(x, a)\psi_{n_2}(y, b)\psi_{n_3}(z, c)$$

where we use the notation

$$\psi_n(x, L) \equiv \sqrt{\frac{2}{L}} \sin \frac{n\pi x}{L}$$

for the eigenfunctions of the operator $-(\hbar^2/2m)\partial^2/\partial x^2$, vanishing on the boundaries $\psi_n(0, L) = \psi_n(L, L) = 0$.[23]

As seen in Eq. (4.74), these eigenfunctions are orthonormal, and thus

$$\int_0^a dx \int_0^b dy \int_0^c dz \, \psi_{n_1 n_2 n_3}^\star(\mathbf{x}) \psi_{n_1' n_2' n_3'}(\mathbf{x}) = \left[\int_0^a dx \, \psi_{n_1}^\star(x, a) \psi_{n_1'}(x, a) \right]$$

$$\times \left[\int_0^b dy \, \psi_{n_2}^\star(y, b) \psi_{n_2'}(y, b) \right] \times \left[\int_0^c dz \, \psi_{n_3}^\star(z, c) \psi_{n_3'}(z, c) \right]$$

$$= \delta_{n_1 n_1'} \delta_{n_2 n_2'} \delta_{n_3 n_3'}$$

(b) Once again the Schrödinger equation for a free particle in a cubical box is separable, and therefore the eigenfunctions obeying periodic boundary conditions are the direct product of the wave functions along each orthogonal direction

$$\psi_{\mathbf{k}}(\mathbf{x}) = \frac{1}{\sqrt{L^3}} e^{i\mathbf{k}\cdot\mathbf{x}} = \phi_{n_1}(x) \, \phi_{n_2}(y) \, \phi_{n_3}(z)$$

Here $\mathbf{k} \equiv (k_{n_1}, k_{n_2}, k_{n_3})$ and $\phi_n(x)$ are the eigenfunctions of the one dimensional problem, obeying periodic boundary conditions, given in Eq. (4.34)

$$\phi_n(x) = \frac{1}{\sqrt{L}} e^{i k_n x}$$

with $k_n = 2\pi n/L$ where $n = 0, \pm 1, \pm 2, \cdots$.

As seen in Eq. (4.35), these eigenfunctions are orthonormal in the interval $[-L/2, L/2]$, and therefore

$$\int_{-L/2}^{L/2} dx \int_{-L/2}^{L/2} dy \int_{-L/2}^{L/2} dz \, \psi_{\mathbf{k}}^\star(\mathbf{x}) \psi_{\mathbf{k}'}(\mathbf{x}) = \delta_{n_x n_x'} \delta_{n_y n_y'} \delta_{n_z n_z'}$$

Problem 4.23 (a) Show that the general solution to the Schrödinger equation for a particle of mass m in the box in Fig. 4.22 in the text is given in terms of the normal-mode solutions in Eqs. (4.115) and (4.116) as

$$\Psi(\mathbf{x}, t) = \sum_{n_1 n_2 n_3} A_{n_1 n_2 n_3} \, \psi_{n_1 n_2 n_3}(\mathbf{x}) \exp\left\{ -\frac{i}{\hbar} E_{n_1 n_2 n_3} t \right\}$$

---

[23]These eigenfunctions are given in Eq. (4.72), although here we have made explicit the dependence upon L.

(b) Use the results in Prob. 4.22 to determine the complex amplitudes $A_{n_1 n_2 n_3}$ in terms of the initial conditions $\Psi(\mathbf{x}, 0)$.

Solution to Problem 4.23

(a) For the particle of mass m in a box we have from Eqs. (4.115)–(4.116)

$$\psi_{n_1, n_2, n_3}(x, y, z) = \left(\frac{8}{abc}\right)^{1/2} \sin\frac{n_1 \pi x}{a} \sin\frac{n_2 \pi y}{b} \sin\frac{n_3 \pi z}{c}$$

$$\frac{2m}{\hbar^2} E_{n_1 n_2 n_3} = \left(\frac{n_1^2}{a^2} + \frac{n_2^2}{b^2} + \frac{n_3^2}{c^2}\right)\pi^2$$

It is then easy to see that the wave function $\Psi(\mathbf{x}, t)$ given above is a solution to the Schrödinger equation, since

$$i\hbar\frac{\partial}{\partial t}\Psi(\mathbf{x}, t) = \sum_{n_1 n_2 n_3} A_{n_1 n_2 n_3} E_{n_1 n_2 n_3} \psi_{n_1 n_2 n_3}(\mathbf{x}) \exp\left\{-\frac{i}{\hbar}E_{n_1 n_2 n_3}t\right\}$$

$$-\frac{\hbar^2}{2m}\nabla^2\Psi(\mathbf{x}, t) = \sum_{n_1 n_2 n_3} A_{n_1 n_2 n_3} E_{n_1 n_2 n_3} \psi_{n_1 n_2 n_3}(\mathbf{x}) \exp\left\{-\frac{i}{\hbar}E_{n_1 n_2 n_3}t\right\}$$

and therefore

$$i\hbar\frac{\partial}{\partial t}\Psi(\mathbf{x}, t) = -\frac{\hbar^2}{2m}\nabla^2\Psi(\mathbf{x}, t)$$

(b) We have

$$\Psi(\mathbf{x}, 0) = \sum_{n_1 n_2 n_3} A_{n_1 n_2 n_3} \psi_{n_1 n_2 n_3}(\mathbf{x})$$

This relation is inverted with the use of the results in Prob. 4.22, and we write

$$\int_0^a dx \int_0^b dy \int_0^c dz\, \psi^\star_{n_1' n_2' n_3'}(\mathbf{x})\Psi(\mathbf{x}, 0)$$

$$= \sum_{n_1 n_2 n_3} A_{n_1 n_2 n_3} \left[\int_0^a dx \int_0^b dy \int_0^c dz\, \psi^\star_{n_1' n_2' n_3'}(\mathbf{x})\psi_{n_1 n_2 n_3}(\mathbf{x})\right]$$

$$= \sum_{n_1 n_2 n_3} A_{n_1 n_2 n_3} \delta_{n_1 n_1'}\delta_{n_2 n_2'}\delta_{n_3 n_3'}$$

$$= A_{n_1' n_2' n_3'}$$

which expresses the coefficients $A_{n_1 n_2 n_3}$ in terms of the wave function at the initial time $\Psi(\mathbf{x}, 0)$. Note that the initial wave function $\Psi(\mathbf{x}, 0)$ must be specified at all points in space in order to determine the wave function $\Psi(\mathbf{x}, t)$ at all subsequent times.

Problem 4.24 (a) Show that the appropriate generalization to three dimensions of the completeness relation in Prob. 4.3 is

$$\frac{1}{(2\pi)^3} \int d^3k \, e^{i\mathbf{k}\cdot(\mathbf{x}-\mathbf{y})} = \delta^{(3)}(\mathbf{x} - \mathbf{y}) \qquad ; \text{ Dirac delta function}$$

where $\int d^3k$ represents a triple integral over all values of the components.[24] In this expression

$$\delta^{(3)}(\mathbf{x} - \mathbf{y}) = 0 \qquad ; \mathbf{x} \neq \mathbf{y}$$
$$\neq 0 \qquad ; \mathbf{x} = \mathbf{y}$$
$$\int d^3y \, q(\mathbf{y}) \delta^{(3)}(\mathbf{x} - \mathbf{y}) = q(\mathbf{x})$$

(b) Use the result in (a) to establish the three-dimensional Fourier transform relations

$$f(\mathbf{x}) = \frac{1}{(2\pi)^{3/2}} \int d^3k \, e^{i\mathbf{k}\cdot\mathbf{x}} \, a(\mathbf{k})$$
$$a(\mathbf{k}) = \frac{1}{(2\pi)^{3/2}} \int d^3x \, e^{-i\mathbf{k}\cdot\mathbf{x}} \, f(\mathbf{x})$$

Solution to Problem 4.24

(a) We may easily generalize the Dirac delta function to three dimensions by observing that

$$\frac{1}{(2\pi)^3} \int d^3k \, e^{i\mathbf{k}\cdot(\mathbf{x}-\mathbf{y})} = \prod_{i=1}^{3} \int \frac{dk_i}{(2\pi)} e^{ik_i(x_i - y_i)}$$

$$= \prod_{i=1}^{3} \delta(x_i - y_i) \equiv \delta^{(3)}(\mathbf{x} - \mathbf{y})$$

The properties of $\delta^{(3)}(\mathbf{x} - \mathbf{y})$ then follow directly from the analogous properties of the one-dimensional Dirac delta function

$$\delta^{(3)}(\mathbf{x} - \mathbf{y}) = 0 \qquad ; \mathbf{x} \neq \mathbf{y}$$
$$\neq 0 \qquad ; \mathbf{x} = \mathbf{y}$$
$$\int d^3y \, q(\mathbf{y}) \delta^{(3)}(\mathbf{x} - \mathbf{y}) = q(\mathbf{x})$$

---

[24] We now simply suppress the multiple integral signs.

(b) We may now use the results in (a) to establish the three-dimensional Fourier transform relations

$$a(\mathbf{k}) = \frac{1}{(2\pi)^{3/2}} \int d^3x \, e^{-i\mathbf{k}\cdot\mathbf{x}} f(\mathbf{x})$$

$$\frac{1}{(2\pi)^{3/2}} \int d^3k \, e^{i\mathbf{k}\cdot\mathbf{x}} a(\mathbf{k}) = \frac{1}{(2\pi)^3} \int d^3k \int d^3y \, e^{i\mathbf{k}\cdot(\mathbf{x}-\mathbf{y})} f(\mathbf{y})$$

$$= \int d^3y \, \delta^{(3)}(\mathbf{x}-\mathbf{y}) f(\mathbf{y}) = f(\mathbf{x})$$

Problem 4.25 It follows from the definition of the angular momentum in Eq. (4.127) that

$$L_x = \frac{1}{i}\left(y\frac{\partial}{\partial z} - z\frac{\partial}{\partial y}\right) \qquad ; \text{ and cyclic permutations of } (x, y, z)$$

(a) Prove the following commutation relations between the components of the angular momentum (see Prob. 4.7)

$$[L_x, L_y] = iL_z \qquad ; \text{ and cyclic permutations of } (x, y, z)$$

(b) The square of the total angular momentum is $\mathbf{L}^2 = L_x^2 + L_y^2 + L_z^2$. Show

$$[\mathbf{L}^2, L_i] = 0 \qquad ; i = (x, y, z)$$

(c) Show that the components of the angular momentum are hermitian operators (see Prob. 4.5).

(d) Use the result in Prob. 4.10(b) to deduce an uncertainty relation between the components of the angular momentum.

Solution to Problem 4.25

(a) We have[25]

$$L_x = \frac{1}{i}\left(y\frac{\partial}{\partial z} - z\frac{\partial}{\partial y}\right)$$

$$L_y = \frac{1}{i}\left(z\frac{\partial}{\partial x} - x\frac{\partial}{\partial z}\right)$$

$$L_z = \frac{1}{i}\left(x\frac{\partial}{\partial y} - y\frac{\partial}{\partial x}\right)$$

---

[25] As usual, the partial derivatives here imply that the other variables in the set (x, y, z) are to be kept constant.

Therefore, when operating on an arbitrary wave function,

$$[L_x, L_y] =$$

$$= -\left\{ \left(y\frac{\partial}{\partial z} - z\frac{\partial}{\partial y} \right) \left(z\frac{\partial}{\partial x} - x\frac{\partial}{\partial z} \right) - \left(z\frac{\partial}{\partial x} - x\frac{\partial}{\partial z} \right) \left(y\frac{\partial}{\partial z} - z\frac{\partial}{\partial y} \right) \right\}$$

$$= -\left\{ y\frac{\partial}{\partial x}\frac{\partial}{\partial z}z - xy\frac{\partial^2}{\partial z^2} - z^2\frac{\partial^2}{\partial x\partial y} + xz\frac{\partial^2}{\partial y\partial z} - yz\frac{\partial^2}{\partial x\partial z} + z^2\frac{\partial^2}{\partial x\partial y} \right.$$

$$\left. + xy\frac{\partial^2}{\partial z^2} - x\frac{\partial}{\partial y}\frac{\partial}{\partial z}z \right\}$$

$$= -\left\{ y\frac{\partial}{\partial x}\frac{\partial}{\partial z}z + xz\frac{\partial^2}{\partial y\partial z} - yz\frac{\partial^2}{\partial x\partial z} - x\frac{\partial}{\partial y}\frac{\partial}{\partial z}z \right\}$$

We may now use the commutator

$$\left[\frac{\partial}{\partial z}, z \right] = 1$$

to express

$$\frac{\partial}{\partial z}z = 1 + z\frac{\partial}{\partial z}$$

Thus we obtain

$$[L_x, L_y] = -\left\{ y\frac{\partial}{\partial x}\left(1 + z\frac{\partial}{\partial z} \right) - yz\frac{\partial^2}{\partial x\partial z} + xz\frac{\partial^2}{\partial y\partial z} - x\frac{\partial}{\partial y}\left(1 + z\frac{\partial}{\partial z} \right) \right\}$$

$$= -y\frac{\partial}{\partial x} + x\frac{\partial}{\partial y} = iL_z$$

In the same way we may prove

$$[L_y, L_z] = iL_x \qquad ; \quad [L_z, L_x] = iL_y$$

These commutation relations can be expressed in a compact form using the Levi-Civita antisymmetric tensor ε_{ijk} with $(i, j, k) = (1, 2, 3)$[26]

$$[L_i, L_j] = i\sum_{k=1}^{3} \varepsilon_{ijk}L_k$$

(b) In proving part (b) we will use the identity[27]

$$[A^2, B] = A[A, B] + [A, B]A$$

---

[26] $\varepsilon_{ijk} = +1$ if (i, j, k) is an even permutation of $(1, 2, 3)$, $\varepsilon_{ijk} = -1$ if (i, j, k) is an odd permutation of $(1, 2, 3)$, and $\varepsilon_{ijk} = 0$ otherwise. Here $(1, 2, 3) \equiv (x, y, z)$

[27] This is a special case of the more general result in Prob. 4.18(b).

which may be verified in a straightforward manner by explicitly writing out the commutators on the r.h.s.

Therefore, since $\mathbf{L}^2 = L_1^2 + L_2^2 + L_3^2$,

$$[\mathbf{L}^2, L_i] = \left[\sum_{j=1}^{3} L_j^2, L_i\right] = \sum_{j=1}^{3} \{L_j [L_j, L_i] + [L_j, L_i] L_j\}$$

$$= i \sum_{j=1}^{3} \sum_{k=1}^{3} \varepsilon_{jik} (L_j L_k + L_k L_j)$$

This expression clearly vanishes, since it is obtained from the contraction of a fully antisymmetric tensor (the Levi-Civita tensor) with a fully symmetric tensor (the operator $L_j L_k + L_k L_j$)

$$[\mathbf{L}^2, L_i] = 0 \qquad ; \; i = (1, 2, 3)$$

(c) As seen in Problem 4.5, an operator is said to be hermitian if it has the property

$$\int_a^b dx \, \Psi^*(x, t) O(p, x) \Psi(x, t) = \int_a^b dx \, [O(p, x) \Psi(x, t)]^* \Psi(x, t)$$

Consider the operator L_x

$$\int d^3x \, \Psi^*(\mathbf{x}, t) L_x \Psi(\mathbf{x}, t) = \frac{1}{i} \int d^3x \, \Psi^*(\mathbf{x}, t) \left(y \frac{\partial}{\partial z} - z \frac{\partial}{\partial y}\right) \Psi(\mathbf{x}, t)$$

If this expression is integrated by parts, we arrive at

$$\int d^3x \, \Psi^*(\mathbf{x}, t) L_x \Psi(\mathbf{x}, t) =$$

$$\frac{1}{i} \int dx dy \, [\Psi^*(\mathbf{x}, t) y \Psi(\mathbf{x}, t)]_{z=-\infty}^{z=\infty} - \frac{1}{i} \int dx dz \, [\Psi^*(\mathbf{x}, t) z \Psi(\mathbf{x}, t)]_{y=-\infty}^{y=\infty}$$

$$- \frac{1}{i} \int d^3x \, \Psi(\mathbf{x}, t) \left(y \frac{\partial}{\partial z} \Psi^*(\mathbf{x}, t) - z \frac{\partial}{\partial y} \Psi^*(\mathbf{x}, t)\right)$$

With the assumption that the wave functions vanish at infinite distances, the surface terms in the above expression cancel, and we are left with

$$\int d^3x \, \Psi^*(\mathbf{x}, t) L_x \Psi(\mathbf{x}, t) = \int d^3x \, [L_x \Psi(\mathbf{x}, t)]^* \Psi(\mathbf{x}, t)$$

Hence we conclude that with these boundary conditions, L_x is an hermitian operator. The proof for L_y and L_z is analogous.

(d) The result in Problem 4.10(b) expresses the mean-square-deviations of two hermitian operators (A, B) in terms of the matrix element of their commutator

$$(\Delta A)^2 (\Delta B)^2 \geq \frac{1}{4} \left| \langle [A, B] \rangle \right|^2$$

With the use of the commutation relations satisfied by the hermitian components of the angular momentum operator in part (a), we then have

$$(\Delta L_i)^2 (\Delta L_j)^2 \geq \frac{1}{4} |\langle L_k \rangle|^2 \qquad ; i \neq j \neq k$$

Problem 4.26 (a) Make a change of variables from cartesian to polar coordinates in two dimensions

$$x = \rho \cos \phi \qquad\qquad ; y = \rho \sin \phi$$

Use the chain rule for differentiating an implicit function $f(x, y) = f[x(\rho, \phi), y(\rho, \phi)] \equiv \bar{f}(\rho, \phi)$ to show that the laplacian in Eq. (4.108) is expressed in polar coordinates as [compare Eqs. (4.130)]

$$\nabla^2 = \frac{1}{\rho} \frac{\partial}{\partial \rho} \rho \frac{\partial}{\partial \rho} + \frac{1}{\rho^2} \frac{\partial^2}{\partial \phi^2}$$

(b) Repeat part (a) for the change from cartesian to spherical coordinates in three dimensions, where (see Fig. 4.26 in the text)

$$x = r \sin \theta \cos \phi \qquad ; y = r \sin \theta \sin \phi \qquad ; z = r \cos \theta$$

Show[28]

$$\nabla^2 = \frac{1}{r^2} \frac{\partial}{\partial r} r^2 \frac{\partial}{\partial r} + \frac{1}{r^2} \left[\frac{1}{\sin \theta} \frac{\partial}{\partial \theta} \sin \theta \frac{\partial}{\partial \theta} + \frac{1}{\sin^2 \theta} \frac{\partial^2}{\partial \phi^2} \right]$$

$$= \frac{1}{r} \frac{\partial^2}{\partial r^2} r + \frac{1}{r^2} \left[\frac{1}{\sin \theta} \frac{\partial}{\partial \theta} \sin \theta \frac{\partial}{\partial \theta} + \frac{1}{\sin^2 \theta} \frac{\partial^2}{\partial \phi^2} \right]$$

(c) Use the fact that the spherical harmonics satisfy the following equation (see, for example, [Fetter and Walecka (2003)])

$$- \left[\frac{1}{\sin \theta} \frac{\partial}{\partial \theta} \sin \theta \frac{\partial}{\partial \theta} + \frac{1}{\sin^2 \theta} \frac{\partial^2}{\partial \phi^2} \right] Y_{lm}(\theta, \phi) = l(l+1) Y_{lm}(\theta, \phi)$$

to derive the radial Eq. (4.142).

---

[28]The algebra here is a little more intense, but worth doing once in your life; for an alternate derivation, see appendix B of [Fetter and Walecka (2003)].

Solution to Problem 4.26

(a) We work from $f(x,y) = \bar{f}(\rho, \phi) = \bar{f}[\rho(x,y), \phi(x,y)]$. With simple algebra, it then follows from the chain rule that

$$\frac{\partial^2 f}{\partial x^2} = \frac{\partial^2 \bar{f}}{\partial \rho^2}\left(\frac{\partial \rho}{\partial x}\right)^2 + 2\frac{\partial^2 \bar{f}}{\partial \rho \partial \phi}\frac{\partial \rho}{\partial x}\frac{\partial \phi}{\partial x} + \frac{\partial^2 \bar{f}}{\partial \phi^2}\left(\frac{\partial \phi}{\partial x}\right)^2 + \frac{\partial \bar{f}}{\partial \rho}\frac{\partial^2 \rho}{\partial x^2} + \frac{\partial \bar{f}}{\partial \phi}\frac{\partial^2 \phi}{\partial x^2}$$

$$\frac{\partial^2 f}{\partial y^2} = \frac{\partial^2 \bar{f}}{\partial \rho^2}\left(\frac{\partial \rho}{\partial y}\right)^2 + 2\frac{\partial^2 \bar{f}}{\partial \rho \partial \phi}\frac{\partial \rho}{\partial y}\frac{\partial \phi}{\partial y} + \frac{\partial^2 \bar{f}}{\partial \phi^2}\left(\frac{\partial \phi}{\partial y}\right)^2 + \frac{\partial \bar{f}}{\partial \rho}\frac{\partial^2 \rho}{\partial y^2} + \frac{\partial \bar{f}}{\partial \phi}\frac{\partial^2 \phi}{\partial y^2}$$

We use the inverse relations

$$\rho = \sqrt{x^2 + y^2} \qquad ; \phi = \arctan\frac{y}{x}$$

to obtain the factors

$$\left(\frac{\partial \rho}{\partial x}\right)^2 + \left(\frac{\partial \rho}{\partial y}\right)^2 = 1 \qquad ; \frac{\partial \rho}{\partial x}\frac{\partial \phi}{\partial x} + \frac{\partial \rho}{\partial y}\frac{\partial \phi}{\partial y} = 0$$

$$\left(\frac{\partial \phi}{\partial x}\right)^2 + \left(\frac{\partial \phi}{\partial y}\right)^2 = \frac{1}{\rho^2} \qquad ; \frac{\partial^2 \phi}{\partial x^2} + \frac{\partial^2 \phi}{\partial y^2} = 0$$

$$\frac{\partial^2 \rho}{\partial x^2} + \frac{\partial^2 \rho}{\partial y^2} = \frac{1}{\rho}$$

Therefore

$$\left(\frac{\partial^2}{\partial x^2} + \frac{\partial^2}{\partial y^2}\right)f = \left(\frac{\partial^2}{\partial \rho^2} + \frac{1}{\rho}\frac{\partial}{\partial \rho} + \frac{1}{\rho^2}\frac{\partial^2}{\partial \phi^2}\right)\bar{f}$$

$$= \left(\frac{1}{\rho}\frac{\partial}{\partial \rho}\rho\frac{\partial}{\partial \rho} + \frac{1}{\rho^2}\frac{\partial^2}{\partial \phi^2}\right)\bar{f}$$

(b) The extension to three dimensions is done in the same way, but involves significantly more algebra. Here $f(x,y,z) = \bar{f}(r, \theta, \phi) = \bar{f}[r(x,y,z), \theta(x,y,z), \phi(x,y,z)]$ and

$$\nabla^2 f = \left(\frac{\partial^2}{\partial x^2} + \frac{\partial^2}{\partial y^2} + \frac{\partial^2}{\partial z^2}\right)f$$

Application of the chain rule then gives

$$
\nabla^2 f =
$$

$$
\frac{\partial^2 \bar{f}}{\partial r^2}\left[\left(\frac{\partial r}{\partial x}\right)^2 + \left(\frac{\partial r}{\partial y}\right)^2 + \left(\frac{\partial r}{\partial z}\right)^2\right] + \frac{\partial^2 \bar{f}}{\partial \theta^2}\left[\left(\frac{\partial \theta}{\partial x}\right)^2 + \left(\frac{\partial \theta}{\partial y}\right)^2 + \left(\frac{\partial \theta}{\partial z}\right)^2\right]
$$

$$
+\frac{\partial^2 \bar{f}}{\partial \phi^2}\left[\left(\frac{\partial \phi}{\partial x}\right)^2 + \left(\frac{\partial \phi}{\partial y}\right)^2 + \left(\frac{\partial \phi}{\partial z}\right)^2\right] + \frac{\partial \bar{f}}{\partial r}\left(\frac{\partial^2 r}{\partial x^2} + \frac{\partial^2 r}{\partial y^2} + \frac{\partial^2 r}{\partial z^2}\right)
$$

$$
+\frac{\partial \bar{f}}{\partial \theta}\left(\frac{\partial^2 \theta}{\partial x^2} + \frac{\partial^2 \theta}{\partial y^2} + \frac{\partial^2 \theta}{\partial z^2}\right) + \frac{\partial \bar{f}}{\partial \phi}\left(\frac{\partial^2 \phi}{\partial x^2} + \frac{\partial^2 \phi}{\partial y^2} + \frac{\partial^2 \phi}{\partial z^2}\right)
$$

$$
+2\frac{\partial^2 \bar{f}}{\partial r\partial \theta}\left(\frac{\partial r}{\partial x}\frac{\partial \theta}{\partial x} + \frac{\partial r}{\partial y}\frac{\partial \theta}{\partial y} + \frac{\partial r}{\partial z}\frac{\partial \theta}{\partial z}\right) + 2\frac{\partial^2 \bar{f}}{\partial r\partial \phi}\left(\frac{\partial r}{\partial x}\frac{\partial \phi}{\partial x} + \frac{\partial r}{\partial y}\frac{\partial \phi}{\partial y} + \frac{\partial r}{\partial z}\frac{\partial \phi}{\partial z}\right)
$$

$$
+2\frac{\partial^2 \bar{f}}{\partial \theta\partial \phi}\left(\frac{\partial \theta}{\partial x}\frac{\partial \phi}{\partial x} + \frac{\partial \theta}{\partial y}\frac{\partial \phi}{\partial y} + \frac{\partial \theta}{\partial z}\frac{\partial \phi}{\partial z}\right)
$$

We use the inverse relations to express (r, θ, ϕ) in terms of (x, y, z)

$$
r = \sqrt{x^2 + y^2 + z^2} \quad ; \; \theta = \arccos\frac{z}{\sqrt{x^2 + y^2 + z^2}} \quad ; \; \phi = \arctan\frac{y}{x}
$$

The various terms in the previous equations are then given by

$$
\left(\frac{\partial r}{\partial x}\right)^2 + \left(\frac{\partial r}{\partial y}\right)^2 + \left(\frac{\partial r}{\partial z}\right)^2 = 1 \qquad ; \; \frac{\partial r}{\partial x}\frac{\partial \theta}{\partial x} + \frac{\partial r}{\partial y}\frac{\partial \theta}{\partial y} + \frac{\partial r}{\partial z}\frac{\partial \theta}{\partial z} = 0
$$

$$
\left(\frac{\partial \theta}{\partial x}\right)^2 + \left(\frac{\partial \theta}{\partial y}\right)^2 + \left(\frac{\partial \theta}{\partial z}\right)^2 = \frac{1}{r^2} \qquad ; \; \frac{\partial r}{\partial x}\frac{\partial \phi}{\partial x} + \frac{\partial r}{\partial y}\frac{\partial \phi}{\partial y} + \frac{\partial r}{\partial z}\frac{\partial \phi}{\partial z} = 0
$$

$$
\left(\frac{\partial \phi}{\partial x}\right)^2 + \left(\frac{\partial \phi}{\partial y}\right)^2 + \left(\frac{\partial \phi}{\partial z}\right)^2 = \frac{1}{r^2 \sin^2 \theta} \quad ; \; \frac{\partial \theta}{\partial x}\frac{\partial \phi}{\partial x} + \frac{\partial \theta}{\partial y}\frac{\partial \phi}{\partial y} + \frac{\partial \theta}{\partial z}\frac{\partial \phi}{\partial z} = 0
$$

$$
\frac{\partial^2 r}{\partial x^2} + \frac{\partial^2 r}{\partial y^2} + \frac{\partial^2 r}{\partial z^2} = \frac{2}{r} \qquad ; \; \frac{\partial^2 \phi}{\partial x^2} + \frac{\partial^2 \phi}{\partial y^2} + \frac{\partial^2 \phi}{\partial z^2} = 0
$$

$$
\frac{\partial^2 \theta}{\partial x^2} + \frac{\partial^2 \theta}{\partial y^2} + \frac{\partial^2 \theta}{\partial z^2} = \frac{\cot \theta}{r^2}
$$

With the use of these expressions, we then obtain

$$
\nabla^2 f = \frac{\partial^2 \bar{f}}{\partial r^2} + \frac{1}{r^2}\frac{\partial^2 \bar{f}}{\partial \theta^2} + \frac{1}{r^2 \sin^2 \theta}\frac{\partial^2 \bar{f}}{\partial \phi^2} + \frac{2}{r}\frac{\partial \bar{f}}{\partial r} + \frac{\cot \theta}{r^2}\frac{\partial \bar{f}}{\partial \theta}
$$

$$
= \left[\frac{1}{r^2}\frac{\partial}{\partial r}r^2\frac{\partial}{\partial r} + \frac{1}{r^2}\left(\frac{1}{\sin \theta}\frac{\partial}{\partial \theta}\sin \theta\frac{\partial}{\partial \theta} + \frac{1}{\sin^2 \theta}\frac{\partial^2}{\partial \phi^2}\right)\right]\bar{f}
$$

$$
= \left[\frac{1}{r}\frac{\partial^2}{\partial r^2}r + \frac{1}{r^2}\left(\frac{1}{\sin \theta}\frac{\partial}{\partial \theta}\sin \theta\frac{\partial}{\partial \theta} + \frac{1}{\sin^2 \theta}\frac{\partial^2}{\partial \phi^2}\right)\right]\bar{f}
$$

This is the laplacian in spherical coordinates in three dimensions.

(c) We assume a central potential $V = V(r)$, and express the wave function in spherical coordinates as

$$\psi(r, \theta, \phi) = \frac{u(r)}{r} Y_{lm}(\theta, \phi)$$

where $Y_{lm}(\theta, \phi)$ are the spherical harmonics. With the use of the results of part (b), the Schrödinger equation is then re-written as

$$\left[-\frac{\hbar^2}{2m_0} \nabla^2 + V(r) \right] \psi(r, \theta, \phi)$$

$$= -\frac{\hbar^2}{2m_0} \left[\frac{1}{r} \frac{d^2 u(r)}{dr^2} Y_{lm}(\theta, \phi) - \frac{u(r)}{r^3} l(l+1) Y_{lm}(\theta, \phi) \right] + V(r) \frac{u(r)}{r} Y_{lm}(\theta, \phi)$$

$$= E \frac{u(r)}{r} Y_{lm}(\theta, \phi)$$

This equation may now be reduced to the form of Eq. (4.142)

$$\frac{d^2 u}{dr^2} + \left\{ \frac{2m_0}{\hbar^2} [E - V(r)] - \frac{l(l+1)}{r^2} \right\} u(r) = 0$$

Problem 4.27 This problem illustrates some applications of the general boundary conditions (1)–(2) in quantum mechanics stated above Eq. (4.66).

(a) Consider the azimuthal solutions in Eq. (4.134). Take a linear combination $\Phi(\phi) = a_1 \phi_{m_1}(\phi) + a_2 \phi_{m_2}(\phi)$ of any two solutions, and demand that the corresponding *probability density* $|\Phi|^2$ be single-valued under the transformation $\phi \to \phi + 2\pi$ (see Fig. 4.25 in the text). Show that $\Delta m = m_2 - m_1 = m$, where $m \equiv$ integer. Show that if one demands that $\Phi = $ constant be an acceptable solution, then *all the m_i must be integers.*[29]

(b) The radial component of the gradient in Eqs. (4.111)–(4.112) is just $\partial/\partial r$. Let $l = 0$, and consider the radial wave function $R(r)$ in Eqs. (4.141) and (4.143) as $r \to 0$. Suppose $R(r) = a$ where a is some complex constant is an acceptable solution. Now consider a linear combination of solutions $a + b/r$. Show that the probability density arising from such a wave function is *finite* when multiplied by the volume element $4\pi r^2 dr$; however, show that the origin acts as a *point source of probability* when the radial probability flux is integrated over the area $4\pi r^2$. Hence conclude that a radial solution that goes as b/r at the origin is too singular in quantum mechanics.

---

[29] This is a very important point, for when we start talking about half-integer angular momenta, we will find that the wave functions themselves are *double-valued* under the transformation $\phi \to \phi + 2\pi$.

Solution to Problem 4.27

(a) Consider the linear combination

$$\Phi(\phi) = \frac{a_1}{\sqrt{2\pi}}e^{im_1\phi} + \frac{a_2}{\sqrt{2\pi}}e^{im_2\phi}$$

and write the corresponding probability density

$$|\Phi(\phi)|^2 = \frac{1}{2\pi}\left[|a_1|^2 + |a_2|^2 + a_1^\star a_2 e^{i(m_2-m_1)\phi} + a_1 a_2^\star e^{-i(m_2-m_1)\phi}\right]$$

Under the transformation $\phi \rightarrow \phi + 2\pi$ the probability density transforms to

$$|\Phi(\phi + 2\pi)|^2 = \frac{1}{2\pi}\left[|a_1|^2 + |a_2|^2 + a_1^\star a_2 e^{i\Delta m(\phi+2\pi)} + a_1 a_2^\star e^{-i\Delta m(\phi+2\pi)}\right]$$

We therefore conclude that the probability density is single-valued only if $\Delta m = m_2 - m_1$ is a integer.

If we want $\Phi = $ constant to be an acceptable solution, then we may apply the previous arguments with $m_1 = 0$ and m_2 arbitrary and conclude that the probability density associated with the linear combination

$$\Phi(\phi) = \frac{a_1}{\sqrt{2\pi}} + \frac{a_2}{\sqrt{2\pi}}e^{im_2\phi}$$

is single–valued only if $\Delta m = m_2$ is an integer. This implies that all the m_i must be integers.

(b) We assume that the radial wave function behaves as $a + b/r$ for $r \rightarrow 0$, with a and b complex constants. The corresponding probability density is then $|a + b/r|^2/4\pi$, where we have used the fact that $l = 0$ and $Y_{00}(\theta, \phi) = 1/\sqrt{4\pi}$. The probability of finding the particle within a small distance r from the origin is then given by

$$\int_0^r \left|a + \frac{b}{r'}\right|^2 r'^2 dr' = \frac{1}{3}|a|^2 r^3 + \frac{1}{2}(a^\star b + ab^\star)r^2 + |b|^2 r$$

This expression is *finite*, and tends to zero as $r \rightarrow 0$.

We now use the same wave function to calculate the radial probability current

$$S_r = \frac{\hbar}{2im}\left[\Psi^\star \frac{\partial}{\partial r}\Psi - \left(\frac{\partial}{\partial r}\Psi\right)^\star \Psi\right]$$

$$= -\frac{\hbar}{2im}(a^\star b - b^\star a)\frac{1}{4\pi r^2}$$

$$= \frac{\hbar}{m}\text{Im}\,(b^\star a)\frac{1}{4\pi r^2}$$

The radial probability flux integrated over a surface $4\pi r^2$ is then constant

$$4\pi r^2 S_r = \frac{\hbar}{m}\mathrm{Im}\,(b^\star a)$$

Since this expression remains finite in the limit $r \to 0$, we conclude that the origin acts as a point source of probability. To avoid this problem, a radial solution that goes as b/r at the origin must be excluded.

Problem 4.28 (a) The analytic expressions for the radial wave functions for the first two s-states $(l = 0)$ in the point Coulomb potential are [Schiff (1968)]

$$R_{10}(r) = 2\left(\frac{Z}{a_0}\right)^{3/2}e^{-Zr/a_0} \quad ; \quad R_{20}(r) = \left(\frac{Z}{2a_0}\right)^{3/2}\left(2 - \frac{Zr}{a_0}\right)e^{-Zr/2a_0}$$

Here $a_0 \equiv \hbar^2/m_0(e^2/4\pi\varepsilon_0)$ is the Bohr radius. Make an accurate to-scale plot of the radial densities $|R_{\bar{n}0}(r)|^2$ for these states in units of Zr/a_0.[30]

(b) The analytic expression for the wave function for the first p-state $(l = 1)$ in the point Coulomb potential is[31]

$$\psi_{21m}(r, \theta, \phi) = R_{21}(r)Y_{1m}(\theta, \phi)$$

$$R_{21}(r) = \left(\frac{Z}{2a_0}\right)^{3/2}\left(\frac{Zr}{a_0\sqrt{3}}\right)e^{-Zr/2a_0}$$

Make good to-scale plots, by any means you find the most informative (density of points, contours, *etc.*), of the probability densities for the three states with $m = 0, \pm 1$.

Solution to Problem 4.28

(a) In Figs. 4.17 and 4.18 we plot the radial probability densities for the first two s-states in the point Coulomb potential. In Fig. 4.17, we plot

$$\frac{a_0}{Z}|R_{10}(r)|^2 r^2 = \left(\frac{2Zr}{a_0}\right)^2 e^{-2Zr/a_0}$$

which reaches a maximum at $r = a_0/Z$.

---

[30]Remember the volume element in spherical coordinates is $d^3x = r^2 dr \sin\theta\, d\theta\, d\phi$.

[31]Note that whereas the newtonian orbit problem in a central potential in three dimensions can always be reduced to motion in a two-dimensional plane (see [Fetter and Walecka (2003)]), this is no longer true in quantum mechanics, where the three-dimensional problem is intrinsically different [compare Eqs. (4.132) and (4.137)]. The s-states with $l = 0$, for example, have a *uniform probability distribution* in both θ and ϕ.

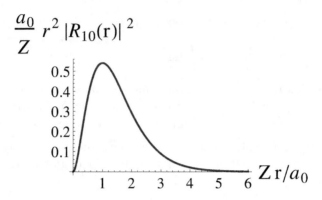

Fig. 4.17   Radial probability density $(a_0/Z)|R_{10}(r)|^2 r^2$.

In Fig. 4.18 we plot

$$\frac{a_0}{Z}|R_{20}(r)|^2 r^2 = \frac{1}{8}e^{-Zr/a_0}\left(2-\frac{Zr}{a_0}\right)^2\left(\frac{Zr}{a_0}\right)^2$$

which reaches a (global) maximum at $r = (3+\sqrt{5})a_0/Z$.

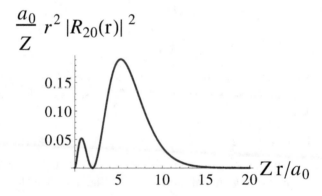

Fig. 4.18   Radial probability density $(a_0/Z)|R_{20}(r)|^2 r^2$.

(b) We start by observing that the radial probability density for the first p-state is

$$|R_{21}(r)|^2 r^2 = \frac{Z^5 r^4 e^{-Zr/a_0}}{24a_0^5}$$

which reaches its maximum at $r_0 = 4a_0/Z$. This must be multiplied by $|Y_{1m}(\theta,\phi)|^2 \sin\theta$ to get the full probability density $\rho(r,\theta,\phi)$, where the probability here is $\rho\,drd\theta d\phi$. We fix $r = r_0$, and in Figs. 4.19 and 4.20 we make a plot of the surface $(a_0/Z)\rho(r_0, \theta, \phi)$ as a function of (θ, ϕ) for the angular functions $|Y_{1,\pm1}(\theta, \phi)|^2 \sin\theta$ and $|Y_{1,0}(\theta, \phi)|^2 \sin\theta$ respectively.

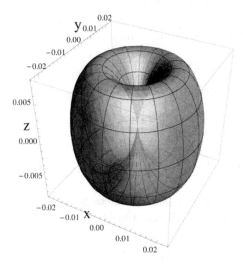

Fig. 4.19 Probability density $(a_0/Z)\rho(r_0, \theta, \phi)$, with $|Y_{1,\pm1}(\theta, \phi)|^2 \sin\theta = (3/8\pi)\sin^3\theta$.

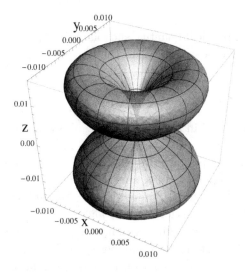

Fig. 4.20 Same as Fig. 4.19, with $|Y_{1,0}(\theta, \phi)|^2 \sin\theta = (3/4\pi)\sin\theta\cos^2\theta$.

Problem 4.29 Assume a collection of non-interacting fermions in their ground state in a large cubical box, of side L, with rigid walls. Use the results in Eqs. (4.115)–(4.116). Work in the limit $L \to \infty$. Write $k_i = (\pi/L)n_i$ and convert the sum over states to an integral over wavenumbers. Show one obtains the *same results* for $(N/V, k_F, E/N)$ as in Eqs. (4.168)–(4.170) where periodic boundary conditions are employed. Discuss.

Solution to Problem 4.29

The wavenumber spacing in this case is

$$\Delta k_i = \frac{\pi}{L}$$

which is half the spacing for periodic boundary conditions of Eq. (4.161). On other hand, since $k_i = (\pi/L)n_i$ with $n_i = 1, 2, \cdots$, the wavenumbers now only take positive values.

In the limit $L \to \infty$, we may repeat the steps followed in Eqs. (4.165)–(4.166) for periodic boundary conditions to obtain

$$\sideset{}{'}\sum_{\mathbf{k}} = \frac{dk_x dk_y dk_z}{(\pi/L)^3} = \frac{V}{\pi^3} d^3k$$

and

$$\sum_{\mathbf{k}} = \frac{V}{\pi^3} \int d^3k \qquad ; L \to \infty \qquad ; k_i > 0$$

The calculation of a physical quantity such as N/V, k_F, and E/N can be represented by means of a sum of the form

$$\sum_{\mathbf{k}} f(|\mathbf{k}|) \to \frac{V}{\pi^3} \int_0^{k_F} d^3k \, f(|\mathbf{k}|)$$

where the integral is obtained in the limit $L \to \infty$, and it is now restricted to the first octant with $(k_x, k_y, k_z) > 0$. Since $f(|\mathbf{k}|)$ is even about the origin in (k_x, k_y, k_z), we can perform the integration over the positive and negative values of the components of $\mathbf{k}$ and divide by a factor of 8, thus recovering the same results as for periodic boundary conditions.

For this reason, the expressions for N/V, k_F and E/N in a box with rigid walls are the *same* as those for periodic boundary conditions in Eqs. (4.168)–(4.170). We thus observe that the local number of modes per unit volume with given $|\mathbf{k}|$ is independent of the far-off boundary conditions imposed on the problem.

Problem 4.30 Consider the ground state of a system of N spin-zero bosons of mass m in a large cubical box, of side L, with rigid walls.

(a) Show the energy of the ground state is

$$E = \frac{\hbar^2}{2m} \frac{3\pi^2}{L^2} N$$

(b) Show the corresponding pressure is

$$P = \frac{1}{N^{2/3}} \left(\frac{\hbar^2 \pi^2}{m} \right) n^{5/3}$$

where $n = N/V$ is the particle density.

(c) Hence conclude that, at fixed particle density, the pressure of this Bose gas goes to zero as the size of the box goes to infinity.

Solution to Problem 4.30

(a) The energies in the box accessible to the bosons are

$$E_{n_1 n_2 n_3} = \frac{\hbar^2 \pi^2}{2mL^2} (n_1^2 + n_2^2 + n_3^2) \qquad ; \ n_i = 1, 2, \cdots, \infty$$

where $n_i = 1, 2, \cdots, \infty$ [see Eq. (4.116)]. In the ground state all the bosons occupy the lowest-energy single-particle state, corresponding to $n_1 = n_2 = n_3 = 1$. The energy of the ground state will therefore be

$$E = \frac{\hbar^2}{2m} \frac{3\pi^2}{L^2} N$$

(b) The pressure is obtained by taking the partial derivative of E with respect to V. We thus need to express E directly in terms of the volume, taking into account the fact that $V = L^3$

$$E = \frac{\hbar^2}{2m} \frac{3\pi^2}{V^{2/3}} N$$

The pressure is then given by

$$P = -\left(\frac{\partial E}{\partial V} \right)_N = \frac{\hbar^2}{2m} 3\pi^2 N \left(\frac{2}{3V^{5/3}} \right) = \frac{1}{N^{2/3}} \left(\frac{\hbar^2 \pi^2}{m} \right) n^{5/3}$$

where the particle density is given by $n = N/V$.

(c) With $N \to \infty$ and $V \to \infty$ at fixed particle density $n = N/V$, the pressure P of the non-interacting Bose gas in its ground state in the box goes to zero as $N^{-2/3}$.

Problem 4.31 Consider a collection of N spin-zero bosons of mass m in a big cubical box of side L satisfying p.b.c. At high temperature, the chemical potential is negative (see Prob. E.3). As the temperature is lowered, the chemical potential increases, and one will eventually reach a point where the chemical potential vanishes. The chemical potential for the Bose gas must then remain zero for $T < T_0$, or the distribution function in the second of Eqs. (4.178) will become singular at finite ε_k. What happens below T_0 is that a finite fraction of the bosons begins to occupy the single state with $\mathbf{k} = 0$.

(a) Show the temperature T_0 at which the chemical potential first vanishes, and hence the temperature for the phase transition in this non-interacting Bose gas, is determined from the equation

$$\frac{N}{V} = \int \frac{d^3k}{(2\pi)^3} \, \frac{1}{e^{\hbar^2 k^2/2mk_B T_0} - 1}$$

(b) Go to dimensionless variables, and show this equation can be rewritten as

$$k_B T_0 = \frac{1}{\gamma^{2/3}} \frac{\hbar^2 n^{2/3}}{2m} \qquad ; \, \gamma \equiv \frac{1}{(2\pi)^2} \int_0^\infty \frac{\sqrt{u}\,du}{e^u - 1}$$

Here $n = N/V$ is the particle density.

(c) Liquid ^{4}He has a mass density $\rho = 0.145\,\text{g/cm}^3$. Compute T_0. Compare with the observed superfluid transition temperature T_λ in Fig. 11.1 in the text.[32]

Solution to Problem 4.31

(a) The single-particle distribution function for Bose systems is given in Eq. (4.178) of the book and reads

$$n_k = \left[\exp\left(\frac{\epsilon_k - \mu}{k_B T}\right) - 1\right]^{-1}$$

We may calculate the total number of bosons, in the limit of a large volume $V = L^3 \to \infty$, as

$$N = \sum_k n_k \to \frac{V}{(2\pi)^3} \int d^3k \left[\exp\left(\frac{\epsilon_k - \mu}{k_B T}\right) - 1\right]^{-1}$$

---

[32]Use $m_{\text{He}^4} \approx 4m_p$.

Call the temperature for which the chemical potential μ first vanishes T_0. One then obtains the equation

$$\frac{N}{V} = \int \frac{d^3k}{(2\pi)^3} \frac{1}{e^{\hbar^2 k^2/2mk_B T_0} - 1}$$

(b) We introduce the dimensionless quantity γ, where $\hbar^2/2mk_B T_0 \equiv (\gamma/n)^{2/3}$, and make a change of variable $k = (n/\gamma)^{1/3}\sqrt{u}$. The equation defining T_0 then becomes

$$n = \int \frac{d^3k}{(2\pi)^3} \frac{1}{e^{(\gamma/n)^{2/3}k^2} - 1} \qquad ; \quad \frac{\hbar^2}{2mk_B T_0} \equiv \left(\frac{\gamma}{n}\right)^{2/3}$$

$$= \frac{n}{4\pi^2\gamma} \int_0^\infty \frac{\sqrt{u}\,du}{e^u - 1}$$

This last equation is solved for γ to obtain

$$\gamma = \frac{1}{(2\pi)^2} \int_0^\infty \frac{\sqrt{u}\,du}{e^u - 1}$$

The integral appearing in this equation is evaluated as before. We make the change of variable $v = e^{-u}$ and write the integral as[33]

$$\gamma = \frac{1}{4\pi^2} \int_0^1 \frac{1}{1-v} \left(\log\frac{1}{v}\right)^{1/2} dv$$

Since $0 \le v \le 1$, we may use the geometric series $1/(1-v) = \sum_{j=0}^\infty v^j$

$$\gamma = \frac{1}{4\pi^2} \sum_{j=0}^\infty \int_0^1 v^j \left(\log\frac{1}{v}\right)^{1/2} dv$$

$$= \frac{1}{4\pi^2} \sum_{j=0}^\infty \int_0^\infty e^{-(j+1)u} \sqrt{u}\,du$$

As we have seen in Prob. 3.2, the gamma function has the integral representation

$$\int_0^\infty e^{-\alpha x} x^r\, dx = \frac{\Gamma(r+1)}{\alpha^{r+1}} \qquad ; \alpha > 0 \qquad ; r > -1$$

and therefore

$$\gamma = \frac{1}{4\pi^2} \sum_{j=0}^\infty \frac{\Gamma(3/2)}{(j+1)^{3/2}} = \frac{\Gamma(3/2)}{4\pi^2} \zeta(3/2)$$

---

[33]This integral is also protected as $v \to 1$.

where $\zeta(s) \equiv \sum_{j=0}^{\infty}(j+1)^{-s}$ is the Riemann zeta function, with $\zeta(3/2) \approx 2.612$.

(c) The approximate numerical value of γ is[34]

$$\gamma \approx 0.0586$$

The transition temperature T_0 in the Bose gas is then given by

$$T_0 = \frac{\hbar^2 n^{2/3}}{\gamma^{2/3} 2mk_B}$$

The particle density of liquid ^{4}He is related to its mass density by

$$\rho = \frac{M}{V} = \frac{4m_p N}{V} = 4m_p n$$

where m_p is the proton mass

$$m_p = 938.3 \frac{\text{Mev}}{c^2} = 1.67 \times 10^{-27} \text{ kg}$$

The particle density is therefore

$$n = \frac{\rho}{4m_p} = 2.17 \times 10^{28} \frac{1}{\text{m}^3}$$

The expression for T_0, with $m_{\text{He}^4} = 4m_p$ and the constants in appendix L, then leads to

$$T_0 \approx 3.1\,^{\circ}\text{K}$$

This is very close to the observed superfluid transition temperature $T_\lambda = 2.17\,^{\circ}\text{K}$ given in Fig. 11.1 in the text.

Problem 4.32 Consider a non-interacting, non-relativistic Fermi gas of spin-1/2 particles in its ground state at a given density $n = N/V$ (electrons in a metal, neutrons in a neutron star, *etc.*). The particles have mass m and magnetic moment μ_0. A uniform magnetic field **B** is applied. Treat the system as two separate Fermi gases, one with magnetic moment aligned with the field, and one with magnetic moment opposed. Parameterize the number of particles of each type as

$$N_\uparrow = \frac{N}{2}(1+\delta) \qquad ; N_\downarrow = \frac{N}{2}(1-\delta)$$

---

[34] Recall $\Gamma(3/2) = \sqrt{\pi}/2$.

(a) What is the contribution to the energy of the system coming from the interaction of the spin magnetic moments with the magnetic field for a given δ?

(b) The kinetic energy of the system must increase to achieve a configuration with finite δ. Let ε_F^0 be the value of the Fermi energy with $\delta = 0$. Show the increase in kinetic energy of the system is given to order δ^2 by[35]

$$\Delta E = \frac{\delta^2}{3}\varepsilon_F^0 N$$

(c) Construct the total change in the energy as a sum of the two contributions in (a) and (b). Minimize with respect to δ to find the new ground state. Show

$$\delta = \frac{3\mu_0 B}{2\varepsilon_F^0}$$

(d) The magnetic spin susceptibility is defined in terms of the magnetic dipole moment per unit volume $\mathbf{M}$ according to $\mathbf{M} = \chi_P \mathbf{B}$. Hence derive the expression of the *Pauli paramagnetic spin susceptibility*

$$\chi_P = \frac{3\mu_0^2}{2\varepsilon_F^0}n$$

Solution to Problem 4.32

(a) The interaction energy of an intrinsic magnetic moment with a magnetic field $\mathbf{B}$ is $H' = -\boldsymbol{\mu} \cdot \mathbf{B}$ [see Eq. (4.157)]. The total interaction energy of the system of spin-1/2 particles is therefore

$$\Delta H' = -\mu_0 B\frac{N(1+\delta)}{2} + \mu_0 B\frac{N(1-\delta)}{2} = -\mu_0 BN\delta$$

where we have fixed the z-direction along $\mathbf{B}$.

(b) Eq. (4.167) relates the number of particles to the maximum wavenumber k_F. In line with the suggestion, we treat the system as made up of two separate Fermi gases and write[36]

$$N_\uparrow = \frac{N(1+\delta)}{2} = \frac{V}{(2\pi)^3}\frac{4\pi}{3}k_{F\uparrow}^3$$

$$N_\downarrow = \frac{N(1-\delta)}{2} = \frac{V}{(2\pi)^3}\frac{4\pi}{3}k_{F\downarrow}^3$$

---

[35] Use the Taylor series expansion $(1+x)^n = 1+nx+n(n-1)x^2/2!+\cdots$, which holds for $|x| < 1$ and any n (integer or non-integer).

[36] Each Fermi gas now has $g = 1$.

These equations can be inverted to give

$$k_{F\uparrow} = \left[6\pi^2 \frac{N}{V} \frac{(1+\delta)}{2} \right]^{1/3}$$

$$k_{F\downarrow} = \left[6\pi^2 \frac{N}{V} \frac{(1-\delta)}{2} \right]^{1/3}$$

The total kinetic energy of the system will then be [see Eq. (4.170)]

$$E = E_\uparrow + E_\downarrow = \frac{3}{5} \frac{N(1+\delta)}{2} \frac{\hbar^2 k_{F\uparrow}^2}{2m} + \frac{3}{5} \frac{N(1-\delta)}{2} \frac{\hbar^2 k_{F\downarrow}^2}{2m}$$

$$= \frac{3}{10} \frac{N\hbar^2}{2m} \left(3\pi^2 \frac{N}{V} \right)^{2/3} \left[(1+\delta)^{5/3} + (1-\delta)^{5/3} \right]$$

For $|\delta| \ll 1$ we may expand the factors $(1 \pm \delta)^{5/3}$ up to order δ^2, as explained in the footnote, to obtain

$$E \approx \frac{3}{10} \frac{N\hbar^2}{2m} \left(3\pi^2 \frac{N}{V} \right)^{2/3} \left[2 + \frac{10}{9} \delta^2 \right]$$

$$= \frac{3}{5} \varepsilon_F^0 N \left(1 + \frac{5}{9} \delta^2 \right)$$

and therefore

$$\Delta E = \frac{\delta^2}{3} \varepsilon_F^0 N$$

(c) The total change in energy is

$$\Delta E_{\text{tot}} = \Delta H' + \Delta E = -\mu_0 B N \delta + \frac{\delta^2}{3} \varepsilon_F^0 N$$

which describes a parabola in δ with a minimum located at[37]

$$\delta = \frac{3\mu_0 B}{2\varepsilon_F^0} \qquad ; \text{ for minimum}$$

If we use this value in ΔE_{tot} we obtain

$$\Delta E_{\text{tot}} = -\frac{3}{4} \frac{N(\mu_0 B)^2}{\varepsilon_F^0}$$

Since $\Delta E_{\text{tot}} < 0$, we conclude that in a constant magnetic field the new ground state of a system of N spin-1/2 particles with mass m and intrinsic

---

[37] Clearly, this result is valid only for $|\delta| \ll 1$, i.e. for $|\mu_0 B|/\varepsilon_F^0 \ll 1$.

magnetic moment μ_0 will have

$$N_\uparrow \approx \left(1 + \frac{3\mu_0 B}{2\varepsilon_F^0}\right)\frac{N}{2} \qquad ; \; N_\downarrow \approx \left(1 - \frac{3\mu_0 B}{2\varepsilon_F^0}\right)\frac{N}{2}$$

(d) The magnetic moment per unit volume of the system is

$$M = \frac{\mu_0 N\delta}{V} = \frac{3\mu_0^2 nB}{2\varepsilon_F^0}$$

Thus the Pauli paramagnetic spin susceptibility is given by

$$\chi_P = \frac{3\mu_0^2}{2\varepsilon_F^0}n$$

Problem 4.33 The spin wave functions for a spin-1/2 fermion in Eq. (4.160) form a two-component column matrix (a "spinor").

(a) Use the results in appendix B to show that these spinors are *orthonormal* in the sense that

$$\underline{\eta}_\lambda^\dagger \underline{\eta}_{\lambda'} = \delta_{\lambda\lambda'} \qquad ; \; (\lambda, \lambda') = (\uparrow, \downarrow)$$

(b) Consider a more general spinor of the form

$$\underline{\eta} = \begin{pmatrix} \alpha \\ \beta \end{pmatrix}$$

where (α, β) are complex numbers. Show the normalization $\underline{\eta}^\dagger \underline{\eta} = 1$ of this spinor implies $|\alpha|^2 + |\beta|^2 = 1$. In this case, one can interpret $|\alpha|^2$ as the *probability* that the particle has spin up along the z-axis and $|\beta|^2$ as the probability that it has spin down. Verify that this interpretation is consistent with the wave functions in (a).

(c) Show the most general hermitian operator in this two-dimensional spin space must be of the form $a1 + \mathbf{b} \cdot \boldsymbol{\sigma}$ where $(1, \boldsymbol{\sigma})$ are the unit and Pauli matrices (see appendix B), and $(a, \mathbf{b})$ are real.

Solution to Problem 4.33

(a) With the use of the expressions in Eqs. (4.160) we have

$$\underline{\eta}_\uparrow^\dagger \underline{\eta}_\uparrow = (1\;0)\begin{pmatrix}1\\0\end{pmatrix} = 1 \qquad ; \; \underline{\eta}_\downarrow^\dagger \underline{\eta}_\downarrow = (0\;1)\begin{pmatrix}0\\1\end{pmatrix} = 1$$

$$\underline{\eta}_\uparrow^\dagger \underline{\eta}_\downarrow = (1\;0)\begin{pmatrix}0\\1\end{pmatrix} = 0 \qquad ; \; \underline{\eta}_\downarrow^\dagger \underline{\eta}_\uparrow = (0\;1)\begin{pmatrix}1\\0\end{pmatrix} = 0$$

This can be compactly written as

$$\underline{\eta}_\lambda^\dagger \underline{\eta}_{\lambda'} = \delta_{\lambda\lambda'} \qquad ; (\lambda, \lambda') = (\uparrow, \downarrow)$$

(b) For a general spinor

$$\underline{\eta} = \begin{pmatrix} \alpha \\ \beta \end{pmatrix}$$

we have

$$\underline{\eta}^\dagger \underline{\eta} = \begin{pmatrix} \alpha^\star & \beta^\star \end{pmatrix} \begin{pmatrix} \alpha \\ \beta \end{pmatrix} = |\alpha|^2 + |\beta|^2 = 1$$

where we have implemented the correct normalization of the spinor.

The expectation value of $S_z = \sigma_z/2$ in this state is therefore

$$\underline{\eta}^\dagger S_z \underline{\eta} = \frac{1}{2} \begin{pmatrix} \alpha^\star & \beta^\star \end{pmatrix} \begin{pmatrix} 1 & 0 \\ 0 & -1 \end{pmatrix} \begin{pmatrix} \alpha \\ \beta \end{pmatrix} = \frac{1}{2} \left(|\alpha|^2 - |\beta|^2 \right)$$

This justifies the interpretation of $|\alpha|^2$ as the *probability* that the particle has spin up along the z-axis, and of $|\beta|^2$ as the probability that it has spin down. In part (a) we either have $\alpha = 1$ and $\beta = 0$, or $\alpha = 0$ and $\beta = 1$. These states have eigenvalues $\pm 1/2$, and correspond to unit probability of having the spin aligned or anti-aligned along the z-axis.

****

(*Aside*) Another example of a spinor with this form is obtained by considering the spin projection along the direction of an arbitrary unit vector $\mathbf{v} = (\sin\theta\cos\phi, \sin\theta\sin\phi, \cos\theta)$. With $\mathbf{S} = \boldsymbol{\sigma}/2$, one has

$$\mathbf{S} \cdot \mathbf{v} = \frac{1}{2} \left[\sin\theta(\cos\phi\,\sigma_x + \sin\phi\,\sigma_y) + \cos\theta\,\sigma_z \right]$$

$$= \frac{1}{2} \left[\sin\theta\cos\phi \begin{pmatrix} 0 & 1 \\ 1 & 0 \end{pmatrix} + \sin\theta\sin\phi \begin{pmatrix} 0 & -i \\ i & 0 \end{pmatrix} + \cos\theta \begin{pmatrix} 1 & 0 \\ 0 & -1 \end{pmatrix} \right]$$

$$= \frac{1}{2} \begin{pmatrix} \cos\theta & \sin\theta\,e^{-i\phi} \\ \sin\theta\,e^{i\phi} & -\cos\theta \end{pmatrix}$$

New spinors $\underline{\eta}_\lambda$ can be introduced which obey the eigenvalue equation

$$\mathbf{S} \cdot \mathbf{v}\,\underline{\eta}_{\pm 1} = \pm\frac{1}{2}\underline{\eta}_{\pm 1}$$

and thus have their spin aligned, or anti-aligned, along $\mathbf{v}$.[38]

---

[38] Here $\boldsymbol{\sigma} \cdot \mathbf{v}\,\underline{\eta}_\lambda = \lambda\,\underline{\eta}_\lambda$.

A little matrix multiplication, and the use of some trigonometric identities, show that the eigenvectors are given by

$$\underline{\eta}_{+1} = e^{i\delta} \begin{pmatrix} e^{-i\phi/2} \cos\theta/2 \\ e^{i\phi/2} \sin\theta/2 \end{pmatrix} \qquad ; \; \underline{\eta}_{-1} = e^{i\delta} \begin{pmatrix} -e^{-i\phi/2} \sin\theta/2 \\ e^{i\phi/2} \cos\theta/2 \end{pmatrix}$$

where δ is an arbitrary phase.

****

(c) The most general 2×2 complex matrix will depend upon 8 parameters, the real and imaginary parts of its four elements. If we impose the condition that the matrix be hermitian, then we reduce the number of independent parameters to four, since the diagonal elements must be real, while the element above the diagonal is the complex conjugate of the element below the diagonal.

On the other hand, the Pauli matrices and the 2×2 identity matrix are a set of 4 linearly independent 2×2 hermitian matrices, and therefore they provide a basis for expressing a general hermitian matrix. We can easily verify this by providing the explicit matrix representation for $a1 + \mathbf{b} \cdot \boldsymbol{\sigma}$

$$a1 + \mathbf{b} \cdot \boldsymbol{\sigma} = \begin{pmatrix} a + b_z & b_x - ib_y \\ b_x + ib_y & a - b_z \end{pmatrix}$$

Chapter 5

Atomic Physics

Problem 5.1 Special relativity (see later) says that there is a magnetic field $c\mathbf{B} = (\mathbf{v}/c) \times \mathbf{E}$ associated with an electric field moving with velocity $\mathbf{v}$.

(a) Show the electric field at the position of an electron in an atom arising from the nucleus is $\mathbf{E} = (Ze_p/4\pi\varepsilon_0 r^3)\,\mathbf{r}$ where $\mathbf{r}$ points from the nucleus to the electron.

(a) Use the fact that the nucleus is moving with velocity $-\mathbf{v}$ with respect to the electron to show that the effective magnetic field at the position of the electron arising from the nuclear current is

$$\mathbf{B}_{\text{eff}} = \frac{1}{m_e c^2}\left(\frac{Ze_p}{4\pi\varepsilon_0 r^3}\right)\mathbf{r} \times (m_e \mathbf{v})$$

(b) Hence derive an expression for V_{so} appearing in Eq. (5.24)

$$V_{\text{so}}(r) = \frac{g_s}{2}\frac{Ze^2}{4\pi\varepsilon_0 \hbar c}\left(\frac{\hbar}{m_e c}\frac{1}{r}\right)^3 m_e c^2$$

(c) Evaluate the expectation value of this quantity [see the second of Eqs. (5.26)] using the radial wave function for the $2p$-state in hydrogen given in Prob. 4.28(b).[1]

Solution to Problem 5.1

(a) We assume the nucleus to be a point charge of value $Q = Ze_p$. The electric field generated from this charge and acting on the electron is then

---

[1] *Note*: The expression in part (b) is a factor of two larger than that obtained through a correct relativistic treatment of the electron provided by the Dirac equation (Prob. 9.3); this is due to the fact that the spinning electron is not moving with a uniform velocity $\mathbf{v}$, but is actually accelerating in its orbit. (There is an additional "Thomas precession".)

simply

$$\mathbf{E} = \frac{Ze_p}{4\pi\varepsilon_0 r^3}\mathbf{r}$$

(b) Since the nucleus moves with a velocity $\mathbf{v}_n = -\mathbf{v}$ with respect to the electron, there is a magnetic field

$$\mathbf{B} = \frac{\mathbf{v}_n}{c^2} \times \mathbf{E} = \frac{1}{m_e c^2}\left(\frac{Ze_p}{4\pi\varepsilon_0 r^3}\right)\mathbf{r} \times (m_e\mathbf{v})$$

at the position of the electron, arising from the nuclear current.

(c) The energy due to the interaction of the intrinsic magnetic moment of the electron with the effective magnetic field is given in Eq. (5.24) and reads

$$H' = -\boldsymbol{\mu}_{\text{spin}} \cdot \mathbf{B}_{\text{eff}}$$

where $\boldsymbol{\mu}_{\text{spin}} = (g_s e\hbar/2m_e)\mathbf{s}$ is the intrinsic magnetic moment of the electron. Therefore, since $e = -e_p$,

$$H' = g_s\frac{e_p\hbar^2}{2m_e}\frac{1}{m_e c^2}\left(\frac{Ze_p}{4\pi\varepsilon_0 r^3}\right)\mathbf{s}\cdot\mathbf{l}$$

where we identify $\mathbf{r} \times (m_e\mathbf{v}) = \hbar\mathbf{l}$ with the angular momentum of the electron. We can now read off the spin-orbit potential from this expression

$$H' = V_{\text{so}}(r)\,\mathbf{s}\cdot\mathbf{l}$$
$$V_{\text{so}}(r) = \frac{g_s}{2}\frac{Ze^2}{4\pi\varepsilon_0\hbar c}\left(\frac{\hbar}{m_e c}\frac{1}{r}\right)^3 m_e c^2$$

(d) The wave function for the first p-state in the point Coulomb potential is given in Prob. 4.28 and reads

$$\psi_{21m}(r, \theta, \phi) = R_{21}(r)Y_{1m}(\theta, \phi)$$
$$R_{21}(r) = \left(\frac{Z}{2a_0}\right)^{3/2}\left(\frac{Zr}{a_0\sqrt{3}}\right)e^{-Zr/2a_0}$$

The expectation value of $V_{\text{so}}(r)$ in the 2p-state in hydrogen-like atoms is then given by

$$\langle V_{\text{so}}(r)\rangle = \int |R_{nl}(r)|^2\, V_{\text{so}}(r)\, r^2 dr$$

Thus[2]

$$\langle V_{\text{so}}(r) \rangle = \frac{g_s}{48} Z^4 \alpha^4 m_e c^2 \int_0^\infty u e^{-u} \, du$$

$$= \frac{g_s}{48} Z^4 \alpha^4 m_e c^2$$

Problem 5.2 Verify Eqs. (5.54).

Solution to Problem 5.2

We use Eqs. (5.37) and (5.53) to obtain $n(r)$ in terms of $\chi(x)$

$$n(r) = \frac{Z^2}{3\pi^2} \left(\frac{2m}{\hbar^2} \frac{e_p^2}{4\pi\varepsilon_0 b a_0} \right)^{3/2} \left[\frac{\chi(x)}{x} \right]^{3/2}$$

$$= \frac{(2b)^{3/2} Z}{3\pi^2} \left(\frac{Z^{1/3}}{b a_0} \right)^3 \left[\frac{\chi(x)}{x} \right]^{3/2}$$

This is precisely the first of Eqs. (5.54), once the definition of b in Eq. (5.48) is employed to give $(2b)^{3/2} = 3\pi/4$

$$n(r) = \frac{Z}{4\pi} \left(\frac{Z^{1/3}}{b a_0} \right)^3 \left[\frac{\chi(x)}{x} \right]^{3/2}$$

The second of Eqs. (5.54) is then obtained directly using this equation and performing the change of variable from r to x with $x = Z^{1/3} r / b a_0$

$$4\pi n(r) r^2 \, dr = Z [\chi(x)]^{3/2} x^{1/2} dx$$

Problem 5.3 Write or obtain a program to integrate a second-order non-linear differential equation and reproduce the results in Fig. 5.11 and Eq. (5.70).

Solution to Problem 5.3

The Thomas-Fermi Eq. (5.52) is

$$\sqrt{x} \frac{d^2}{dx^2} \chi(x) = [\chi(x)]^{3/2} \qquad ; \text{ Thomas-Fermi equation}$$

$$\chi(0) = 1 \qquad ; \text{ nucleus at origin}$$

$$\chi(\infty) = 0 \qquad ; \text{ neutral atom}$$

---

[2]Recall $a_0 = \hbar^2/m_e(e^2/4\pi\varepsilon_0) = \hbar/\alpha m_e c$.

Before trying to solve the T-F equation numerically, it is useful to analyze the behavior of its solution close to $x = 0$. For $x \to 0$ we expect that $\chi(x) \approx 1 + \chi'(0)x + \beta x^r + \cdots$, where we have used $\chi(0) = 1$ and $r > 1$. Thus $\chi'(x) \approx \chi'(0) + \beta r x^{r-1} + \cdots$, and $\chi''(x) \approx \beta r(r-1)x^{r-2} + \cdots$.

The second boundary condition in Eqs. (5.52) is specified at $x = \infty$. Therefore we should look for a value of $\chi'(0)$ for which $\chi(\infty) = 0$. If we substitute the expression for $\chi''(x)$ in the T-F equation for $x \to 0$, we obtain the equation

$$\beta r(r-1)x^{r-3/2} = 1$$

which implies $r = 3/2$ and $\beta = 4/3$. Thus we see that $\chi(x)$ is not analytic in x near $x = 0$, and $\lim_{x \to 0} \chi''(x) = \infty$.

If we wish to solve the equation numerically, we need to start at a point $x = \epsilon$, with $0 < \epsilon \ll 1$, and use the small x behavior of $\chi(x)$ discussed above to obtain the correct boundary conditions

$$\chi(\epsilon) \approx 1 + \chi'(0)\epsilon + \cdots$$
$$\chi'(\epsilon) \approx \chi'(0) + 2\sqrt{\epsilon} + \cdots$$

The Thomas-Fermi equation can now be solved numerically starting at $x = \epsilon$, using the initial conditions given in the equations above, which involve the free parameter $\chi'(0)$. The results are shown in Fig. 5.1.[3]

For a generic value of $\chi'(0)$ the solution will not have the correct boundary condition at $x = \infty$, either blowing up as $x \to \infty$, or vanishing at some finite x. As we see in Fig. 5.1, for $\chi'(0) = -1.5$ the solution blows up at $x \gg 1$, while for $\chi'(0) = -1.6$, the solution vanishes at $x \approx 3.5$. Through an application of the bisection method on an interval containing the exact value of $\chi'(0)$, one can iteratively restrict the interval and obtain a better value. The solid line in the figure displays the solution of the Thomas-Fermi equation corresponding to $\chi'(0) = -1.58807$, which provides a very precise approximation to the exact solution.

The numerical solution corresponding to $\chi'(0) = -1.58807$ can now be used to calculate C_{TF}, defined in Eq (5.68) of the text, to give $C_{\text{TF}} = 20.9$ eV.

As stated in the text, this calculation represents one of the first applications of modern computers to a problem in physics [Feynman, Metropolis, and Teller (1949)].

---

[3] The differential equation is solved numerically using the NDSolve program in Mathematica. Here we take $\epsilon = 10^{-10}$.

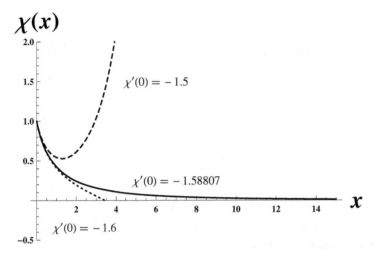

Fig. 5.1 Numerical solution to the Thomas-Fermi equation using three different values for $\chi'(0)$.

Problem 5.4 For an electron with $s = 1/2$, one has $j = l \pm 1/2$. Assume $g_s = 2$.

(a) Show that the Landé g-factor takes the form

$$g_{\text{Landé}} = \left(1 \pm \frac{1}{2l+1}\right) \qquad ; j = l \pm \frac{1}{2}$$

(b) Show that if $l = 0$, the answer is $g_{\text{Landé}} = 2$. Interpret this result.

Solution to Problem 5.4

(a) We just need to substitute $s = 1/2$, $g_s = 2$ and $j = l \pm 1/2$ in Eq. (5.20) for the Landé g-factor. The result for $j = l + 1/2$ is then

$$g_{\text{Landé}} = \frac{1}{2(l+1/2)(l+3/2)} \{[(l+1/2)(l+3/2) + l(l+1) - 3/4] +$$

$$2[(l+1/2)(l+3/2) + 3/4 - l(l+1)]\}$$

$$= \frac{1}{2(l+1/2)(l+3/2)} [2l^2 + 4l + 3/2 + (l+3/2)]$$

$$= 1 + \frac{1}{2l+1} \qquad ; j = l + \frac{1}{2}$$

The result for $j = l - 1/2$ is

$$g_{\text{Landé}} = \frac{1}{2(l - 1/2)(l + 1/2)} \{[(l - 1/2)(l + 1/2) + l(l + 1) - 3/4] +$$
$$2[(l - 1/2)(l + 1/2) + 3/4 - l(l + 1)]\}$$
$$= \frac{1}{2(l - 1/2)(l + 1/2)} [2l^2 - 1/2 - (l - 1/2)]$$
$$= 1 - \frac{1}{2l + 1} \qquad\qquad ; j = l - \frac{1}{2}$$

Therefore

$$g_{\text{Landé}} = 1 \pm \frac{1}{2l + 1}$$

(b) For $l = 0$, one has $j = l + 1/2 = 1/2$ [see Eq. (5.7)], and therefore from the first formula in part (a)

$$g_{\text{Landé}} = 2$$

In this case the g-factor simply amounts to g_s since there is no orbital magnetic moment.

Problem 5.5 If the atom is neutral in T-F theory, it must contain exactly Z electrons. Start from Eqs. (5.52) and (5.54) and prove

$$\int d^3r\, n(r) = Z$$

Solution to Problem 5.5

We need to calculate

$$\int d^3r\, n(r) = \int_0^\infty 4\pi r^2 dr\, \frac{Z}{4\pi} \left(\frac{Z^{1/3}}{ba_0}\right)^3 \left[\frac{\chi(x)}{x}\right]^{3/2}$$
$$= Z \int_0^\infty dx \sqrt{x}\, \chi(x)^{3/2}$$

Since $\chi(x)$ obeys the Thomas-Fermi equation, we may use $\chi(x)^{3/2} = \sqrt{x}\, \chi''(x)$ to obtain

$$\int d^3r\, n(r) = Z \int_0^\infty dx\, x\chi''(x)$$
$$= Z[x\chi'(x)]_0^\infty - Z \int_0^\infty dx\, \chi'(x)$$

where the second line follows from an integration by parts. The boundary terms vanish, and therefore

$$\int d^3r\, n(r) = -Z \int_0^\infty dx\, \chi'(x)$$

$$= Z\chi(0)$$

$$= Z$$

Problem 5.6[4] The unphysical nature of the long-range tail on the electron distribution can be avoided if one studies an *ionized* atom with z electrons removed. In this case, the shielding function will vanish at some finite radius x_R.

(a) For the ionized atom, show the T-F equation remains valid with the electrostatic potential inside the atom given by

$$\Phi(r) - \frac{ze_p}{4\pi\varepsilon_0 R} = \frac{Ze_p}{4\pi\varepsilon_0 r}\chi(x) \qquad ; r \le R$$

(b) Show the radius of the atom is determined from

$$\chi(x_R) = 0 \qquad ; x_R = \frac{Z^{1/3}R}{ba_0}$$

(c) Show the degree of ionization is determined by (compare Prob. 5.5)

$$\int_0^R 4\pi r^2 n(r)\, dr = Z \int_0^{x_R} \chi(x)^{3/2} x^{1/2} dx = Z - z$$

Solution to Problem 5.6

(a) We are considering an ionized atom of radius R, where z electrons have been removed. For $r \ge R$ the electric field of the atom may be calculated using Eq. (5.39) as

$$\mathcal{E}_r = \frac{ze_p}{4\pi\epsilon_0 r^2} \qquad ; r \ge R$$

and the corresponding electrostatic potential of the atom is

$$\Phi(r) = \frac{ze_p}{4\pi\epsilon_0 r} \qquad ; r \ge R$$

---

[4]Note there were two Probs. 5.5 in the book, hence the re-labeling of Probs. 5.6-5.9.

The integration of Eq. (5.36) must be performed choosing the constant of integration so that $\Phi(R) = ze_p/4\pi\epsilon_0 R$. Thus[5]

$$\Phi(r) = \frac{\hbar^2}{2m|e|} \left(3\pi^2\right)^{2/3} n(r)^{2/3} + \frac{ze_p}{4\pi\epsilon_0 R} \qquad ; r \leq R$$

For $r \leq R$, Eq. (5.42) therefore gets modified to

$$\frac{1}{r}\frac{\partial^2}{\partial r^2}(r\Phi) = \kappa \left[\Phi(r) - \frac{ze_p}{4\pi\epsilon_0 R}\right]^{3/2}$$

As suggested in the problem, we define the difference as

$$\Phi(r) - \frac{ze_p}{4\pi\varepsilon_0 R} \equiv \left(\frac{Ze_p}{4\pi\varepsilon_0 r}\right)\chi(x)$$

as in Eq. (5.53). The additional constant term leaves $\partial^2(r\Phi)/\partial r^2$ unchanged, and substitution of this relation in the previous relation then reproduces the dimensionless Thomas-Fermi Eq. (5.52) exactly as in the text.

(b) At $r = R$, we have $\Phi(R) = ze_p/4\pi\epsilon_0 R$, which implies that

$$\chi(x_R) = 0 \qquad ; x_R = \frac{Z^{1/3}R}{ba_0}$$

(c) We may express the degree of ionization of the atom directly in terms of the shielding function, using the results of parts (a) and (b). Start from the result in (a) that

$$n(r) = \left(\frac{2m|e|}{\hbar^2}\right)^{3/2}\frac{1}{3\pi^2}\left[\Phi(r) - \frac{ze_p}{4\pi\epsilon_0 R}\right]^{3/2}$$

Then use the last result from (a) to show

$$\int_0^R 4\pi r^2 n(r)\, dr = \int_0^R 4\pi r^2 \left(\frac{2m|e|}{\hbar^2}\right)^{3/2}\frac{1}{3\pi^2}\left[\Phi(r) - \frac{ze_p}{4\pi\epsilon_0 R}\right]^{3/2}$$

$$= \left(\frac{2m|e|}{\hbar^2}\right)^{3/2}\frac{4}{3\pi}\int_0^R r^2 \left[\frac{Z|e|}{4\pi\epsilon_0 r}\chi(x)\right]^{3/2} dr$$

Now combine the constants

$$\frac{4}{3\pi}\left(\frac{2me^2 Z}{4\pi\varepsilon_0\hbar^2}\right)^{3/2} = \left(\frac{Z}{ba_0}\right)^{3/2} = Z\left(\frac{Z^{1/3}}{ba_0}\right)^{3/2}$$

---

[5]Notice that $n(R) = 0$.

It follows that

$$\int_0^R 4\pi r^2 n(r)\, dr = Z \int_0^{x_R} \chi(x)^{3/2} x^{1/2} dx$$

$$= Z \int_0^{x_R} x\, \chi''(x) dx$$

$$= Z\left[x_R \chi'(x_R) + 1\right]$$

where the last line again follows from a partial integration, as in the solution to Prob. 5.5. This is just the total number of electrons left in the ionized atom

$$\int_0^R 4\pi r^2 n(r)\, dr = Z - z$$

Therefore we have a relation between the slope of the T-F function at the radius of the ionized atom $\chi'(x_R)$ and the degree of ionization of the atom

$$x_R \chi'(x_R) = -\frac{z}{Z}$$

The Thomas-Fermi Eq. (5.52) must now be solved for the ionized atom with the boundary conditions $\chi(0) = 1$ and $\chi(x_R) = 0$, and the degree of ionization is then determined from the resulting slope $x_R \chi'(x_R) = -z/Z$ at the radius x_R. The actual radius is related to x_R by $R = (ba_0/Z^{1/3})x_R$.

Problem 5.7 Express the coefficients (c_0, c_1) in the T-F density functional $F(n)$ in Eq. (5.69) in terms of the physical constants of the atomic system.

Solution to Problem 5.7

We may use Eq. (5.54) in the text to express $\chi(x)$ directly in terms of the density

$$Z[\chi(x)]^{3/2} x^{1/2} dx = 4\pi n(r) r^2 dr$$

$$\frac{\chi(x)}{x} = \left[\frac{4\pi n(r)}{Z}\right]^{2/3} \left(\frac{ba_0}{Z^{1/3}}\right)^2$$

If these expressions are substituted into Eq. (5.66), we obtain

$$E = \int_0^\infty \left[\frac{\hbar^2}{20m}[3\pi^2 n(r)]^{2/3} - \frac{1}{2}\left(\frac{Ze_p^2}{4\pi\epsilon_0 r}\right)\right] 4\pi r^2 n(r) dr$$

The kinetic energy of the Fermi gas of electrons contributes to the first term, and there is a contribution to both terms coming from the electrostatic self-energy of the atom in the T-F approximation.

This result can be re-expressed as a series in $n(r)$

$$E = \frac{(3\pi^2)^{2/3}\hbar^2}{20m} \int n(r)^{5/3} \left[1 + \frac{10m}{(3\pi^2)^{2/3}\hbar^2} \left(-\frac{Ze_p^2}{4\pi\epsilon_0 r} \right) n(r)^{-2/3} \right] d^3r$$

from which we can read off the constants c_0 and c_1 of Eq. (5.69)

$$c_0 = (3\pi^2)^{2/3} \frac{\hbar^2}{20m} \qquad ; c_1 = \frac{1}{2c_0} = \frac{1}{(3\pi^2)^{2/3}} \frac{10m}{\hbar^2}$$

Problem 5.8 (a) Give the anticipated electron configurations for the following elements: $_{55}$Cs, $_{70}$Yb, $_{85}$At, $_{86}$Rn ;

(b) Discuss the expected chemical properties of these elements;

(c) Go to the website in Fig. 5.18 in the text, click on the elements, and check your answers to parts (a) and (b).

Solution to Problem 5.8

(a) The expected electron configurations can be read off from Fig. 5.15 in the text. Start from the core configuration of the inert gas $_{54}$Xe and build on this

$_{54}$Xe : $(1s)^2(2s)^2(2p)^6(3s)^2(3p)^6(4s)^2(3d)^{10}(4p)^6(5s)^2(4d)^{10}(5p)^6$

$_{55}$Cs : $\cdots (6s)^1$

$_{70}$Yb : $\cdots (6s)^2(4f)^{14}$

$_{85}$At : $\cdots (6s)^2(4f)^{14}(5d)^{10}(6p)^5$

$_{86}$Rn : $\cdots (6s)^2(4f)^{14}(5d)^{10}(6p)^6$

(b) The expected chemical properties then follow:

- The element $_{55}$Cs has a single valence $6s$ electron with the $_{54}$Xe core, and it should therefore be an alkali metal;
- The element $_{70}$Yb has a pair of $6s$ electrons and a filled $4f$ inner shell. It should be a divalent metal;
- The element $_{86}$Rn with filled shells should be another inert gas;
- The element $_{85}$At has a hole in the $6p$ shell in the inert gas $_{86}$Rn core. It should be a halogen.

(c) These are indeed the observed properties of these elements.

Problem 5.9 Identify the next alkali, halogen, and inert gas in Fig. 5.17 in the text. Compare with Fig. 5.18.

Solution to Problem 5.9

From Fig. 5.15 in the text:

- The next alkali metal after the inert gas $_{86}$Rn should be element 87 with one additional $7s$ valence electron;
- The next inert gas after $_{86}$Rn should have the additional electron configuration $(7s)^2(5f)^{14}(6d)^{10}(7p)^6$. With 32 additional electrons, it should be element 118;
- The next halogen should have a hole in the $7p$ shell, and it should thus be element 117.

In Fig. 5.18 in the text, $_{87}$Fr is tabulated as an alkali, and its properties are known. Although included in this periodic table, it is not clear that the inert gas $_{118}$Uuo and halogen $_{117}$Uus have ever been observed.

Chapter 6

Nuclear Physics

Problem 6.1 (a) The hamiltonian for two non-relativistic particles interacting through a potential that only depends on their relative coordinate is

$$H(\mathbf{r}_1, \mathbf{r}_2, \mathbf{p}_1, \mathbf{p}_2) = \frac{\mathbf{p}_1^2}{2m_1} + \frac{\mathbf{p}_2^2}{2m_2} + V(|\mathbf{r}_2 - \mathbf{r}_1|)$$

Introduce the following change of variables

$$\mathbf{R} = \frac{m_1\mathbf{r}_1 + m_2\mathbf{r}_2}{m_1 + m_2} \qquad ; \ \mathbf{r} = \mathbf{r}_2 - \mathbf{r}_1$$

$$\mathbf{P} = \mathbf{p}_1 + \mathbf{p}_2 \qquad ; \ \mathbf{p} = \frac{m_1\mathbf{p}_2 - m_2\mathbf{p}_1}{m_1 + m_2}$$

Show the hamiltonian takes the form

$$H = \frac{\mathbf{P}^2}{2M} + \frac{\mathbf{p}^2}{2\mu} + V(r) \qquad ; \ \frac{1}{\mu} \equiv \frac{1}{m_1} + \frac{1}{m_2}$$

Here $M = m_1 + m_2$ is the total mass, and μ is the *reduced mass*.

(b) Show that the commutation relations in quantum mechanics that follow from the definitions in part (a) are[1]

$$[p_i, x_j] = \frac{\hbar}{i}\delta_{ij} \qquad ; \ [P_i, X_j] = \frac{\hbar}{i}\delta_{ij}$$

$$[p_i, X_j] = [P_i, x_j] = 0$$

Hence, justify the following replacements in part (a)

$$p_i \to \frac{\hbar}{i}\frac{\partial}{\partial x_i} \qquad ; \ P_i \to \frac{\hbar}{i}\frac{\partial}{\partial X_i}$$

---

[1]See Prob. 4.7. Here the operators for the two particles are independent, and they commute with each other.

[*Note*: These relations can be checked with an explicit change of variables in the laplacians involved. We leave this as an exercise for the dedicated reader.]

(c) Show the solution to the stationary-state Schrödinger equation for two particles in a large box with periodic boundary conditions then *factors* into $\Psi(\mathbf{r}_1, \mathbf{r}_2) = \psi_{\mathbf{K}}(\mathbf{R})\phi(\mathbf{r})$ where the C-M wave function is that of Eq. (4.118). What is the remaining Schrödinger equation for $\phi(\mathbf{r})$?

Solution to Problem 6.1

(a) Upon inverting the equation defining $\mathbf{P}$ and $\mathbf{p}$, we have

$$\mathbf{p}_1 = \frac{m_1}{m_1 + m_2}\, \mathbf{P} - \mathbf{p} \qquad ; \; \mathbf{p}_2 = \frac{m_2}{m_1 + m_2}\, \mathbf{P} + \mathbf{p}$$

The kinetic-energy term in the hamiltonian may then be expressed directly in terms of $\mathbf{P}$ and $\mathbf{p}$ using these relations

$$T = \frac{\mathbf{p}_1^2}{2m_1} + \frac{\mathbf{p}_2^2}{2m_2} = \frac{\mathbf{P}^2}{2(m_1 + m_2)} + \frac{\mathbf{p}^2}{2}\left(\frac{1}{m_1} + \frac{1}{m_2}\right)$$

$$= \frac{\mathbf{P}^2}{2M} + \frac{\mathbf{p}^2}{2\mu}$$

Here M is the total mass and μ the reduced mass defined above. Since the potential term in the hamiltonian is clearly a function of the relative distance, the hamiltonian takes the form specified in the problem.

(b) The pairs of vectors $(\mathbf{p}_1, \mathbf{r}_1)$ and $(\mathbf{p}_2, \mathbf{r}_2)$ represent canonically conjugate variables that fulfill the commutation relations

$$[p_{1i}, r_{1j}] = \frac{\hbar}{i}\delta_{ij} \qquad ; \; [p_{2i}, r_{2j}] = \frac{\hbar}{i}\delta_{ij}$$

while variables corresponding to different particles commute

$$[p_{1i}, r_{2j}] = [p_{2i}, r_{1j}] = 0$$

We may now derive the commutation relations among the new variables, using the properties stated above

$$[p_i, r_j] = \frac{m_1}{M}\,[p_{2i}, r_{2j}] + \frac{m_2}{M}\,[p_{1i}, r_{1j}] = \frac{\hbar}{i}\delta_{ij}$$

$$[P_i, R_j] = \frac{m_1}{M}\,[p_{1i}, r_{1j}] + \frac{m_2}{M}\,[p_{2i}, r_{2j}] = \frac{\hbar}{i}\delta_{ij}$$

and

$$[P_i, r_j] = [p_{2i}, r_{2j}] - [p_{1i}, r_{1j}] = 0$$

$$[p_i, R_j] = \frac{m_1 m_2}{M^2}[p_{2i}, r_{2j}] - \frac{m_1 m_2}{M^2}[p_{1i}, r_{1j}] = 0$$

These commutation relations imply that $(\mathbf{p}, \mathbf{r})$ and $(\mathbf{P}, \mathbf{R})$ are also pairs of canonically conjugate variables. The replacements

$$p_i \to \frac{\hbar}{i}\frac{\partial}{\partial x_i} \qquad ; P_i \to \frac{\hbar}{i}\frac{\partial}{\partial X_i}$$

then provide an explicit representation of the total and relative momentum operators $(\mathbf{P}, \mathbf{p})$ in the coordinate space of the total and relative coordinates $(\mathbf{R}, \mathbf{r})$, respectively.

(c) In part (a) we expressed the hamiltonian in terms of the new variables, $(\mathbf{P}, \mathbf{R}, \mathbf{p}, \mathbf{r})$. In part (b) we have shown that it is possible to provide an explicit representation of the momenta $(\mathbf{P}, \mathbf{p})$ in the coordinate space of $(\mathbf{R}, \mathbf{r})$. This means that the hamiltonian operator may be represented explicitly in coordinate space as

$$H = -\frac{\hbar^2}{2M}\nabla_R^2 + \left[-\frac{\hbar^2}{2\mu}\nabla_r^2 + V(r)\right] \equiv H_1 + H_2$$

where H_1 depends uniquely on $(\mathbf{P}, \mathbf{R})$ and H_2 on $(\mathbf{p}, \mathbf{r})$.

Suppose now that $\psi(\mathbf{R})$ and $\phi(\mathbf{r})$ are eigenfunctions of H_1 and H_2, so that

$$H_1\psi(\mathbf{R}) = E_1\psi(\mathbf{R}) \qquad ; H_2\phi(\mathbf{r}) = E_2\phi(\mathbf{r})$$

Then $\psi(\mathbf{R})\phi(\mathbf{r})$ is an eigenfunction of the total hamiltonian, since

$$H\psi(\mathbf{R})\phi(\mathbf{r}) = (H_1 + H_2)\psi(\mathbf{R})\phi(\mathbf{r}) = (E_1 + E_2)\psi(\mathbf{R})\phi(\mathbf{r})$$

In particular, we observe that H_1 is the hamiltonian of a free particle of mass M. If we assume a large box with periodic boundary conditions, the wave function $\psi(\mathbf{R})$ must then have the form of Eq. (4.118)

$$\psi_{\mathbf{K}}(\mathbf{R}) = \frac{1}{\sqrt{L^3}}e^{i\mathbf{K}\cdot\mathbf{R}}$$

$$\mathbf{K} = \frac{2\pi}{L}(n_x, n_y, n_z) \qquad ; n_i = 0, \pm 1, \pm 2, \cdots$$

Thus we conclude that the center-of-mass of the system has a definite momentum $\mathbf{P} = \hbar\mathbf{K}$.

Finally, the remaining part of the total wave function, which depends only upon the relative distance **r**, obeys the stationary-state Schrödinger equation

$$\left[-\frac{\hbar^2}{2\mu}\nabla_r^2 + V(r) \right]\phi(\mathbf{r}) = E_2\phi(\mathbf{r})$$

This is a familiar one-body Schrödinger equation in the relative coordinate **r**, with the reduced mass μ. In this case, it is not possible to obtain the eigenfunctions and eigenvalues without specifying the potential $V(r)$.

Problem 6.2 The binding energy of the deuteron is $|E| = 2.22\,\text{MeV}$.

(a) Given the potential in Eq. (6.21) and the second of Eqs. (6.29), and this value of the binding energy, solve the eigenvalue Eq. (6.24) numerically to determine the new V_0;

(b) Determine and plot the corresponding wave function $u_{10}(r)$;

(c) Plot the corresponding probability density. Discuss.

Solution to Problem 6.2

(a) Before solving the eigenvalue equation, we need to assign values to the physical parameters

$$\hbar c \approx 197.3\,\text{MeV-F} \qquad ;\ m_pc^2 = 938.3\,\text{MeV}$$

The mass of the neutron is slightly larger than the mass of the proton, and for the present purposes it can be assumed $m_n = m_p$; in this case $\mu \approx m_p/2$. Thus

$$\frac{2\mu}{\hbar^2} \approx \frac{m_pc^2}{(\hbar c)^2} = 0.0241\,\text{MeV}^{-1}\text{-F}^{-2}$$

With $R = 2\,\text{F}$ from Eq. (6.29), this gives

$$\sqrt{\frac{2\mu}{\hbar^2}}\,R \approx 0.310\,\text{MeV}^{-1/2} \quad ;\ R = 2\,\text{F}$$

The eigenvalue Eq (6.24) is

$$\kappa\cot\kappa R = -\gamma \qquad ;\ \kappa^2 \equiv \frac{2\mu}{\hbar^2}(V_0 - |E|)$$

$$;\ \gamma^2 \equiv \frac{2\mu}{\hbar^2}|E|$$

It now reads

$$\sqrt{V_0 - 2.22} \cot\left[0.31\sqrt{V_0 - 2.22}\right] = -1.49$$

where $|E|$ and V_0 are expressed in units of MeV.

Notice that the cotangent is singular at $V_0 \approx 2.22 + 102.7\, n^2$, with $n = 0, \pm 1, \pm 2, \cdots$, although for $n = 0$ the eigenvalue equation is not singular. Since physically we are interested in $\sqrt{V_0 - 2.22} > 0$, the relevant singularities correspond to $n = 1, 2, \cdots$. As we see from Fig. 6.1, the problem admits multiple solutions, each falling in a region between two singularities. The first solution, which is represented by a circle in the l.h.s. of the plot, may be calculated numerically to give

$$V_0 = 36.6\,\text{MeV}$$

This is the solution that we were looking for. We may easily see that the remaining solutions, which correspond to larger values of V_0 where the two curves in the plot cross, are excited states in deeper wells.[2]

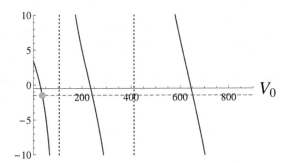

Fig. 6.1 Graphical solution of the eigenvalue Eq. (6.24) for the parameters in Prob. 6.2. The horizontal dashed line corresponds to $\gamma = 1.49$, and the solid lines correspond to $\sqrt{V_0 - 2.22} \cot[0.31\sqrt{V_0 - 2.22}]$. The solution we are looking for is identified by the circle on the l.h.s. of the plot.

(b) The general form of the wave function is given in Eq. (6.22) in the text, where the constants $\mathcal{B}$ and $\mathcal{D}$ must vanish to ensure the boundary condition at the origin and the normalizability of the solution. Thus

$$u_I(r) = \mathcal{A} \sin \kappa r \qquad ; 0 \leq r \leq R$$
$$u_{II}(r) = \mathcal{C} e^{-\gamma r} \qquad ; r > R$$

---

[2]Larger values of V_0 imply larger values of κ and therefore more oscillations in the region I of Fig. 6.5 in the text.

The continuity of the wave function at $r = R$ allows us to express $\mathcal{A}$ in terms of $\mathcal{C}$. Finally, the normalization of the wave function $\int_0^\infty dr\, |u(r)|^2 = 1$ allows us to determine $\mathcal{C}$. This gives[3]

$$\mathcal{C} = 1.54\,\mathcal{A} \qquad ; \mathcal{A} = 0.562\,\mathrm{F}^{-1/2}$$

The resulting wave function $u(r)$ is plotted in Fig. 6.2.

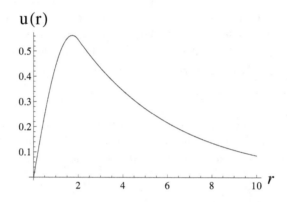

Fig. 6.2 S-wave bound-state wave function $u(r)$ in the attractive square-well potential in Prob. 6.2. Here r is in Fermis, $u(r)$ in $(\mathrm{Fermi})^{-1/2}$, and $R = 2\,\mathrm{F}$.

(c) The probability density for the state that we have calculated is displayed in Fig. 6.3.

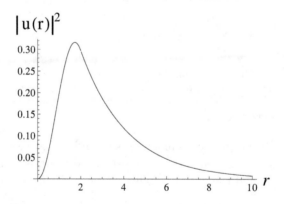

Fig. 6.3 Probability density $|u(r)|^2$ corresponding to the s-wave bound-state wave function $u(r)$ shown in Fig. 6.2.

---

[3]The algebra here is detailed in Prob. 6.3. Recall the full wave function is $\psi_{100}(\mathbf{r}) = [u(r)/r]Y_{00}(\theta, \phi) = u(r)/\sqrt{4\pi}\,r$.

In particular, the probability of finding the particle outside the well ($r > R$) is rather large, since $\int_R^\infty |u(r)|^2 dr = 0.642$. Thus we see that it is more likely to find the particle in the classically-forbidden region, rather than in the classically-allowed region.

Problem 6.3 Determine the correct normalization for the wave function in Fig. 6.6 in the text for small, non-zero γ. Show that it reduces to Eq. (6.33) as $\gamma \to 0$.

Solution to Problem 6.3

For $|E| \to 0$, or in dimensionless form $\gamma R \to 0$, Eq. (6.24) implies

$$\kappa R = R\sqrt{\frac{2\mu V_0}{\hbar^2} - \gamma^2} \to \frac{\pi}{2} \qquad ; \gamma R \to 0$$

From Eqs. (6.22) with $\mathcal{B} = \mathcal{D} = 0$, the full normalization integral is,

$$\int_0^\infty u_{10}^2(r)dr = \frac{\mathcal{A}^2}{2}\left[R - \frac{\sin 2\kappa R}{2\kappa}\right] + \frac{\mathcal{C}^2 e^{-2\gamma R}}{2\gamma} = 1$$

This is re-written as

$$\mathcal{A}^2\left[\frac{R}{2} - \frac{\sin\left(2R\sqrt{2\mu V_0/\hbar^2 - \gamma^2}\right)}{4\sqrt{2\mu V_0/\hbar^2 - \gamma^2}}\right] + \frac{\mathcal{C}^2 e^{-2\gamma R}}{2\gamma} = 1$$

We can use the continuity of the wave function at R to relate $\mathcal{A}$ to $\mathcal{C}$

$$\mathcal{A} = \mathcal{C}e^{-\gamma R}\csc(\kappa R)$$

Thus the correct normalization condition is

$$\csc^2\left(R\sqrt{\frac{2\mu V_0}{\hbar^2} - \gamma^2}\right)\left[\frac{R}{2} - \frac{\sin\left(2R\sqrt{2\mu V_0/\hbar^2 - \gamma^2}\right)}{4\sqrt{2\mu V_0/\hbar^2 - \gamma^2}}\right] + \frac{1}{2\gamma} = \frac{e^{2\gamma R}}{\mathcal{C}^2}$$

After noticing that the first term is just $R/2$ as $\gamma R \to 0$, we can write this as

$$\frac{R}{2} + \frac{1}{2\gamma} = \frac{e^{2\gamma R}}{\mathcal{C}^2} \qquad ; \gamma R \to 0$$

For $\gamma R \to 0$, this is re-written as

$$\mathcal{C}^2 = 2\gamma \qquad ; \gamma R \to 0$$

We conclude from the above that

$$\int_0^\infty u_{10}^2(r)dr \approx \frac{C^2 e^{-2\gamma R}}{2\gamma} = 1 \qquad ; \gamma R \to 0$$

thus reproducing Eq. (6.33).

Problem 6.4 Suppose the potential in Eq. (6.21) is modified to contain an infinite repulsion ("hard core") at short distances so that it is augmented with the relation $V = +\infty$ for $r \le c < R$.[4]

(a) How is the eigenvalue Eq. (6.24) changed?

(b) Suppose $c = 0.4\,\mathrm{F}$ and $R = 2\,\mathrm{F}$, two fairly realistic values. How deep must V_0 be to produce just one bound state at zero energy?

Solution to Problem 6.4

(a) The potential is now

$$V(r) = \begin{cases} \infty & ; \ 0 \le r \le c \\ -V_0 & ; \ c < r \le R \\ 0 & ; \ R < r \end{cases}$$

The general solution of Eq. (6.22) must obey the boundary condition $u_I(c) = 0$, which implies the wave function in region I takes the form

$$u_I(r) = \mathcal{A} \sin \kappa(r - c)$$

The constant $\mathcal{D}$ must also vanish in this case because of the requirement of normalizability of the wave function, so that in region II

$$u_{II}(r) = \mathcal{C} e^{-\gamma r}$$

The new eigenvalue equation is once again obtained by equating the logarithmic derivative of the solutions $u_I(r)$ and $u_{II}(r)$ at $r = R$

$$\kappa \cot \kappa(R - c) = -\gamma$$

which correctly reduces to Eq. (6.24) for $c = 0$.

(b) For a bound state at zero energy, the solution to the eigenvalue equation becomes

$$\kappa(R - c) = (R - c)\sqrt{\frac{2\mu V_0}{\hbar^2}} = \frac{\pi}{2} \qquad ; \ |E| \to 0$$

---

[4]The nucleon-nucleon potential exhibits such a short-range repulsion.

With $\mu = m_p/2$ and

$$R - c = 1.6\,\mathrm{F} \qquad ; \qquad \frac{\hbar^2}{2m_p} = 20.7\,\mathrm{MeV\text{-}F^2}$$

this gives

$$V_0 = \frac{\hbar^2\pi^2}{4m_p(R-c)^2} \approx 39.9\,\mathrm{MeV}$$

Problem 6.5 The parity operator P in quantum mechanics is defined to reflect the spatial coordinate $P\Psi(\mathbf{r}, t) \equiv \eta_P\Psi(-\mathbf{r}, t)$, where, for the present purposes, $\eta_P = +1$. Suppose one has a state that at a time t_0 is an *eigenstate* of parity so that $P\Psi(\mathbf{r}, t_0) \equiv \Psi(-\mathbf{r}, t_0) = \pi\Psi(\mathbf{r}, t_0)$, where π is the eigenvalue. Solution to the time-dependent Schrödinger equation involves repeated application of the hamiltonian as a function of time. Prove that if the hamiltonian *commutes* with the parity operator (*i.e.* $[P, H] = 0$), then $\Psi(\mathbf{r}, t)$ remains an eigenstate of parity for all time.

Solution to Problem 6.5

Assume the hamiltonian H is independent of time. The solution to the time-dependent Schrödinger equation may then be obtained in the form

$$\Psi(\mathbf{r}, t) = e^{-iH(t-t_0)/\hbar}\Psi(\mathbf{r}, t_0)$$

If the parity operator commutes with the hamiltonian, then it also commutes with any power of the hamiltonian

$$[P, H^n] = 0 \qquad ; \, n = 1, 2, \cdots$$

As a result, the parity operator commutes with the time-evolution operator

$$[P, e^{-iH(t-t_0)/\hbar}] = 0$$

and we have

$$P\Psi(\mathbf{r}, t) = Pe^{-iH(t-t_0)/\hbar}\Psi(\mathbf{r}, t_0) = e^{-iH(t-t_0)/\hbar}P\Psi(\mathbf{r}, t_0)$$

Thus if $\Psi(\mathbf{r}, t)$ is an eigenstate of P at $t = t_0$

$$P\Psi(\mathbf{r}, t_0) = \pi\Psi(\mathbf{r}, t_0)$$

then $\Psi(\mathbf{r}, t)$ remains an eigenstate of parity for all time

$$P\Psi(\mathbf{r}, t) = \pi\Psi(\mathbf{r}, t)$$

Problem 6.6 Prove the following theorem: Of two isobars of given B, with Z differing by one, at least one is unstable under the β-decay interactions in Eqs. (6.2)–(6.3). Assume massless neutrinos.

Solution to Problem 6.6

The theorem follows from energetics. Let (Z, N) denote the nucleus with proton number Z, neutron number N, baryon number $B = Z + N$, and ground-state energy $M_{Z,N}c^2$. Let $e^{\pm}$ denote positrons and electrons with rest energy $m_e c^2$, and assume the neutrinos are massless. Consider the three nuclear β-decay interactions

$$(Z, N) \to (Z + 1, N - 1) + e^- + \bar{\nu}_e \qquad ; \ \beta^-\text{-decay}$$
$$(Z + 1, N - 1) + e^- \to (Z, N) + \nu_e \qquad ; \ \beta^-\text{-capture}$$
$$(Z + 1, N - 1) \to (Z, N) + e^+ + \nu_e \qquad ; \ \beta^+\text{-decay}$$

We now observe the following:

- If $M_{Z,N} > M_{Z+1,N-1} + m_e$, the first reaction will proceed;
- If $M_{Z,N} < M_{Z+1,N-1} + m_e$, the second reaction will proceed;
- Thus one of the nuclei (Z, N) or $(Z + 1, N - 1)$ is unstable against the β-decay interaction;
- Furthermore, if $M_{Z+1,N-1} > M_{Z,N} + m_e$, the third reaction will also take place.[5]

Problem 6.7 Suppose a radial wave function $\psi(r)$ is of the form $\psi(r) = u(r)/r$. Show that the boundary conditions that (ψ, ψ') be continuous at finite r are equivalent to (u, u') being continuous.

Solution to Problem 6.7

We have

$$\psi(r) = \frac{u(r)}{r}$$
$$\psi'(r) = \frac{u'(r)}{r} - \frac{u(r)}{r^2}$$

First, we notice that the continuity of $\psi(r)$ implies the continuity of $u(r)$, since from the first equation $u(r) = r\psi(r)$. On the other hand, we also have

---

[5]These arguments can be converted from nuclear to atomic masses by adding Z electrons to the first two reactions, and $Z + 1$ electrons to the third.

from the second equation

$$u'(r) = r\psi'(r) + \frac{u(r)}{r}$$

Since $\psi'(r)$ is continuous by hypothesis, and we have just shown that $u(r)$ is also continuous, we have that $u'(r)$ is continuous as well.

Problem 6.8 Use the atomic mass tables [Lawrence Berkeley Laboratory (2003)], and consider the following reactions in the center-of-mass system:

$$^4_2\text{He} + {}^{12}_6\text{C} \rightarrow {}^{16}_8\text{O} + \gamma$$
$$^{28}_{14}\text{Si} \rightarrow {}^{14}_7\text{N} + {}^{14}_7\text{N}$$
$$^{238}_{92}\text{U} \rightarrow {}^{104}_{42}\text{Mo} + {}^{134}_{50}\text{Sn}$$
$$^{12}_6\text{C}^*(7.6\,\text{MeV}) \rightarrow {}^4_2\text{He} + {}^4_2\text{He} + {}^4_2\text{He}$$

Here γ is a massless photon, and $^{12}_6\text{C}^*(7.6\,\text{MeV})$ represents an excited state at the indicated energy.

Determine which of these nuclear reactions is exothermic, and, if so, determine the energy release.

Solution to Problem 6.8

We consider the process

$$^4_2\text{He} + {}^{12}_6\text{C} \rightarrow {}^{16}_8\text{O} + \gamma$$

With the use of the atomic mass tables [Lawrence Berkeley Laboratory (2003)], we have

$$M(^4_2\text{He})c^2 = 3{,}728.4\,\text{MeV}$$
$$M(^{12}_6\text{C})c^2 = 11{,}177.9\,\text{MeV}$$
$$M(^{16}_8\text{O})c^2 = 14{,}899.2\,\text{MeV}$$

We see that

$$Q = M(^{16}_8\text{O})c^2 - M(^4_2\text{He})c^2 - M(^{12}_6\text{C})c^2 \approx -7.1\,\text{MeV}$$

Since $Q < 0$ the process is exothermic, and the reaction can proceed as indicated.

We next consider the process

$$^{28}_{14}\text{Si} \rightarrow {}^{14}_7\text{N} + {}^{14}_7\text{N}$$

In this case we have

$$M(^{28}_{14}\text{Si})c^2 = 26,060.3\,\text{MeV}$$
$$M(^{14}_{7}\text{N})c^2 = 13,043.8\,\text{MeV}$$

and

$$Q = 2M(^{14}_{7}\text{N})c^2 - M(^{28}_{14}\text{Si})c^2 = 27.3\,\text{MeV}$$

Since $Q > 0$ the process is endothermic, and it will not go unless additional energy is supplied.

We now consider the process

$$^{238}_{92}\text{U} \rightarrow {}^{104}_{42}\text{Mo} + {}^{134}_{50}\text{Sn}$$

In this case we have

$$M(^{238}_{92}\text{U})c^2 = 221,743\,\text{MeV}$$
$$M(^{104}_{42}\text{Mo})c^2 = 96,795\,\text{MeV}$$
$$M(^{134}_{50}\text{Sn})c^2 = 124,753\,\text{MeV}$$

and

$$Q = M(^{104}_{42}\text{Mo})c^2 + M(^{134}_{50}\text{Sn})c^2 - M(^{238}_{92}\text{U})c^2 = -195\,\text{MeV}$$

Since $Q < 0$ the process is exothermic, and this fission reaction can take place.

Finally, consider the process

$$^{12}_{6}\text{C}^\star(7.6\,\text{MeV}) \rightarrow {}^{4}_{2}\text{He} + {}^{4}_{2}\text{He} + {}^{4}_{2}\text{He}$$

In this case we have

$$M^\star(^{12}_{6}\text{C})c^2 = 11,177.9\,\text{MeV} + 7.6\,\text{MeV} = 11,185.5\,\text{MeV}$$
$$M(^{4}_{2}\text{He})c^2 = 3,728.4\,\text{MeV}$$

and

$$Q = 3M(^{4}_{2}\text{He})c^2 - M^\star(^{12}_{6}\text{C})c^2 = -0.3\,\text{MeV}$$

The process is exothermic, and the $J^\pi = 0^+$ excited state in $^{12}_{6}\text{C}$ at this energy can indeed decay into three alpha-particles.[6]

---

[6]The inverse reaction $3\alpha \rightarrow {}^{12}_{6}\text{C}^\star(0^+) \rightarrow {}^{12}_{6}\text{C} + e^+ + e^-$ provides one of the paths for the formation of heavier elements in stars.

Problem 6.9 This problem examines the Coulomb interaction energy of two charges e_p uniformly distributed over a sphere (drop) of radius R. One makes use of Gauss' law, which states that the electric field at a radius r arising from a uniform spherical distribution of charge comes from the charge *inside* of the sphere of radius r, all of that charge acting as a point source at the origin (see chapter 5).

(a) Show that the electric field at a radius r inside the drop coming from the first particle is then given by

$$E_r = \frac{e_p}{4\pi\varepsilon_0} \frac{r}{R^3} \qquad ; r \leq R$$

(b) The field is derivable from a potential as $E_r = -\partial\Phi/\partial r$, and at the radius of the drop, the potential is known to be $e_p/4\pi\varepsilon_0 R$. Hence show that inside of the drop the potential is given by

$$\Phi = \frac{e_p}{4\pi\varepsilon_0 R} \left(\frac{3}{2} - \frac{1}{2}\frac{r^2}{R^2} \right) \qquad ; r \leq R$$

(c) Show the interaction energy of that part of the second charge in the shell $4\pi r^2\, dr$ with the first charge is given by

$$dW = \frac{e_p^2}{4\pi\varepsilon_0} \frac{3}{R^4} \left(\frac{3}{2} - \frac{1}{2}\frac{r^2}{R^2} \right) r^2\, dr$$

(d) Show that the total interaction energy of the pair of charges is

$$W = \frac{6}{5} \frac{e_p^2}{4\pi\varepsilon_0} \frac{1}{R}$$

Hence conclude that the interaction of $Z(Z-1)/2$ pairs of charges is given by the first of Eqs. (6.53).

Solution to Problem 6.9

(a) According to Gauss' law, the flux of the electric field across a closed surface yields q/ε_0, where q is the charge enclosed in the surface. For a uniformly charged sphere, with total charge e_p and radius R, the charge enclosed in a smaller sphere of radius $r \leq R$ with the same center is $q = e_p(r/R)^3$. Thus

$$4\pi r^2 E_r = \frac{e_p}{\varepsilon_0} \left(\frac{r}{R} \right)^3 \qquad ; r \leq R$$

and therefore

$$E_r = \frac{e_p}{4\pi\varepsilon_0}\frac{r}{R^3} \qquad ; r \le R$$

(b) This radial electric field is related to the electrostatic potential by

$$E_r = -\frac{d\Phi(r)}{dr}$$

Upon integrating this relation, we obtain

$$\Phi(r) = -\frac{e_p}{8\pi\varepsilon_0 R^3}r^2 + c$$

where c is a constant of integration. Since

$$\Phi(R) = \frac{e_p}{4\pi\varepsilon_0 R}$$

we find

$$c = \frac{3}{2}\frac{e_p}{4\pi\varepsilon_0 R}$$

We thus obtain the electrostatic potential inside the charge distribution as

$$\Phi = \frac{e_p}{4\pi\varepsilon_0 R}\left(\frac{3}{2} - \frac{1}{2}\frac{r^2}{R^2}\right) \qquad ; r \le R$$

(c) Consider now a second charge similarly uniformly distributed over the sphere. The interaction energy of that part of the second charge in the shell $4\pi r^2\, dr$ with the first charge is given by

$$dW = \Phi dq$$

where dq is the electric charge contained in the shell.[7] Since $q = e_p(r/R)^3$ from the above, we have

$$dq = \frac{3e_p r^2 dr}{R^3}$$

Therefore

$$dW = \frac{e_p^2}{4\pi\varepsilon_0}\frac{3}{R^4}\left(\frac{3}{2} - \frac{1}{2}\frac{r^2}{R^2}\right)r^2\, dr$$

---

[7]We assume that the total charge e_p is distributed uniformly inside the sphere.

(d) Integration of this relation for dW over the sphere gives

$$W = \int_0^R dW = \frac{e_p^2}{4\pi\varepsilon_0} \frac{3}{R^4} \left(\frac{1}{2} - \frac{1}{10}\right) R^3$$

$$= \frac{6}{5} \frac{e_p^2}{4\pi\varepsilon_0} \frac{1}{R}$$

For a system of Z charges there are $Z(Z-1)/2$ pairs of charges, each making a contribution to the total interaction energy of this amount. Hence the total interaction energy is given by Eq. (6.53)

$$E_C = \frac{3}{5} \frac{Z(Z-1)e_p^2}{4\pi\varepsilon_0 R}$$

Problem 6.10 Make use of the following relation from Eqs. (6.53)[8]

$$a_3 = \frac{3}{5} \frac{e^2}{4\pi\varepsilon_0 r_{0c}}$$

Show that the empirical value of a_3 in Table 6.2 yields the value of r_{0c} in Eq. (6.58).

Solution to Problem 6.10

It is useful to express r_{0c} in terms of the fine-structure constant, $\alpha = e^2/(4\pi\varepsilon_0 \hbar c) \approx 1/137$:

$$r_{0c} = \frac{3}{5} \frac{\alpha\hbar c}{a_3}$$

We have already seen that $\hbar c \approx 197.3 \, \text{MeV-F}$, so the substitution of the empirical value of a_3 from Table 6.2 in the text

$$(a_3)_{\text{exp}} = 0.710 \, \text{MeV}$$

then leads to the empirical value of r_{0c} in Eq. (6.58)

$$r_{0c} = 1.22 \, \text{F}$$

Problem 6.11 (a) Use the semi-empirical mass formula (SEMF) in Eq. (6.57) and Table 6.2 in the text to compute the energy release in the third process in Prob. 6.8

$$^{238}_{92}\text{U} \rightarrow \, ^{104}_{42}\text{Mo} + \, ^{134}_{50}\text{Sn}$$

---

[8] It is assumed there that the nucleus is heavy enough so that $Z(Z-1) \approx Z^2$.

(b) Compare with the answer obtained in Prob. 6.8 from the observed atomic masses.

Solution to Problem 6.11

(a) With the use of the semi-empirical mass formula, we have for the energy release

$$E(104, 42) + E(134, 50) - E(238, 92) = -176 \,\text{MeV}$$

where the nuclei are denoted by $E(B, Z)$.

(b) The result of part (a) should be compared with the value

$$Q = -195 \,\text{MeV}$$

calculated in Problem 6.8. The SEMF thus provides a rather accurate result.

Problem 6.12 Consider the SEMF in the case of an odd nucleus ($\lambda = 0$).

(a) Minimize with respect to Z at fixed B and show that the β-stable nuclei have a $Z(B)$ of

$$Z(B) = \frac{B}{2 + (a_3 B^{2/3})/2a_4} \qquad ; \text{at minimum}$$

This determines a "line of stability." Write $B = N + Z$, solve self-consistently for Z at fixed N, and make a plot of $Z(N)$ with Z as the ordinate and N as the abscissa.

(b) For a self-bound system in its ground state where the pressure and entropy vanish, the neutron chemical potential at fixed Z is obtained from Eq. (D.10) as $\mu = dE/dN$. Write $B = N + Z$, and compute the neutron chemical potential $\mu_n(Z, N)$ at fixed Z from the SEMF. A nucleus will be unbound with respect to neutron emission when the neutron chemical potential first vanishes $\mu_n(Z, N) = 0$. Solve this equation for $Z(N)$, and determine the "neutron drip line". Plot this curve on the graph in part (a).

(c) Repeat part (b) for the "proton drip line".

(d) Compare the results in parts (a)–(c) with the chart of the nuclides in [National Nuclear Data Center (2007)].[9]

---

[9]Note that the simple 5-parameter SEMF discussed here is designed to describe nuclei along the line of stability and with small deviations about it. For a more accurate description of the entire energy surface, see the discussion of effective field theory in [Walecka (2004)].

Solution to Problem 6.12

(a) For an odd nucleus with $\lambda = 0$, the semi-empirical mass formula is given by Eq. (6.57) as

$$E(B, Z) = -a_1 B + a_2 B^{2/3} + a_3 \frac{Z^2}{B^{1/3}} + a_4 \frac{(B - 2Z)^2}{B}$$

where $B = N + Z$ is the total number of baryons. Numerical values of the coefficients $(a_1, \cdots, a_4)$ are given in Table 6.2 in the text. This is a parabola in Z with a minimum in Z at a given B located at

$$\frac{\partial E(B, Z)}{\partial Z} = 2a_3 \frac{Z}{B^{1/3}} - 4a_4 \frac{(B - 2Z)}{B} = 0$$

which implies

$$Z_{\min}(B) = \frac{B}{2 + (a_3 B^{2/3})/2a_4}$$

Since $B = N + Z$, one can look for the solution to the self-consistency equation $Z = Z_{\min}(N + Z)$ at fixed N, which provides the line of stability $Z(N)$ plotted in Fig. 6.4.

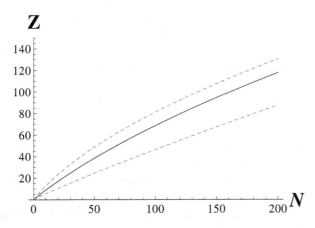

Fig. 6.4 Line of stability $Z(N)$ as a function of N. The neutron and proton drip lines are indicated by the dotted curves.

(b) Rewrite the semi-empirical mass formula in terms of N and Z

$$E(N, Z) = -a_1(N + Z) + a_2(N + Z)^{2/3} + a_3 \frac{Z^2}{(N + Z)^{1/3}} + a_4 \frac{(N - Z)^2}{N + Z}$$

The neutron chemical potential at fixed Z is then given by

$$\mu_N(N, Z) = \frac{\partial E(N, Z)}{\partial N}$$

$$= -a_1 + \frac{2a_2}{3}\frac{1}{(N+Z)^{1/3}} - \frac{a_3}{3}\frac{Z^2}{(N+Z)^{4/3}}$$

$$+ a_4 \left[\frac{(N+3Z)(N-Z)}{(N+Z)^2}\right]$$

The value of N for which this first vanishes defines the "neutron drip line"

$$\mu_N(N_{\text{drip}}, Z) = 0 \qquad ; \text{ neutron drip line}$$

At this value of N, neutrons can be removed from the system without changing its energy. The neutron drip line is indicated with the bottom dotted line in Fig. 6.4.

(c) In a similar fashion, the proton chemical potential at fixed N is

$$\mu_P(N, Z) = \frac{\partial E(N, Z)}{\partial Z}$$

$$= -a_1 + \frac{2a_2}{3}\frac{1}{(N+Z)^{1/3}} + \frac{a_3}{3}\frac{Z(5Z+6N)}{(N+Z)^{4/3}}$$

$$+ a_4 \left[\frac{(Z+3N)(Z-N)}{(N+Z)^2}\right]$$

The value of Z for which this first vanishes defines the "proton drip line"

$$\mu_P(N, Z_{\text{drip}}) = 0 \qquad ; \text{ proton drip line}$$

This curve is also indicated with the top dotted line in Fig. 6.4.

(d) The results in Fig. 6.4 can be compared with the chart of the nuclides in [National Nuclear Data Center (2007)]. The calculated *line* of stability is pretty accurate.[10] For $N = 150$, the long-lived nucleus $^{246}_{96}\text{Cm}$ exists, while the predicted value for the stable $Z_{\min}(246) = 95$. The calculated *region* of particle-stable nuclei is qualitatively correct. For $N = 100$, for example, the observed nuclei run from $^{159}_{59}\text{Pr}$ to $^{182}_{82}\text{Pb}$, a range of $\Delta Z = 23$, while the calculated range is $\Delta Z \approx 30$. For $Z = 82$, the observed nuclei run from $^{178}_{82}\text{Pb}$ to $^{220}_{82}\text{Pb}$, a range of $\Delta N = 42$ while the calculated range is $\Delta N \approx 70$. Clearly things get worse as one gets away from the line of stability, and a more sophisticated symmetry energy is required. Shell effects, neglected in

---

[10]After all, the parameters are fit to this!

the SEMF, also play a role in these stability arguments, as does α-emission for heavy nuclei.

Problem 6.13 Consider a uniformly charged sphere of radius R and charge Z.

(a) Compute the form factor in Eq. (6.69) by explicitly carrying out the angular and radial integrals. Show

$$F(q) = Z\left[\frac{3j_1(qR)}{qR}\right] \qquad ; \text{ uniform sphere}$$

$$j_1(x) \equiv \frac{\sin x}{x^2} - \frac{\cos x}{x}$$

(b) Plot this result as a function of qR. Discuss.

Solution to Problem 6.13

(a) Let ρ_0 be the constant charge density inside the sphere and call θ the angle formed by the vectors $\mathbf{q}$ and $\mathbf{r}$ so that $\mathbf{q}\cdot\mathbf{r} = qr\cos\theta$. In spherical coordinates the volume element is expressed as $d^3r = r^2 dr\, d\phi\, d\cos\theta$. Since the integrand in Eq. (6.69) does not depend upon ϕ, we have

$$\begin{aligned}
F(\mathbf{q}) &= 2\pi\rho_0 \int_{-1}^{+1} d\cos\theta \int_0^R r^2 dr\, e^{iqr\cos\theta} \\
&= \frac{4\pi\rho_0}{q} \int_0^R r dr\, \sin qr \\
&= \frac{4\pi R^3 \rho_0}{qR}\left[\frac{\sin qR}{(qR)^2} - \frac{\cos qR}{qR}\right]
\end{aligned}$$

The spherical Bessel function $j_1(x)$ is defined as $j_1(x) = \sin x/x^2 - \cos x/x$, while $\rho_0(4\pi R^3/3) = \rho_0 V = Z$, and therefore

$$F(q) = Z\left[\frac{3j_1(qR)}{qR}\right]$$

(b) In Fig. 6.5 we have plotted $|F(q)/Z|^2$ as a function of qR. The form factor has an oscillating behavior with an infinite sequence of maxima and minima, while the function decays as $1/(qR)^2$ for $qR \gg 1$. In particular, the first zero of $F(q)$ occurs at $qR \approx 1.43\pi$ [notice the similarity with Eq. (6.64)]; this tells us that one needs to reach high momenta to resolve a small charge distribution. We may use this formula to estimate the momentum $\hbar q$ needed to resolve a typical nuclear distance $R \approx 1\text{F}$

$$q \approx \frac{1.43\pi}{R}$$

As we have already seen, $\hbar c = 197.3$ MeV-F, and therefore at the first minimum

$$\hbar q \approx 886 \frac{\text{MeV}}{c} \qquad ; \text{ for } R = 1\,\text{F}$$

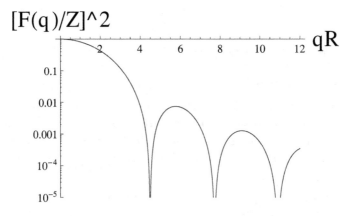

Fig. 6.5 $|F(q)/Z|^2$ for a uniform distribution of charge inside a sphere of given radius. The first minimum occurs at $qR \approx 1.43\pi$.

Problem 6.14 (a) Make use of Fig. 6.23 to predict the spins and parities of the following nuclei

$$^{17}_{9}\text{F} \qquad ; \, ^{61}_{29}\text{Cu} \qquad ; \, ^{105}_{46}\text{Pd} \qquad ; \, ^{121}_{49}\text{In} \qquad ; \, ^{211}_{87}\text{Fr}$$

Compare with experiment [National Nuclear Data Center (2007)].

(b) Compute the magnetic moments of these nuclei and compare with the experimental results [Lawrence Berkeley Laboratory (2005)].

Solution to Problem 6.14

(a) The even group of neutrons or protons couples to $J^\pi = 0^+$, and therefore the spin and parity of the nucleus in its ground state come from the odd group of nucleons. With a single valence nucleon above closed shells, the spin and parity come from that valence nucleon. Furthermore, in a partially-filled shell, even pairs couple to $J^\pi = 0^+$ and the spin and parity then come from the remaining valence nucleon. For a hole in a shell, the spin and parity are those of the unfilled state. It then follows from filling the levels in Fig. 6.23 in the text that

- $^{17}_{9}$F has 9 protons, and a $(1d_{5/2})_p$ configuration with $J^\pi = 5/2^+$;
- $^{61}_{29}$Cu has 29 protons, and a $(2p_{3/2})_p$ configuration with $J^\pi = 3/2^-$;
- $^{105}_{46}$Pd has 59 neutrons, and a $(2d_{5/2})_n$ configuration with $J^\pi = 5/2^+$;
- $^{121}_{49}$In has 49 protons, and a $(1g_{9/2})_p^{-1}$ configuration with $J^\pi = 9/2^+$;
- $^{211}_{87}$Fr has 87 protons, and a $(1h_{9/2})_p$ configuration of with $J^\pi = 9/2^-$.

The results are summarized and compared with experiment in Table 6.1.

Table 6.1   Configurations and magnetic moments for the nuclei in Prob. 6.14. For the experimental results see [Lawrence Berkeley Laboratory (2005)].

Nucleus	Configuration	$(J^\pi)_{th}$	$(J^\pi)_{exp}$	$(\mu/\mu_N)_{th}$	$(\mu/\mu_N)_{exp}$
$^{17}_{9}$F	$(1d_{5/2})_p$	$5/2^+$	$5/2^+$	4.79	4.72
$^{61}_{29}$Cu	$(2p_{3/2})_p$	$3/2^-$	$3/2^-$	3.79	2.14
$^{105}_{46}$Pd	$(2d_{5/2})_n$	$5/2^+$	$5/2^+$	-1.91	-.64
$^{121}_{49}$In	$(1g_{9/2})_p^{-1}$	$9/2^+$	$9/2^+$	6.79	5.50
$^{211}_{87}$Fr	$(1h_{9/2})_p$	$9/2^-$	$9/2^-$	2.63	4.00

(b) The magnetic moments are predicted by the Schmidt-line calculations in Eqs. (6.101) and (6.94); they are shown in Table 6.1.[11] The experimental results for the magnetic moments are also indicated. Not bad.

Problem 6.15 Repeat Prob. 6.14 for $^{133}_{54}$Xe. Discuss.

Solution to Problem 6.15

Here the situation is a little more complex than those discussed above. The nucleus $^{133}_{54}$Xe has 79 neutrons. The shell-model assignment from Fig. 6.23 in the text is $(1h_{11/2})_n^{-3}$, three holes in the $(1h_{11/2})$ shell; however, when there is a close-lying high-spin state, we might expect that the $(3s_{1/2})_n^2$ pair is actually promoted to enhance pairing overlap, leaving a $(1h_{11/2})_n^{-1}$ configuration with $J^\pi = 11/2^-$. For the same reason, a $(2d_{3/2})_n$ might be promoted to *fill* the $1h_{11/2}$ shell, leaving a $(2d_{3/2})_n^{-1}$ configuration and a $J^\pi = 3/2^+$ ground state. This possibility is compared with the observed ground state of $^{133}_{54}$Xe in the first row of Table 6.2.

---

[11] A hole also acts as a single valence nucleon as far as the magnetic moment is concerned, since the remaining pairs in the shell are coupled to $J^\pi = 0^+$. (If you need a proof, see [Fetter and Walecka (2003a)].)

By these same arguments, we would then expect to see the $(1h_{11/2})_n^{-1}$ state appear as a low-lying excitation (a *hole excitation*) in this nucleus. This is indeed the case, as indicated in the second line in Table 6.2.

Table 6.2   Configurations and magnetic moments for the nucleus in Prob. 6.15. For the experimental results see [Lawrence Berkeley Laboratory (2005)].

Nucleus	Configuration	$(J^\pi)_{\text{th}}$	$(J^\pi)_{\text{exp}}$	$(\mu/\mu_N)_{\text{th}}$	$(\mu/\mu_N)_{\text{exp}}$
$^{133}_{54}$Xe	$(2d_{3/2})_n^{-1}$	$3/2^+$	$3/2^+$	1.15	.81
$^{133}_{54}$Xe* (233 KeV)	$(1h_{11/2})_n^{-1}$	$11/2^-$	$11/2^-$	-1.91	-1.08

Problem 6.16 Consider the scattering of light from a single slit where the location of the first diffraction minimum is given for small angles by Eq. (6.66).

(a) Show that this expression can be written in the more familiar form

$$\theta = \lambda/a \qquad ; \text{ first diffraction minimum}$$

(b) Hence conclude that if one wants a diffraction pattern to show up on a laboratory screen at small, but finite angles, one wants to use light whose wavelength is comparable to the size of the slit, or $\lambda = O(a)$.

Solution to Problem 6.16

(a) In the limit of small angles ($\theta \ll 1$), Eq. (6.66) may be approximated as

$$k a \theta \approx 2\pi \qquad ; \text{ first diffraction minimum}$$

$$\theta \approx \frac{2\pi}{ka} = \frac{\lambda}{a}$$

where we have used $k = 2\pi/\lambda$ in the last relation.

(b) The result of part (a) implies that one obtains a diffraction minimum at a small but finite angle when the wavelength of the light is comparable to the size of the slit, $\lambda = O(a)$.

Problem 6.17 Assume that the neutron and proton densities in nuclear matter are driven apart by the Coulomb interaction of the protons. Write

$$N = \frac{B}{2}(1+\delta) \qquad ; \; Z = \frac{B}{2}(1-\delta)$$

It follows, as in Prob. 4.32, that the kinetic energy of nuclear matter at fixed baryon density, to leading order in δ, is increased by an amount

$$\frac{\Delta T}{B} = \frac{\varepsilon_{F0}}{3}\delta^2$$

Here ε_{F0} is the Fermi energy for symmetric nuclear matter.

(a) Derive this result;

(b) Show this implies a *symmetry energy coefficient* in the semi-empirical mass formula of

$$a_4 = \frac{\varepsilon_{F0}}{3}$$

(c) Use $k_{F0} = 1.42\,\mathrm{F}^{-1}$ from Table 6.3, and compare this result with the observed value of a_4 in Table 6.2;

(c) How would you improve this calculation?

Solution to Problem 6.17

(a) As in Problem 4.32 we use Eq. (4.167) of the text, to relate the number of particles to the maximum wavenumber k_F. We treat neutrons and protons as two separate spin-1/2 Fermi gases and write

$$N = \frac{B(1+\delta)}{2} = \frac{2V}{(2\pi)^3}\frac{4\pi k_{\mathrm{FN}}^3}{3}$$

$$Z = \frac{B(1-\delta)}{2} = \frac{2V}{(2\pi)^3}\frac{4\pi k_{\mathrm{FP}}^3}{3}$$

which can be inverted to give

$$k_{\mathrm{FN}} = \left[3\pi^2\frac{B}{V}\frac{(1+\delta)}{2}\right]^{1/3}$$

$$k_{\mathrm{FP}} = \left[3\pi^2\frac{B}{V}\frac{(1-\delta)}{2}\right]^{1/3}$$

The total kinetic energy of the system will then be [see Eq. (4.170)]

$$T = T_{\mathrm{N}} + T_{\mathrm{P}} = \frac{3}{5}\frac{B(1+\delta)}{2}\frac{\hbar^2 k_{\mathrm{FN}}^2}{2m} + \frac{3}{5}\frac{B(1-\delta)}{2}\frac{\hbar^2 k_{\mathrm{FP}}^2}{2m}$$

$$= \frac{3}{10}\frac{\hbar^2}{2m}\left(\frac{3\pi^2 B}{2 V}\right)^{2/3} B\left[(1+\delta)^{5/3} + (1-\delta)^{5/3}\right]$$

For $|\delta| \ll 1$ we may expand the factors $(1 \pm \delta)^{5/3}$ up to order δ^2, as explained in the footnote in Prob. 4.32, with the result

$$T \approx \frac{3}{10} \frac{\hbar^2}{2m} \left(\frac{3\pi^2}{2} \frac{B}{V} \right)^{2/3} B \left[2 + \frac{10}{9} \delta^2 \right]$$

$$= \frac{3}{10} \varepsilon_F^0 B \left(2 + \frac{10}{9} \delta^2 \right)$$

and therefore

$$\frac{\Delta T}{B} = \frac{\varepsilon_{F0}}{3} \delta^2$$

(b) The symmetry energy term in the semi-empirical mass formula of Eq. (6.54) may be expressed directly in terms of δ as

$$E_{\text{sym}} = a_4 \frac{(N-Z)^2}{B}$$

$$= a_4 B \delta^2$$

A comparison with the result of part (a) then yields

$$a_4 = \frac{\varepsilon_{F0}}{3}$$

(c) With the given value of $k_{F0} = 1.42\,\text{F}^{-1}$, the Fermi energy of symmetric nuclear matter is

$$\varepsilon_{F0} = \frac{\hbar^2 k_{F0}^2}{2m_N} \approx 41.8\,\text{MeV}$$

and the result in part (b) then gives

$$a_4 \approx 13.9\,\text{MeV}$$

This is substantially below the observed value given in Table 6.2 in the text

$$(a_4)_{\text{exp}} = 23.7\,\text{MeV}$$

Clearly, although the right idea, attempting to make a quantitative calculation of the symmetry energy without taking into account the nuclear interactions is too naive.

Problem 6.18 Explicitly evaluate the magnetic moment in Eqs. (6.99)–(6.100), and derive the expression for the Schmidt lines in Eq. (6.101).

Solution to Problem 6.18

If Eq. (6.99) is inserted in Eq. (6.100), one obtains

$$\frac{\mu}{\mu_N} = \frac{1}{2(j+1)} \{ j(j+1) + l(l+1) - s(s+1) + \\ 2\lambda \left[j(j+1) + s(s+1) - l(l+1) \right] \}$$

We first consider the case where $j = l + 1/2$ and $s = 1/2$. In this case we have

$$j(j+1) + l(l+1) - s(s+1) = 2l(l+3/2)$$
$$j(j+1) + s(s+1) - l(l+1) = l + 3/2$$

Substitution in the above gives

$$\frac{\mu}{\mu_N} = l + \lambda = j - \frac{1}{2} + \lambda$$

which corresponds to the first of Eqs. (6.101).

We now consider the case where $j = l - 1/2$ and $s = 1/2$. In this case we have

$$j(j+1) + l(l+1) - s(s+1) = 2(l-1/2)(l+1)$$
$$j(j+1) + s(s+1) - l(l+1) = -l + 1/2$$

We then obtain

$$\frac{\mu}{\mu_N} = \left(l - \frac{1}{2} \right) \frac{l+1-\lambda}{l+1/2}$$
$$= j + \frac{j}{j+1} \left(\frac{1}{2} - \lambda \right)$$

which corresponds to the second of Eqs. (6.101).

Chapter 7

Particle Physics

Problem 7.1 Work through appendix A.1 of [Walecka (2004)]. Use the Feynman rules for scalar meson exchange given there to derive the Yukawa potential in Eq. (7.29).[1]

Solution to Problem 7.1

We follow the discussion of the meson-exchange potential contained in appendix A of [Walecka (2004)]. The interaction lagrangian density for scalar meson exchange is

$$\mathcal{L}_I = g_s \bar{\psi}\psi\phi$$

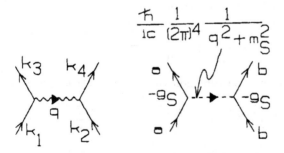

Fig. 7.1   Feynman rules for scalar meson exchange.

---

[1]That the meson responsible for the long-range part of the nuclear force, the pion, should be a *pseudoscalar* rather than a scalar meson was, in fact, predicted by Pauli on the basis of observed properties of the deuteron [Pauli (1948)].

With the use of the Feynman rules given in Fig. 7.1, the S-matrix may be written to lowest order as

$$S_{fi} = \frac{(2\pi)^4}{\Omega^2} \delta^{(4)}(k_1 + k_2 - k_3 - k_4) \left(\frac{-i}{\hbar c}\right)^2 (-g_s)^2 \frac{\hbar}{ic} \frac{1}{q^2 + m_s^2}$$
$$\times \bar{u}_a(k_3) u_a(k_1) \, \bar{u}_b(k_4) u_b(k_2)$$

where Ω is the volume of a large cubical box and the Dirac delta function expresses the conservation of energy and momentum.

This expression must be compared with the analogous expression for the S-matrix corresponding to a non-relativistic interaction hamiltonian

$$\int d^3x \, \mathcal{H}_I(\mathbf{x}) = \int d^3x \, d^3y \, \hat{\psi}_a^\dagger(\mathbf{x}) \hat{\psi}_b^\dagger(\mathbf{y}) V_{\text{eff}}(\mathbf{x} - \mathbf{y}) \hat{\psi}_b(\mathbf{y}) \hat{\psi}_a(\mathbf{x})$$

where $(\hat{\psi}, \hat{\psi}^\dagger)$ are field operators.[2] The first-order S-matrix in this case is

$$S_{fi} = \frac{(2\pi)^4}{\Omega^2} \delta^{(4)}(k_1 + k_2 - k_3 - k_4) \left(\frac{-i}{\hbar c}\right) \tilde{V}_{\text{eff}}(\mathbf{q})$$

where $\mathbf{q} = \mathbf{k}_1 - \mathbf{k}_3 = \mathbf{k}_4 - \mathbf{k}_2$ is the three-momentum transfer, and $\tilde{V}_{\text{eff}}(\mathbf{q})$ is the Fourier transform of the potential. In principle, $\tilde{V}_{\text{eff}}(\mathbf{q})$ may have a spin dependence although we assume here that this is not the case (for this reason, the expression does not depend on the spin indices a and b).

In the limit of large fermion mass ($M \to \infty$), we may treat the fermions as non-relativistic in the first expression for S_{fi} and use $q_0 = O(1/M)$ and $\bar{u}u \to \delta_{ss'}$. A comparison of the two expressions for S_{fi} then gives

$$\tilde{V}_{\text{eff}}(\mathbf{q}) = -\frac{g_s^2}{c^2} \frac{1}{\mathbf{q}^2 + m_s^2}$$

The Fourier transform of this expression yields the effective potential[3]

$$V_{\text{eff}}(r) = \int e^{i\mathbf{q}\cdot\mathbf{r}} \, \tilde{V}_{\text{eff}}(\mathbf{q}) \frac{d^3q}{(2\pi)^3} = -\frac{g_s^2}{4\pi c^2} \frac{e^{-m_s r}}{r}$$

This is the celebrated attractive Yukawa interaction.

Problem 7.2 (a) When the laboratory is taken to be a very large cubical box of volume $\Omega = L^3$, and periodic boundary conditions are applied, the

---

[2] See chapter 12 in the text. We get ahead of ourselves here, but the result is of sufficient importance to do so. It is assumed that the fermions are distinguishable.

[3] For the required (inverse) Fourier transform, see the footnote on p. 388 of the text.

final integral giving overall momentum conservation is of the form

$$\int_\Omega d^3x \, e^{i(\mathbf{K}_i - \mathbf{K}_f) \cdot \mathbf{x}} = \Omega \delta_{\mathbf{K}_f, \mathbf{K}_i}$$

where $\delta_{\mathbf{K}_f, \mathbf{K}_i}$ is a Kronecker delta. Show that in the limit $\Omega \to \infty$, this relation becomes (compare Prob. 4.24)

$$\int d^3x \, e^{i(\mathbf{K}_i - \mathbf{K}_f) \cdot \mathbf{x}} = (2\pi)^3 \, \delta^{(3)}(\mathbf{K}_i - \mathbf{K}_f)$$

where $\delta^{(3)}(\mathbf{K}_i - \mathbf{K}_f)$ is now a Dirac delta function. Hence justify the following replacement in this limit

$$\Omega \, \delta_{\mathbf{K}_f, \mathbf{K}_i} \to (2\pi)^3 \, \delta^{(3)}(\mathbf{K}_i - \mathbf{K}_f) \qquad ; \Omega \to \infty$$

(b) Use the second of Eqs. (7.45) and the results in part (a) to show that in the limits $T \to \infty$, and then $\Omega \to \infty$, the S-matrix in Eq. (7.36) takes the form

$$S_{fi} = (2\pi)^4 \, \delta^{(4)}(K_f - K_i) \frac{1}{\Omega^{(n+2)/2}} \left[\frac{-iT_{fi}}{\hbar c} \right] \qquad ; T \to \infty$$

$$\Omega \to \infty$$

Here

$$\delta^{(4)}(K_f - K_i) \equiv \delta^{(3)}(\mathbf{K}_f - \mathbf{K}_i)\delta(K_{f0} - K_{i0})$$
$$K_{f0} - K_{i0} \equiv (W_f - W_i)/\hbar c$$

Solution to Problem 7.2

(a) Consider $\sum_{\mathbf{K}_f}$ of the first of the equations in part (a)

$$\sum_{\mathbf{K}_f} \int_\Omega d^3x \, e^{i(\mathbf{K}_i - \mathbf{K}_f) \cdot \mathbf{x}} = \Omega \sum_{\mathbf{K}_f} \delta_{\mathbf{K}_f, \mathbf{K}_i} = \Omega$$

In a limit of a large box of volume $\Omega = L^3$, the sum over wavenumbers may be replaced by an integral. As seen in Eq. (4.165), in the limit $L \to \infty$ the sum over a small region in $\mathbf{k}$ space (denoted with a prime), may be approximated as

$$\sum_{\mathbf{k}}' = \frac{\Omega}{(2\pi)^3} d^3k \qquad ; L \to \infty$$

In this limit we may approximate the sum with an integral, and the above becomes

$$\sum_{\mathbf{K}_f} \int_\Omega d^3x\, e^{i(\mathbf{K}_i - \mathbf{K}_f)\cdot\mathbf{x}} \to \frac{\Omega}{(2\pi)^3} \int d^3K_f \int_\Omega d^3x\, e^{i(\mathbf{K}_i - \mathbf{K}_f)\cdot\mathbf{x}} = \Omega$$

which implies

$$\frac{1}{(2\pi)^3} \int d^3K_f \left(\int_\Omega d^3x\, e^{i(\mathbf{K}_i - \mathbf{K}_f)\cdot\mathbf{x}} \right) = 1$$

This equation is satisfied if

$$\int_\Omega d^3x\, e^{i(\mathbf{K}_i - \mathbf{K}_f)\cdot\mathbf{x}} = (2\pi)^3\, \delta^{(3)}(\mathbf{K}_i - \mathbf{K}_f) \qquad ; \Omega \to \infty$$

This justifies the replacement in part (a)

$$\Omega\delta_{\mathbf{K}_f,\mathbf{K}_i} \to (2\pi)^3\, \delta^{(3)}(\mathbf{K}_i - \mathbf{K}_f) \qquad ; \Omega \to \infty$$

in the limit $\Omega \to \infty$.

(b) We use the definition of $\Delta_T(W_f - W_i)$ in Eq. (7.39)

$$\Delta_T(W_f - W_i) \equiv \int_{-T/2}^{T/2} dt\, e^{i(W_f - W_i)t/\hbar}$$

This integral corresponds to the one-dimensional case of the integral treated in part (a), and in the limit $T \to \infty$ we have from Eq. (7.45)

$$\Delta_T(W_f - W_i) \to 2\pi\hbar\, \delta\,(W_f - W_i) \qquad ; T \to \infty$$

We can make use of the properties of the Dirac delta function to write[4]

$$\hbar\delta(W_f - W_i) = \delta\left(\frac{1}{\hbar}(W_f - W_i)\right) = \frac{1}{c}\delta\left(\frac{1}{\hbar c}(W_f - W_i)\right)$$

and therefore

$$\Delta_T(W_f - W_i) \to \frac{2\pi}{c}\delta(K_{f0} - K_{i0}) \qquad ; T \to \infty$$

where $K_{f0} - K_{i0} \equiv (W_f - W_i)/\hbar c$.

---

[4]See the footnote on p. 336 of the book.

We can now use this result, together with that of part (a), to re-write Eq. (7.36) in this limit as

$$S_{fi} = (2\pi)^4 \, \delta^{(4)}(K_f - K_i) \frac{1}{\Omega^{(n+2)/2}} \left[\frac{-iT_{fi}}{\hbar c} \right] \qquad ; \, T \to \infty$$

$$\Omega \to \infty$$

Problem 7.3 Two particles in a large box of volume Ω with parallel velocities $\mathbf{v}(1)$ and $\mathbf{v}(2)$ interact to produce a final state f. Explain why the appropriate incident flux to be used in Eq. (7.49) is

$$I_{\text{inc}} = \frac{1}{\Omega} |\mathbf{v}(1) - \mathbf{v}(2)|$$

Solution to Problem 7.3

The incident particle flux is the number of incident particles per unit transverse area per unit time. In the case of two particles in a large box of volume Ω interacting to produce a final state f, we may first work in the laboratory frame, where one of the particles, is at rest. We call $\mathbf{v}_0(1)$ and $\mathbf{v}_0(2) = 0$, the velocities of the two particles.

In this frame, the wave function of the incident particle is a plane wave, $\psi_{\text{inc}}(\mathbf{x}) = (1/\sqrt{\Omega}) e^{i\mathbf{k}_1^{(0)} \cdot \mathbf{x}}$, and it provides an incident probability flux of

$$I_{\text{inc}}^{(\text{LAB})} = \left| \frac{\hbar}{2im_1} \left[\psi_{\text{inc}}^\star \boldsymbol{\nabla} \psi_{\text{inc}} - (\boldsymbol{\nabla} \psi_{\text{inc}})^\star \psi_{\text{inc}} \right] \right|$$

$$= \frac{1}{\Omega} \frac{\hbar k_1^{(0)}}{m_1} = \frac{1}{\Omega} |\mathbf{v}_0(1)|$$

A Galilean transformation along $\mathbf{v}_0(1)$ may now be used to go to a frame where the parallel velocities of the two particles are $\mathbf{v}(1)$ and $\mathbf{v}(2)$. In this case $\mathbf{v}_0(1)$ transforms into the relative velocity of the particles and[5]

$$I_{\text{inc}} = \frac{1}{\Omega} |\mathbf{v}(1) - \mathbf{v}(2)|$$

Problem 7.4 Reactions can proceed with decreasing rates through the strong, electromagnetic, or weak interactions, or they can be absolutely forbidden. This is governed by appropriate conservation laws (see text). Give the leading interaction through which the following reactions can take

---

[5] We here assume non-relativistic particles, but the result is more general.

place (assume $H_{\mathrm{UW}} = 0$):

$$p + n \rightarrow n + n + \pi^+$$
$$\mu^+ \rightarrow e^+ + \nu_e + \nu_\mu$$
$$\pi^- + p \rightarrow n + \pi^0 + \pi^+$$
$$\Xi^- \rightarrow \Sigma^0 + e^- + \bar{\nu}_e$$
$$\gamma + p \rightarrow n + \pi^+$$
$$p + \bar{p} \rightarrow p + \rho^- + n + \bar{n}$$

Solution to Problem 7.4

- $p + n \rightarrow n + n + \pi^+$ can proceed through the *strong* interaction;
- $\mu^+ \rightarrow e^+ + \nu_e + \nu_\mu$ is *forbidden* by lepton conservation;[6]
- $\pi^- + p \rightarrow n + \pi^0 + \pi^+$ is *forbidden* by charge conservation;
- $\Xi^- \rightarrow \Sigma^0 + e^- + \bar{\nu}_e$ can proceed through the *weak* interaction;
- $\gamma + p \rightarrow n + \pi^+$ can proceed through the *electromagnetic* interaction;
- $p + \bar{p} \rightarrow p + \rho^- + n + \bar{n}$ is *forbidden* by baryon conservation.

Problem 7.5 Assume $(1s)$ spatial states:

(a) Give the quark configurations for all the states in the $S = -1$ and $S = -2$ sectors in Fig. 7.1 in the text;

(b) Give the quark configurations for all the mesons in the $S = \pm 1$ sectors in Fig. 7.2 in the text.

Solution to Problem 7.5

We assume $(1s)$ spatial states:

(a) The quark configurations for the strangeness $S = -1$ and $S = -2$ baryons in Fig. 7.1 in the text are then shown in Table 7.1. All have quark spins coupled to give $J = 1/2$.

Table 7.1 Quark configurations for the baryons with $S = -1$ and $S = -2$ in Fig. 7.1 in the text.

Baryon	S	Configuration	(j^π, t)
Λ	-1	(uds)	$(1/2^+, 0)$
Σ^+	-1	(uus)	$(1/2^+, 1)$
Σ^0	-1	(uds)	$(1/2^+, 1)$
Σ^-	-1	(dds)	$(1/2^+, 1)$
Ξ^0	-2	(uss)	$(1/2^+, 1/2)$
Ξ^-	-2	(dss)	$(1/2^+, 1/2)$

---

[6]This assumes $H_{\mathrm{UW}} = 0$.

(b) The quark configurations for the mesons with strangeness $S = -1$ and $S = +1$ in Fig. 7.2 in the text are shown in Table 7.2. All have quark spins coupled to give $J = 0$.

Table 7.2 Quark configurations for the mesons with $S = \pm 1$ in Fig. 7.2 in the text.

Meson	S	Configuration	(j^π, t)
K^-	-1	$(\bar{u}s)$	$(0^-, 1/2)$
$\bar{K}^0$	-1	$(\bar{d}s)$	$(0^-, 1/2)$
K^0	$+1$	$(d\bar{s})$	$(0^-, 1/2)$
K^+	$+1$	$(u\bar{s})$	$(0^-, 1/2)$

Problem 7.6 What hadrons can be made from (scu) and (scd) quarks? Give all quantum numbers.

Solution to Problem 7.6

The quantum numbers of the baryons made from the quark configurations (scu) and (scd) with the quarks in $(1s)$ spatial states are shown in Table 7.3.

Table 7.3 Baryons made from the quark configurations (scu) and (scd) with the quarks in $(1s)$ spatial states.

Configuration	j^π	B	Q	T	T_3	S	C
(scu)	$1/2^+$	1	1	1/2	$+1/2$	-1	1
(scd)	$1/2^+$	1	0	1/2	$-1/2$	-1	1
(scu)	$3/2^+$	1	1	1/2	$+1/2$	-1	1
(scd)	$3/2^+$	1	0	1/2	$-1/2$	-1	1

Problem 7.7 Show that if the four-momentum transfers of interest in Eq. (7.70) are much smaller than the inverse Compton wavelengths of the massive weak vector bosons, then the amplitude for the process in Fig. 7.21 in the text reduces to

$$\frac{-iT_{fi}}{\hbar c} = \frac{ig_W^2}{\hbar c M_W^2} j_W j_W$$

This is the same amplitude one would get with an *effective point coupling*, and a corresponding amplitude[7]

$$\frac{-iT_{fi}}{\hbar c} = \frac{-i}{\hbar c} \left[-\frac{G_F}{\sqrt{2}} j_W j_W \right]$$

---

[7] Compare Prob. I.2.

Hence make the identification of the Fermi constant in Eq. (7.71).

Solution to Problem 7.7

From the Feynman rules accompanying Fig. 7.21 in the text, the T-matrix for the weak scattering process in the figure is given by

$$\frac{-iT_{fi}}{\hbar c} = \left(\frac{ig_W}{\hbar c}\right)^2 \frac{\hbar c}{i} j_W \frac{1}{q^2 + M_W^2} j_W$$

$$= \frac{ig_W^2}{\hbar c} j_W \frac{1}{q^2 + M_W^2} j_W$$

Here q^2 is the square of the four-momentum-transfer in Eq. (7.70), and $M_W = m_W c/\hbar$ is the inverse Compton wavelength of the intermediate vector boson. As stated in the problem, if $q^2 \ll M_W^2$, this is the same amplitude one would get from an effective point coupling

$$\frac{-iT_{fi}}{\hbar c} = \frac{-i}{\hbar c} \left[-\frac{G_F}{\sqrt{2}} j_W j_W \right] \qquad ; q^2 \ll M_W^2$$

where the Fermi constant is defined by

$$\frac{G_F}{\sqrt{2}} \equiv \frac{g_W^2}{M_W^2}$$

This effective theory provides a remarkably successful description of low-energy weak interactions. The dimensional nature of G_F, coupled with the present argument, makes a thoroughly convincing case for the existence of a massive weak vector boson. Unfortunately, this argument provides no guidance as to just what M_W should be.[8] It was only after sufficient energy was achieved, coupled with appropriate detectors, that the Z^0 and $W^\pm$ were discovered in the experiments at CERN.

Problem 7.8 Consider the scattering of a massless neutrino by a heavy nucleus. Assume the point coupling of Prob. 7.7, and replace the interaction hamiltonian of Prob. I.1 by the contact interaction

$$H_1 = -\frac{G_F}{\sqrt{2}} \int d^3x \, j_W(\mathbf{x}_p) \delta^{(3)}(\mathbf{x}_p - \mathbf{x}) J_W(\mathbf{x})$$

Assume the situation is analogous to that in Prob. I.1, and repeat the arguments given there.

---

[8] One of the authors (jdw) spent the 1963-64 year at CERN participating in the search for the weak vector boson. It was an exciting time. Ultimately, the search turned out to be at too low an energy.

(a) Show that in this case

$$\langle \phi_f | H_1 | \phi_i \rangle = -\frac{G_F}{\sqrt{2}\,\Omega} F_W(\mathbf{q}) \qquad ; \mathbf{q} = \mathbf{k}_i - \mathbf{k}_f$$

$$F_W(\mathbf{q}) = \int d^3 x \, e^{i\mathbf{q}\cdot\mathbf{x}} J_W(\mathbf{x}) \qquad ; \text{form factor}$$

(b) Assume relativistic kinematics for the scattering particle with $\varepsilon = \hbar k c$, and show

$$\int dk \, \delta(E_f - E_i) = \frac{1}{\hbar c} \qquad ; k \equiv k_i = k_f$$

(c) Show the incident flux in the laboratory frame is $I_{\text{inc}} = c/\Omega$.

(d) Hence show the differential cross section is given by

$$\frac{d\sigma}{d\Omega_k} = \frac{1}{8\pi^2} \frac{G_F^2 \, k^2}{(\hbar c)^2} |F_W(\mathbf{q})|^2 \qquad ; \mathbf{q} = \mathbf{k}_i - \mathbf{k}_f$$

Solution to Problem 7.8

(a) The matrix element of H_1 taken between plane-wave states is given by

$$\langle \phi_f | H_1 | \phi_i \rangle = \frac{1}{\Omega} \int d^3 x_p \, e^{i(\mathbf{k}_i - \mathbf{k}_f)\cdot\mathbf{x}_p} H_1$$

With $\mathbf{q} \equiv \mathbf{k}_i - \mathbf{k}_f$, and the H_1 given above, this becomes[9]

$$\begin{aligned}
\langle \phi_f | H_1 | \phi_i \rangle &= = -\frac{G_F}{\Omega\sqrt{2}} \int d^3 x_p \int d^3 x \, e^{i\mathbf{q}\cdot\mathbf{x}_p} \delta^{(3)}(\mathbf{x}_p - \mathbf{x}) J_W(\mathbf{x}) \\
&= -\frac{G_F}{\Omega\sqrt{2}} \int d^3 x \, e^{i\mathbf{q}\cdot\mathbf{x}} J_W(\mathbf{x}) \\
&= -\frac{G_F}{\Omega\sqrt{2}} F_W(\mathbf{q})
\end{aligned}$$

(b) We assume relativistic kinematics with massless neutrinos. The neutrino energy is $\varepsilon = \hbar k c$, and therefore $d\varepsilon = \hbar c \, dk$. Thus

$$\int dk \, \delta(E_f - E_i) = \frac{1}{\hbar c} \int d\varepsilon \, \delta(\varepsilon - E_i) = \frac{1}{\hbar c}$$

---

[9]To make the situation analogous to that in Prob. I.1, we here suppress the Dirac structure and assume a weak current of unit amplitude $j_W(\mathbf{x}_p) = 1$.

(c) As we have seen before, the incident probability flux of a particle moving with velocity v in the laboratory frame is $I_{\text{inc}} = v/\Omega$. For a massless neutrino $v = c$, and therefore

$$I_{\text{inc}} = \frac{c}{\Omega}$$

(d) We use Eqs. (I.35)–(I.36) to write the differential cross section

$$d\sigma = \frac{R_{fi}}{I_{\text{inc}}} \frac{\Omega d^3 k}{(2\pi)^3}$$

where R_{fi} is the transition rate into one state as given in Eq. (I.32)

$$R_{fi} = \frac{2\pi}{\hbar} |\langle \phi_f | H_1 | \phi_i \rangle|^2 \, \delta(E_f - E_i)$$

With the use of the results obtained in the parts (a)–(c), we have

$$\frac{d\sigma}{d\Omega_k} = \frac{2\pi}{\hbar} \left(\frac{G_F}{\Omega\sqrt{2}} \right)^2 |F_W(\mathbf{q})|^2 \, \delta(E_f - E_i) \frac{\Omega k^2 dk}{(2\pi)^3} \frac{\Omega}{c}$$

$$= \frac{1}{8\pi^2} \frac{G_F^2 k^2}{(\hbar c)^2} |F_W(\mathbf{q})|^2$$

which is the stated answer.

Problem 7.9 Consider a wave function with two components ζ_i with $i = (1, 2)$, describing, for example, a spin-1/2 particle [see Eq. (4.160)].[10] Define the following transformation on this wave function by

$$\zeta_i' = \sum_{j=1}^{2} r(\boldsymbol{\omega})_{ij} \zeta_j$$

$$\underline{r}(\boldsymbol{\omega}) = \exp\left\{ \frac{i}{2} \boldsymbol{\sigma} \cdot \boldsymbol{\omega} \right\} \qquad ; \, \boldsymbol{\omega} = (\omega_1, \omega_2, \omega_3)$$

Here $\boldsymbol{\sigma} = (\sigma_1, \sigma_2, \sigma_3)$ are the Pauli matrices (which we do not underline), and $\boldsymbol{\omega}$ characterizes some spatial rotation.

(a) Expand the exponential, and show the 2×2 matrix $\underline{r}(\boldsymbol{\omega})$ can be re-written as (here $\underline{1}$ is the unit 2×2 matrix, and $\mathbf{n}$ is a unit vector)

$$\underline{r}(\boldsymbol{\omega}) = \underline{1} \cos \frac{\omega}{2} + i\mathbf{n} \cdot \boldsymbol{\sigma} \sin \frac{\omega}{2} \qquad ; \, \boldsymbol{\omega} \equiv \omega \mathbf{n}$$

---

[10] Analogous arguments hold in the internal isospin space.

(b) Show that $\underline{r}(\boldsymbol{\omega})$ is both unitary and unimodular (*i.e.* it has unit determinant)

$$\underline{r}(\boldsymbol{\omega})^\dagger \underline{r}(\boldsymbol{\omega}) = 1 \qquad ; \ \det \underline{r}(\boldsymbol{\omega}) = 1$$

(c) Show that the product $\underline{r}(\boldsymbol{\omega}_1)\underline{r}(\boldsymbol{\omega}_2)$ is again a unitary, unimodular matrix.[11]

(d) Show that *any* unitary unimodular matrix can be written in the form given above for *some* choice of $\boldsymbol{\omega}$. Hence show that the transformations form a three-parameter, continuous *group*[12]

$$\underline{r}(\boldsymbol{\omega}_1)\underline{r}(\boldsymbol{\omega}_2) = \underline{r}(\boldsymbol{\omega}_3)$$

This is the group $SU(2)$.

(e) Consider a two-particle direct-product state $\psi_{ij}(1,2) \equiv \zeta_i(1)\zeta_j(2)$ that transforms as

$$\psi'_{i'j'}(1,2) = \sum_{i=1}^{2}\sum_{j=1}^{2} r(\boldsymbol{\omega})_{i'i}\, r(\boldsymbol{\omega})_{j'j}\, \psi_{ij}(1,2)$$

Consider linear combinations with symmetrized and antisymmetrized indices

$$\psi_{ij}^{S}(1,2) \equiv \frac{1}{\sqrt{2}}\left[\psi_{ij}(1,2) + \psi_{ji}(1,2)\right]$$

$$\psi_{ij}^{A}(1,2) \equiv \frac{1}{\sqrt{2}}\left[\psi_{ij}(1,2) - \psi_{ji}(1,2)\right]$$

Show that these states are transformed among themselves by the above transformation, and hence each provides a basis for an *irreducible representation* of the group.[13] How many states of each type are there? What total spin do they correspond to?

Solution to Problem 7.9

We use the series representation of the exponential to write $\underline{r}(\boldsymbol{\omega}) =$

---

[11] *Hint:* Recall Prob. B.3 and use the fact that for matrices $\det \underline{A}\,\underline{B} = \det \underline{A} \det \underline{B}$.
[12] Note that the infinitesimal rotations add here just as they should $\boldsymbol{\omega}_3 = \boldsymbol{\omega}_1 + \boldsymbol{\omega}_2$.
[13] For an introduction to group theory, see [Hamermesh (1989)].

$\exp\{i\boldsymbol{\sigma}\cdot\boldsymbol{\omega}/2\}$ as

$$\underline{r}(\boldsymbol{\omega}) = \sum_{r=0}^{\infty}\frac{1}{r!}\left(\frac{i}{2}\right)^{r}(\boldsymbol{\sigma}\cdot\boldsymbol{\omega})^{r}$$

$$= \sum_{r=0}^{\infty}\left[\left(\frac{i}{2}\right)^{2r}\frac{(\boldsymbol{\sigma}\cdot\boldsymbol{\omega})^{2r}}{(2r)!} + \left(\frac{i}{2}\right)^{2r+1}\frac{(\boldsymbol{\sigma}\cdot\boldsymbol{\omega})^{2r+1}}{(2r+1)!}\right]$$

where in the last expression we have separated the terms in the series containing even and odd powers of $\boldsymbol{\sigma}\cdot\boldsymbol{\omega}$.

We may now use Eq. (B.13) to write

$$(\boldsymbol{\sigma}\cdot\boldsymbol{\omega})^{2} = \sum_{i=1}^{3}\sum_{j=1}^{3}\sigma_i\sigma_j\omega_i\omega_j = \frac{1}{2}\sum_{i=1}^{3}\sum_{j=1}^{3}(\sigma_i\sigma_j + \sigma_j\sigma_i)\omega_i\omega_j$$

$$= \underline{1}\sum_{i=1}^{3}\omega_i\omega_i = \omega^2\underline{1}$$

Thus

$$(\boldsymbol{\sigma}\cdot\boldsymbol{\omega})^{2r} = \omega^{2r}\underline{1} \qquad ; \quad (\boldsymbol{\sigma}\cdot\boldsymbol{\omega})^{2r+1} = \omega^{2r}(\boldsymbol{\sigma}\cdot\boldsymbol{\omega})$$

and

$$\underline{r}(\boldsymbol{\omega}) = \sum_{r=0}^{\infty}\omega^{2r}\left[\frac{1}{(2r)!}\left(\frac{i}{2}\right)^{2r}\underline{1} + \left(\frac{i}{2}\right)^{2r+1}\frac{(\boldsymbol{\sigma}\cdot\boldsymbol{\omega})}{(2r+1)!}\right]$$

$$= \underline{1}\cos\frac{\omega}{2} + i\mathbf{n}\cdot\boldsymbol{\sigma}\sin\frac{\omega}{2}$$

where $\mathbf{n} \equiv \boldsymbol{\omega}/\omega$ is a unit vector.

(b) Let us first prove the unitarity of $\underline{r}(\boldsymbol{\omega})$

$$\underline{r}(\boldsymbol{\omega})^{\dagger}\underline{r}(\boldsymbol{\omega}) = \left(\underline{1}\cos\frac{\omega}{2} - i\mathbf{n}\cdot\boldsymbol{\sigma}\sin\frac{\omega}{2}\right)\left(\underline{1}\cos\frac{\omega}{2} + i\mathbf{n}\cdot\boldsymbol{\sigma}\sin\frac{\omega}{2}\right)$$

$$= \underline{1}\cos^2\frac{\omega}{2} + (\mathbf{n}\cdot\boldsymbol{\sigma})^2\sin^2\frac{\omega}{2}$$

$$= \underline{1}\left[\cos^2\frac{\omega}{2} + \sin^2\frac{\omega}{2}\right] = \underline{1}$$

The same result is obtained in a simpler fashion using the representation $\underline{r}(\boldsymbol{\omega}) = \exp\{i\boldsymbol{\sigma}\cdot\boldsymbol{\omega}/2\}$

$$\underline{r}(\boldsymbol{\omega})^{\dagger}\underline{r}(\boldsymbol{\omega}) = e^{-i\boldsymbol{\sigma}\cdot\boldsymbol{\omega}/2}\,e^{i\boldsymbol{\sigma}\cdot\boldsymbol{\omega}/2} = \underline{1}$$

To show the matrices are unimodular, we may use the representation of the Pauli matrices given in Eq. (B.12) to obtain the explicit representation of $\underline{r}(\boldsymbol{\omega})$

$$\underline{r}(\boldsymbol{\omega}) = \begin{bmatrix} \cos\omega/2 + i(\omega_z/\omega)\sin\omega/2 & i[(\omega_x - i\omega_y)/\omega]\sin\omega/2 \\ i[(\omega_x + i\omega_y)/\omega]\sin\omega/2 & \cos\omega/2 - i(\omega_z/\omega)\sin\omega/2 \end{bmatrix}$$

Thus we see that

$$\det\underline{r}(\boldsymbol{\omega}) = \cos^2\frac{\omega}{2} + \left[\left(\frac{\omega_z}{\omega}\right)^2 + \left(\frac{\omega_x - i\omega_y}{\omega}\right)\left(\frac{\omega_x + i\omega_y}{\omega}\right)\right]\sin^2\frac{\omega}{2}$$

$$= \cos^2\frac{\omega}{2} + \sin^2\frac{\omega}{2} = 1$$

(c) Let us first prove the unitarity relation for the product $\underline{r}(\boldsymbol{\omega}_1)\underline{r}(\boldsymbol{\omega}_2)$[14]

$$[\underline{r}(\boldsymbol{\omega}_1)\underline{r}(\boldsymbol{\omega}_2)]^\dagger[\underline{r}(\boldsymbol{\omega}_1)\underline{r}(\boldsymbol{\omega}_2)] = \underline{r}(\boldsymbol{\omega}_2)^\dagger\underline{r}(\boldsymbol{\omega}_1)^\dagger\underline{r}(\boldsymbol{\omega}_1)\underline{r}(\boldsymbol{\omega}_2)$$
$$= \underline{r}(\boldsymbol{\omega}_2)^\dagger\underline{r}(\boldsymbol{\omega}_2) = \underline{1}$$

To prove that $\underline{r}(\boldsymbol{\omega}_1)\underline{r}(\boldsymbol{\omega}_2)$ is unimodular, we use the property of determinants that $\det\underline{A}\,\underline{B} = \det\underline{A}\det\underline{B}$ to write

$$\det\underline{r}(\boldsymbol{\omega}_1)\underline{r}(\boldsymbol{\omega}_2) = \det\underline{r}(\boldsymbol{\omega}_1)\,\det\underline{r}(\boldsymbol{\omega}_2) = 1$$

(d) The most general 2×2 complex matrix is characterized by 8 real parameters (the real and imaginary parts of its 4 complex elements). Unitarity provides 4 conditions, and the fact that the determinant is 1 supplies a fifth condition. Hence, the most general 2×2 unitary, unimodular matrix has 3 real parameters. The result in part (b) provides a general representation for such a matrix

$$\underline{r}(\boldsymbol{\omega}) = \begin{bmatrix} \cos\omega/2 + i(\omega_z/\omega)\sin\omega/2 & i[(\omega_x - i\omega_y)/\omega]\sin\omega/2 \\ i[(\omega_x + i\omega_y)/\omega]\sin\omega/2 & \cos\omega/2 - i(\omega_z/\omega)\sin\omega/2 \end{bmatrix}$$

where $\boldsymbol{\omega} = (\omega_x, \omega_y, \omega_z)$ provides the 3 parameters. As we have seen, this is the same as

$$\underline{r}(\boldsymbol{\omega}) = \exp\left\{i\boldsymbol{\sigma}\cdot\boldsymbol{\omega}/2\right\}$$

(e) The two-particle states $\psi_{ij}(1,2) \equiv \zeta_i(1)\zeta_j(2)$ span a space of dimension 4, corresponding to all possible combinations of the values of the third component of the spin of the two particles. The states $\psi_{ij}^S(1,2)$ and $\psi_{ij}^A(1,2)$ are linear combinations of these 4 states, and they provide a different basis

---

[14] Recall Prob. B.3.

for the space[15]

$$\psi_{ij}^S(1,2) = \begin{cases} \zeta_\uparrow(1)\zeta_\uparrow(2) \\ [\zeta_\uparrow(1)\zeta_\downarrow(2) + \zeta_\downarrow(1)\zeta_\uparrow(2)]/\sqrt{2} \\ \zeta_\downarrow(1)\zeta_\downarrow(2) \end{cases}$$

$$\psi_{ij}^A(1,2) = [\zeta_\uparrow(1)\zeta_\downarrow(2) - \zeta_\downarrow(1)\zeta_\uparrow(2)]/\sqrt{2}$$

It is readily verified that the bilinear transformation in part (e) preserves the *symmetry* of the state under interchange of the indices (i, j). That is, if the initial state is symmetric (antisymmetric) under the interchange of (i, j), then so will the transformed state be symmetric (antisymmetric) under the interchange of (i', j'). Thus the states $\psi^S(1, 2)$ get mixed among themselves by the transformation, while the state $\psi^A(1, 2)$ remains invariant.[16]

Alternatively, we may here also define the total spin operator

$$\mathbf{S} \equiv \frac{1}{2}[\boldsymbol{\sigma}(1)\underline{1}(2) + \underline{1}(1)\boldsymbol{\sigma}(2)]$$

and observe that the two-particle states transform as

$$\psi'(1,2) = e^{i\mathbf{S}\cdot\boldsymbol{\omega}}\,\psi(1,2)$$

We observe that the operators $\mathbf{S}^2$ and $\mathbf{S}\cdot\boldsymbol{\omega}$ commute

$$\left[\mathbf{S}^2, \mathbf{S}\cdot\boldsymbol{\omega}\right] = 0$$

Therefore the two-particle states which are eigenstates of $\mathbf{S}^2$ will transform under the action of $e^{i\mathbf{S}\cdot\boldsymbol{\omega}}$ into a linear combination of states with the same eigenvalue of $\mathbf{S}^2$.

It is easy to verify that $\psi^A(1, 2)$ is an eigenstate of $\mathbf{S}^2$ with eigenvalue of the form $S(S + 1)$ where $S = 0$, while the $\psi^S(1, 2)$ are eigenstates with eigenvalue $S = 1$.

---

[15]These states are also normalized to 1.

[16]This is the simplest example of the reduction of the direct-product states into bases for irreducible representations of the group, here SU(2).

Chapter 8

Special Relativity

Problem 8.1 (a) Make an expansion for small v/c, and verify the expression for the difference in optical pathlength in the second of Eqs. (8.7);

(b) Verify the expression for the inverse Lorentz transformation in Eq. (8.9);

(c) Verify that the Lorentz transformation in Eqs. (8.8) leaves the quadratic form in Eq. (8.10) invariant;

(d) Verify that the three-dimensional Lorentz transformation in Eqs. (8.24) leaves the quadratic form in Eq. (8.25) invariant.

Solution to Problem 8.1

(a) The difference in optical pathlength is given in Eq. (8.7) and reads

$$
\begin{aligned}
\Delta &= \frac{2\pi L}{\lambda} \left[\frac{1}{1 - v/c} + \frac{1}{1 + v/c} - \frac{2}{\sqrt{1 - v^2/c^2}} \right] \\
&= \frac{2\pi L}{\lambda} \left[\frac{2}{1 - v^2/c^2} - \frac{2}{\sqrt{1 - v^2/c^2}} \right] \\
&\approx \frac{4\pi L}{\lambda} \left[1 + \frac{v^2}{c^2} - \left(1 + \frac{v^2}{2c^2} \right) \right] + O\left[\frac{v^4}{c^4} \right] \\
&= \frac{2\pi L}{\lambda} \left(\frac{v}{c} \right)^2 + O\left[\frac{v^4}{c^4} \right]
\end{aligned}
$$

which agrees with the second of Eqs. (8.7).

(b) In order to check that the expressions in Eqs. (8.9) correspond to the inverse Lorentz transformation, we may substitute Eqs. (8.8) into Eqs. (8.9) and verify that one obtains an identity. For instance, if we use Eqs. (8.8) in the r.h.s. of the first Eq. (8.9), we obtain the l.h.s. of the same equation

$$\frac{x - Vt}{\sqrt{1 - V^2/c^2}} = \frac{x' + Vt'}{1 - V^2/c^2} - \frac{V(t' + Vx'/c^2)}{1 - V^2/c^2}$$

$$= x'$$

A similar result holds for t'.

(c) We have

$$x^2 - c^2t^2 = \frac{(x' + Vt')^2}{1 - V^2/c^2} - c^2\frac{(t' + Vx'/c^2)^2}{1 - V^2/c^2}$$

$$= \frac{x'^2(1 - V^2/c^2) - t'^2(c^2 - V^2)}{1 - V^2/c^2}$$

$$= x'^2 - c^2t'^2$$

(d) The demonstration of Eq. (8.25) follows directly from the demonstration carried out in part (c), the only difference being the presence of a contribution $y^2 + z^2$ in the quadratic form; however, the Lorentz transformation of Eq. (8.24) leaves the y and z directions invariant, and therefore the proof follows.

Problem 8.2 In deriving the formula for the Lorentz contraction of a meter stick, only the first of Eqs. (8.14) is used. Explain the corresponding physical content of the second of Eqs. (8.14).

Solution to Problem 8.2

The length of the meter stick in the lab is determined by a simultaneous measurement a given time t at the two ends. Since the corresponding time t' in the rest frame of the meter stick depends on position, the times of the measurement at the two ends of the stick will be different in the rest frame.

Problem 8.3 Start from Eqs. (8.40) and (8.42).

(a) Verify by matrix multiplication that $\underline{x}' = \underline{a}\,\underline{x}$ gives the proper expressions for the Lorentz transformation in the z-direction of the four-vector x_μ (see Fig. 8.12 in the text);

(b) Verify by matrix multiplication that $\underline{a}^T\underline{a} = \underline{a}\,\underline{a}^T = \underline{1}$.

Solution to Problem 8.3

(a) It is readily verified that matrix multiplication provides the proper expression for the Lorentz transformation in the z-direction

$$\underline{a}\,\underline{x} = \begin{bmatrix} 1 & 0 & 0 & 0 \\ 0 & 1 & 0 & 0 \\ 0 & 0 & \dfrac{1}{\sqrt{1-v^2/c^2}} & \dfrac{iv/c}{\sqrt{1-v^2/c^2}} \\ 0 & 0 & \dfrac{-iv/c}{\sqrt{1-v^2/c^2}} & \dfrac{1}{\sqrt{1-v^2/c^2}} \end{bmatrix} \begin{pmatrix} x \\ y \\ z \\ ict \end{pmatrix} = \begin{pmatrix} x \\ y \\ \dfrac{z-vt}{\sqrt{1-v^2/c^2}} \\ \dfrac{ic(t-zv/c^2)}{\sqrt{1-v^2/c^2}} \end{pmatrix} = \begin{pmatrix} x' \\ y' \\ z' \\ ict' \end{pmatrix}$$

(b) We verify this property by direct matrix multiplication

$$\underline{a}^T \underline{a} = \underline{a}\,\underline{a}^T = \begin{bmatrix} 1 & 0 & 0 & 0 \\ 0 & 1 & 0 & 0 \\ 0 & 0 & \dfrac{1}{\sqrt{1-v^2/c^2}} & \dfrac{-iv/c}{\sqrt{1-v^2/c^2}} \\ 0 & 0 & \dfrac{iv/c}{\sqrt{1-v^2/c^2}} & \dfrac{1}{\sqrt{1-v^2/c^2}} \end{bmatrix} \begin{bmatrix} 1 & 0 & 0 & 0 \\ 0 & 1 & 0 & 0 \\ 0 & 0 & \dfrac{1}{\sqrt{1-v^2/c^2}} & \dfrac{iv/c}{\sqrt{1-v^2/c^2}} \\ 0 & 0 & \dfrac{-iv/c}{\sqrt{1-v^2/c^2}} & \dfrac{1}{\sqrt{1-v^2/c^2}} \end{bmatrix}$$

$$= \begin{bmatrix} 1 & 0 & 0 & 0 \\ 0 & 1 & 0 & 0 \\ 0 & 0 & 1 & 0 \\ 0 & 0 & 0 & 1 \end{bmatrix}$$

We may explicitly write the components of these equations as in Eq. (8.32)

$$\sum_{\mu=1}^{4} a_{\mu\nu} a_{\mu\rho} = \sum_{\mu=1}^{4} a_{\nu\mu} a_{\rho\mu} = \delta_{\nu\rho}$$

Problem 8.4 (a) Prove that the scalar product of two four-vectors in Eq. (8.44) is invariant under Lorentz transformations;

(b) Use this result to prove that the wave operator

$$\left(\frac{\partial}{\partial x_\mu}\right)^2 = \frac{\partial^2}{\partial x^2} - \frac{1}{c^2}\frac{\partial^2}{\partial t^2} = \Box \qquad ; \text{D'Alembertian}$$

is invariant under Lorentz transformations;

(c) The electromagnetic current in special relativity is defined by

$$e j_\mu \equiv e\left(\frac{1}{c}\mathbf{j}, i\rho\right) \qquad ; \text{electromagnetic current}$$

where the electric charge e is now explicitly exhibited. This current transforms as a four-vector under Lorentz transformations.

Establish the invariant *continuity equation* for the electromagnetic current[1]

$$\frac{\partial j_\mu}{\partial x_\mu} = \boldsymbol{\nabla} \cdot \mathbf{j} + \frac{\partial \rho}{\partial t} = 0 \qquad \text{; continuity equation}$$

Solution to Problem 8.4

(a) Let u_μ and w_μ be two four-vectors. Their scalar product is

$$u \cdot w = \sum_{\mu=1}^{4} u_\mu w_\mu$$

This scalar product is invariant under a Lorentz transformation

$$u' \cdot w' = \sum_{\mu=1}^{4}\sum_{\nu=1}^{4}\sum_{\rho=1}^{4} a_{\mu\nu} u_\nu a_{\mu\rho} w_\rho = \sum_{\nu=1}^{4}\sum_{\rho=1}^{4} \delta_{\nu\rho} u_\nu w_\rho = u \cdot w$$

(b) The gradient with respect to x_μ is a four-vector [see Eq. (8.49)]. Since the D'Alembertian operator is obtained as the scalar product

$$\Box = \sum_{\mu=1}^{4} \frac{\partial}{\partial x_\mu}\frac{\partial}{\partial x_\mu} = \nabla^2 - \frac{1}{c^2}\frac{\partial^2}{\partial t^2} \qquad \text{; invariant}$$

it is invariant under a Lorentz transformation.

(c) Since both $\partial/\partial x_\mu = (\boldsymbol{\nabla},\, \partial/\partial ict)$ and $j_\mu = (\mathbf{j}/c,\, i\rho)$ are four-vectors, their scalar product is a Lorentz-invariant quantity

$$\sum_{\mu=1}^{4} \frac{\partial j_\mu}{\partial x_\mu} = \frac{1}{c}\left(\boldsymbol{\nabla}\cdot\mathbf{j} + \frac{\partial \rho}{\partial t}\right) \qquad \text{; invariant}$$

In the frame where the charge is at rest, the spatial components of the four-vector j_μ vanish, and the conservation of charge implies $\partial \rho/\partial t = 0$. Therefore, in this frame, $\partial j_\mu/\partial x_\mu = 0$. However, $\partial j_\mu/\partial x_\mu$ is a Lorentz invariant and therefore

$$\frac{\partial j_\mu}{\partial x_\mu} = 0$$

in all frames.

---

[1] As an example, consider an electron fluid with electron density n and velocity field $\mathbf{v}$. The charge density is $e\rho = en$ and the current is $e\mathbf{j} = en\mathbf{v}$. The continuity equation then reduces to the familiar statement of electron number conservation $\partial n/\partial t + \boldsymbol{\nabla}\cdot(n\mathbf{v}) = 0$.

Problem 8.5 An electron of energy 50 GeV and rest mass $m_0c^2 = 0.5$ MeV travels down an accelerator pipe of length 1 km. How long does the pipe appear to be in the rest frame of the electron (in cm)?[2]

Solution to Problem 8.5

From the ratio between the rest energy and the total energy of this electron we can obtain $\sqrt{1 - v^2/c^2}$ [see Eqs. (8.58)]

$$\frac{E_{\rm rest}}{E} = \sqrt{1 - \frac{v^2}{c^2}} = \frac{0.5\,{\rm MeV}}{50 \times 10^3\,{\rm MeV}} = 10^{-5}$$

We therefore conclude that the accelerator pipe will appear to the electron in its rest frame to be just 1 cm long

$$l_{\rm rest} = \sqrt{1 - \frac{v^2}{c^2}}\, l = 10^{-5} \times 10^5\,{\rm cm} = 1\,{\rm cm}$$

Problem 8.6 A light signal, and a neutrino with energy $E = 2$ MeV and rest mass $m_0c^2 = 2$ eV, are emitted simultaneously from a supernova which is at a distance 10^4 light-years from earth.
(a) What is the difference in arrival times at the earth?
(b) How long does the trip take in the neutrino's rest frame?
(c) What is the distance to earth as viewed in the neutrino's rest frame?

Solution to Problem 8.6

(a) As in the previous problem, we use the ratio between the rest energy of the neutrino and its total energy to obtain its velocity

$$\frac{E_{\rm rest}}{E} = \sqrt{1 - \frac{v^2}{c^2}} = \frac{2\,{\rm eV}}{2 \times 10^6\,{\rm eV}} = 10^{-6}$$

and therefore

$$v = c\sqrt{1 - 10^{-12}}$$

Call l the distance of the supernova from earth. The difference in the arrival times at earth is then given by[3]

$$\Delta t = \frac{l}{v} - \frac{l}{c} = \frac{l}{c}\left(\frac{1}{\sqrt{1 - 10^{-12}}} - 1\right) \approx 10^{-12}\frac{l}{2c}$$

---

[2] Although not required here, we note that in doing numerical calculations when v/c is very close to 1, it is often useful to write $v/c = 1 - \varepsilon$ and then $\sqrt{1 - v^2/c^2} = \sqrt{(1 - v/c)(1 + v/c)} = \sqrt{2\varepsilon}\,[1 + O(\varepsilon)]$.

[3] Recall the footnote in Prob. 4.32.

One light-year is the distance traveled by light in a year

$$1 \, \text{yr} = 3.15 \times 10^7 \, \text{sec}$$
$$1 \, \text{light-yr} = 3.15c \times 10^7 \, \text{sec}$$

Hence

$$l = 10^4 \, \text{light-yr} = 3.15c \times 10^{11} \, \text{sec}$$
$$\Delta t = 0.157 \, \text{sec}$$

(b) Equation (8.20) provides the relation between the time intervals measured in the laboratory frame (t) and in the rest frame of the particle (τ). There is a time dilation of the first with respect to the second and therefore

$$\tau = \frac{l}{v}\sqrt{1 - \frac{v^2}{c^2}} \approx \frac{l}{c}\sqrt{1 - \frac{v^2}{c^2}} = 10^{-2} \, \text{yr}$$

which is just 3.65 days.

(c) From part (a) of the problem, we know that the distance to earth viewed in the neutrino's rest frame will be 10^{-6} smaller than its actual value

$$l_{\text{rest}} = l\sqrt{1 - \frac{v^2}{c^2}} = 10^{-2} \, \text{light-yr}$$

Problem 8.7 In special relativity, a tensor of rank two, three, *etc.* is defined by how it transforms under a Lorentz transformation

$$T'_{\mu\nu\ldots} = a_{\mu\mu'}a_{\nu\nu'}\cdots T_{\mu'\nu'\ldots} \qquad ; \text{tensor}$$

A four-vector is then a tensor of rank one.

(a) Prove that $\delta_{\mu\nu}$ transforms as a second-rank tensor (*Hint*: Define it to have its usual value in all frames, and then show that it satisfies the tensor transformation law.)

(b) The completely antisymmetric tensor in 4-D is defined by

$$\varepsilon_{\alpha\beta\gamma\delta} = \pm 1 \qquad ; (\alpha\beta\gamma\delta) \text{ an even (odd) permutation of (1234)}$$
$$= 0 \qquad ; \text{otherwise}$$

Show that this quantity transforms as a fourth-rank tensor under Lorentz transformations with $\det \underline{a} = 1$. (*Hint*: Make use of the following definition of the determinant $a_{\mu\mu'}a_{\nu\nu'}a_{\rho\rho'}a_{\sigma\sigma'}\varepsilon_{\mu'\nu'\rho'\sigma'} = \varepsilon_{\mu\nu\rho\sigma} \det \underline{a}$.)

Solution to Problem 8.7

(a) We write[4]

$$\delta'_{\mu\nu} = a_{\mu\mu'} a_{\nu\nu'} \delta_{\mu'\nu'} = a_{\mu\mu'} a_{\nu\mu'} = \delta_{\mu\nu}$$

which proves that $\delta_{\mu\nu}$ is a rank-two tensor.

(b) For $\det \underline{a} = 1$, the definition of determinant reads

$$\varepsilon'_{\mu\nu\rho\sigma} = a_{\mu\mu'} a_{\nu\nu'} a_{\rho\rho'} a_{\sigma\sigma'} \varepsilon_{\mu'\nu'\rho'\sigma'} = \varepsilon_{\mu\nu\rho\sigma} \det \underline{a} = \varepsilon_{\mu\nu\rho\sigma}$$

which proves that $\varepsilon_{\mu\nu\rho\sigma}$ is a rank-four tensor.

Problem 8.8 The electromagnetic field tensor is defined by

$$\underline{F} \equiv \begin{bmatrix} 0 & cB_3 & -cB_2 & -iE_1 \\ -cB_3 & 0 & cB_1 & -iE_2 \\ cB_2 & -cB_1 & 0 & -iE_3 \\ iE_1 & iE_2 & iE_3 & 0 \end{bmatrix} \qquad ; \text{ field tensor}$$

$$F_{\mu\nu} = [\underline{F}]_{\mu\nu} \qquad\qquad\qquad ; (\mu, \nu) = (1, 2, 3, 4)$$

(a) Show that Maxwell's Eqs. (2.90) in SI units can be written in the following form

$$\frac{\partial}{\partial x_\nu} F_{\mu\nu} = \frac{e}{\varepsilon_0} j_\mu \qquad ; \text{ Maxwell's equations}$$

$$\varepsilon_{\mu\nu\rho\sigma} \frac{\partial}{\partial x_\sigma} F_{\nu\rho} = 0 \qquad \mu = 1, 2, 3, 4$$

Here ej_μ is the electromagnetic current in Prob. 8.4(c).[5]

(b) Show that the statement that Maxwell's equations hold in any inertial frame is equivalent to the statement that $F_{\mu\nu}$ transforms as a second-rank tensor under Lorentz transformations.

Solution to Problem 8.8

First, we notice that the electromagnetic field tensor is antisymmetric $F_{\mu\nu} = -F_{\nu\mu}$. Let us then explicitly work out the first component of the

---

[4] Note that we are now using the convention that repeated Greek indicies are summed from 1 to 4.

[5] Note that if one defines a *dual* tensor by $G_{\mu\nu} \equiv \varepsilon_{\mu\nu\rho\sigma} F_{\rho\sigma}$, then the second set of Maxwell's equations can be written as $\partial G_{\mu\nu}/\partial x_\nu = 0$.

four-vector $\partial F_{\mu\nu}/\partial x_\nu$

$$\frac{\partial}{\partial x_\nu}F_{1\nu} = \frac{\partial}{\partial x_1}F_{11} + \frac{\partial}{\partial x_2}F_{12} + \frac{\partial}{\partial x_3}F_{13} + \frac{\partial}{\partial x_4}F_{14}$$

$$= c\frac{\partial}{\partial x_2}B_3 - c\frac{\partial}{\partial x_3}B_2 - \frac{1}{c}\frac{\partial}{\partial t}E_1$$

We use Maxwell's equations to write

$$\frac{\partial B_3}{\partial x_2} - \frac{\partial B_2}{\partial x_3} = (\mathbf{\nabla} \times \mathbf{B})_1 = \left(\mu_0 e\mathbf{j} + \frac{1}{c^2}\frac{\partial \mathbf{E}}{\partial t}\right)_1$$

where the electromagnetic charge e is explicitly exhibited. Recall

$$j_\mu = \left(\frac{1}{c}\mathbf{j},\, i\rho\right) \qquad ; \mu_0\varepsilon_0 = \frac{1}{c^2}$$

Therefore we have

$$\frac{\partial}{\partial x_\nu}F_{1\nu} = \frac{e}{\varepsilon_0}j_1$$

In a similar way we obtain

$$\frac{\partial}{\partial x_\nu}F_{2\nu} = \frac{e}{\varepsilon_0}j_2$$

$$\frac{\partial}{\partial x_\nu}F_{3\nu} = \frac{e}{\varepsilon_0}j_3$$

for the remaining spatial components.

Let us now consider the time component, using the first of Maxwell's equations

$$\frac{\partial}{\partial x_\nu}F_{4\nu} = \frac{\partial}{\partial x_1}F_{41} + \frac{\partial}{\partial x_2}F_{42} + \frac{\partial}{\partial x_3}F_{43} + \frac{\partial}{\partial x_4}F_{44}$$

$$= i\frac{\partial}{\partial x_1}E_1 + i\frac{\partial}{\partial x_2}E_2 + i\frac{\partial}{\partial x_3}E_3$$

$$= i\mathbf{\nabla} \cdot \mathbf{E}$$

$$= i\frac{e}{\varepsilon_0}\rho$$

Therefore we obtain the equation for the four-vector $\partial F_{\mu\nu}/\partial x_\nu$

$$\frac{\partial}{\partial x_\nu}F_{\mu\nu} = \frac{e}{\varepsilon_0}j_\mu$$

Consider, next, the first component of the four-vector $\varepsilon_{\mu\nu\rho\sigma}(\partial F_{\nu\rho}/\partial x_\sigma)$, writing its components explicitly as

$$\varepsilon_{1\nu\rho\sigma}\frac{\partial}{\partial x_\sigma}F_{\nu\rho} = \frac{\partial}{\partial x_2}(F_{34} - F_{43}) + \frac{\partial}{\partial x_3}(F_{42} - F_{24}) + \frac{\partial}{\partial x_4}(F_{23} - F_{32})$$

$$= -2i\frac{\partial}{\partial x_2}E_3 + 2i\frac{\partial}{\partial x_3}E_2 + \frac{2}{i}\frac{\partial}{\partial t}B_1$$

$$= -2i\left(\mathbf{\nabla}\times\mathbf{E} + \frac{\partial\mathbf{B}}{\partial t}\right)_1$$

$$= 0$$

Here the last relation follows from the third of Eqs. (2.90). A similar result holds for the other two spatial components.

We may now calculate the time component

$$\varepsilon_{4\nu\rho\sigma}\frac{\partial}{\partial x_\sigma}F_{\nu\rho} = \frac{\partial}{\partial x_1}(F_{32} - F_{23}) + \frac{\partial}{\partial x_2}(F_{13} - F_{31}) + \frac{\partial}{\partial x_3}(F_{21} - F_{12})$$

$$= -2c\left(\frac{\partial}{\partial x_1}B_1 + \frac{\partial}{\partial x_2}B_2 + \frac{\partial}{\partial x_3}B_3\right)$$

$$= -2c\,\mathbf{\nabla}\cdot\mathbf{B}$$

$$= 0$$

where we have used the second of Eqs. (2.90). Hence, for the four-vector $\varepsilon_{\mu\nu\rho\sigma}(\partial F_{\nu\rho}/\partial x_\sigma)$

$$\varepsilon_{\mu\nu\rho\sigma}\frac{\partial}{\partial x_\sigma}F_{\nu\rho} = 0$$

(b) Assume the quantities $\partial F_{\mu\nu}/\partial x_\nu$ and $\varepsilon_{\mu\nu\rho\sigma}(\partial F_{\nu\rho}/\partial x_\sigma)$ do indeed transform as four-vectors. This will be true if $F_{\mu\nu}$ transforms as a second-rank tensor. Inertial frames are now related by means of a Lorentz transformation. Since we have proven in part (a) that Maxwell's laws may be expressed in terms of four-vectors, their expressions in any two inertial frames are precisely related by a Lorentz transformation.[6] We may therefore conclude that the statement that Maxwell's equations hold in any inertial frame is equivalent to the statement that $F_{\mu\nu}$ transforms as a second-rank tensor under Lorentz transformations.

Problem 8.9 The electromagnetic field tensor is antisymmetric in its two indices $F_{\nu\mu} = -F_{\mu\nu}$. It can be expressed in terms of the *vector potential*

---

[6]The relations are *covariant* (see the footnote on p. 171).

A_μ according to

$$\frac{1}{c}F_{\mu\nu} \equiv \frac{\partial A_\nu}{\partial x_\mu} - \frac{\partial A_\mu}{\partial x_\nu} \quad ; \; A_\mu = \left(\mathbf{A}, \frac{i}{c}\Phi\right)$$

(a) Show that the second set of Maxwell's equations in Prob. 8.8(a) is now satisfied identically;

(b) Compare with the expression for $F_{\mu\nu}$ in Prob. 8.8, and show that the relation of A_μ to the electric and magnetic fields is given by[7]

$$\mathbf{B} = \mathbf{\nabla} \times \mathbf{A}$$

$$\mathbf{E} = -\mathbf{\nabla}\Phi - \frac{\partial \mathbf{A}}{\partial t}$$

(c) Show that the electromagnetic fields are invariant under a *gauge transformation* of the vector potential

$$A_\mu \to A_\mu + \frac{\partial \Lambda}{\partial x_\mu} \quad ; \text{ gauge transformation}$$

where Λ is a differentiable scalar function of x_μ. [Note Prob. F.3(a)].

Solution to Problem 8.9

(a) We may substitute the expression for $F_{\mu\nu}$ in terms of the vector potential A_μ into the second set of Maxwell's equation to obtain[8]

$$\varepsilon_{\mu\nu\rho\sigma}\frac{\partial}{\partial x_\sigma}F_{\nu\rho} = c\varepsilon_{\mu\nu\rho\sigma}\frac{\partial}{\partial x_\sigma}\left(\frac{\partial A_\rho}{\partial x_\nu} - \frac{\partial A_\nu}{\partial x_\rho}\right)$$

The expression that we need to evaluate contains the contraction of the completely antisymmetric rank-four tensor $\varepsilon_{\mu\nu\rho\sigma}$ with the rank-three tensor $\partial^2 A_\rho/\partial x_\sigma\partial x_\nu$, which is symmetric in two of its indices since the order of partial derivatives can always be interchanged. It is easy to verify that the contraction of an antisymmetric tensor with a symmetric one must vanish. Explicitly, for the first term,

$$\varepsilon_{\mu\nu\rho\sigma}\frac{\partial^2 A_\rho}{\partial x_\sigma\partial x_\nu} = \varepsilon_{\mu\sigma\rho\nu}\frac{\partial^2 A_\rho}{\partial x_\nu\partial x_\sigma} = -\varepsilon_{\mu\nu\rho\sigma}\frac{\partial^2 A_\rho}{\partial x_\sigma\partial x_\nu}$$

---

[7] In advanced quantum mechanics, Maxwell's Eqs. (2.90) are frequently written in rationalized cgs [Heaviside-Lorentz (H-L)] units where $\varepsilon_0 = 1$, $\mu_0 = 1/c^2$, $c\mathbf{B} \to \mathbf{B}$, $\mathbf{j}/c \to \mathbf{j}$, $e\rho$ is measured in esu, and ej in emu, where 1 emu = 1 esu/c (see appendix K). To get from SI units to H-L units in Probs. 8.4(c), 8.8, and 8.9 make the following replacements: $\varepsilon_0 \to 1$, $c\mathbf{B} \to \mathbf{B}$, $\mathbf{j}/c \to \mathbf{j}$, $c\mathbf{A} \to \mathbf{A}$.

[8] As usual, we are adopting the convention that repeated Greek indices are summed from 1 to 4.

where we have first re-labeled the indices $\nu \rightleftharpoons \sigma$, and then used the antisymmetry of $\varepsilon_{\mu\nu\rho\sigma}$ under the interchange $\nu \rightleftharpoons \sigma$. Since the l.h.s. of the equation equals its negative, we conclude that this quantity vanishes.

(b) Let us express $\boldsymbol{\nabla} \times \mathbf{A}$ directly in terms of $F_{\mu\nu}$

$$\boldsymbol{\nabla} \times \mathbf{A} = \left[\left(\frac{\partial A_3}{\partial x_2} - \frac{\partial A_2}{\partial x_3}\right), \left(\frac{\partial A_1}{\partial x_3} - \frac{\partial A_3}{\partial x_1}\right), \left(\frac{\partial A_2}{\partial x_1} - \frac{\partial A_1}{\partial x_2}\right)\right]$$

$$= \frac{1}{c}\left(F_{23}, F_{31}, F_{12}\right) = \left(B_1, B_2, B_3\right)$$

$$= \mathbf{B}$$

Similarly

$$-\boldsymbol{\nabla}\Phi - \frac{\partial \mathbf{A}}{\partial t} = ic\left[\left(\frac{\partial A_4}{\partial x_1} - \frac{\partial A_1}{\partial x_4}\right), \left(\frac{\partial A_4}{\partial x_2} - \frac{\partial A_2}{\partial x_4}\right), \left(\frac{\partial A_4}{\partial x_3} - \frac{\partial A_3}{\partial x_4}\right)\right]$$

$$= i\left(F_{14}, F_{24}, F_{34}\right) = \left(E_1, E_2, E_3\right)$$

$$= \mathbf{E}$$

(c) Under the gauge transformation we have

$$F_{\mu\nu} \rightarrow \frac{\partial}{\partial x_\mu}\left[A_\nu + \frac{\partial \Lambda}{\partial x_\nu}\right] - \frac{\partial}{\partial x_\nu}\left[A_\mu + \frac{\partial \Lambda}{\partial x_\mu}\right]$$

$$= F_{\mu\nu} + \left(\frac{\partial^2 \Lambda}{\partial x_\mu \partial x_\nu} - \frac{\partial^2 \Lambda}{\partial x_\nu \partial x_\mu}\right) = F_{\mu\nu}$$

Here the term depending on the second partial derivative of Λ vanishes because of its invariance under the interchange of the order of partial differentiation.

Problem 8.10 One can attempt to write a covariant equation of motion for a charged particle in an electromagnetic field as follows

$$\frac{dp_\mu}{d\tau} = \frac{q}{c}F_{\mu\nu}u_\nu \qquad ; \mu = 1, 2, 3, 4$$

Here τ is the proper time, (u_μ, p_μ) are the four-velocity and four-momentum of Eqs. (8.54) and (8.57), and $F_{\mu\nu}$ is the electromagnetic field tensor of Prob. 8.8.

(a) Given that $F_{\mu\nu}$ is a second-rank tensor, show that this is a relation between four-vectors;[9]

---

[9]More precisely, if $u' \cdot v' = u \cdot v$, then an expression is Lorentz *invariant*. If $u'_\mu = v'_\mu$ when $u_\mu = v_\mu$, then the expression is Lorentz *covariant* — it takes the same form in any Lorentz frame.

(b) Show that the spatial parts of this relation give the relativistic form of Newton's second law for a particle of charge q acted upon by the Lorentz force of Eq. (2.83)

$$\frac{d\mathbf{p}}{dt} = q\,(\mathbf{E} + \mathbf{v} \times \mathbf{B}) \qquad ; \text{ Newton's second law}$$

(c) Show that with $p_\mu \equiv (\mathbf{p}, iE_p/c)$, the fourth component of the above gives the correct relativistic power equation $dE_p/dt = q\mathbf{v} \cdot \mathbf{E}$. Derive this result from the relativistic relation between energy and momentum in Eq. (8.59) and the result in part (b).

Solution to Problem 8.10

(a) The proper time τ is a scalar (since it is invariant under Lorentz transformations), u_μ and p_μ are four vectors, while $F_{\mu\nu}$ is a rank-two tensor. The contraction of a rank-two tensor with an arbitrary four-vector generates a four-vector, while the multiplication or division of a four-vector by a scalar does not change its rank. We therefore conclude that both the l.h.s. and the r.h.s. of the equation of motion are four-vectors.

(b) Recall

$$dt = \frac{d\tau}{\sqrt{1 - v^2/c^2}}$$

$$u_\mu = \frac{1}{\sqrt{1 - v^2/c^2}}\,(\mathbf{v}, ic)$$

$$p_\mu = m_0 u_\mu = \left(\mathbf{p}, \frac{i}{c}E_p\right)$$

Let us evaluate the spatial components of $F_{\mu\nu}u_\nu$. We explicitly work out the first component

$$F_{1\nu}u_\nu = F_{12}u_2 + F_{13}u_3 + F_{14}u_4$$

$$= \frac{c}{\sqrt{1 - v^2/c^2}}\,[B_3 v_2 - B_2 v_3 + E_1]$$

$$= \frac{c}{\sqrt{1 - v^2/c^2}}\,[\mathbf{E} + \mathbf{v} \times \mathbf{B}]_1$$

The remaining two spatial components can be obtained in the same fashion, and they read

$$F_{2\nu}u_{\nu} = \frac{c}{\sqrt{1 - v^2/c^2}}\,[\mathbf{E} + \mathbf{v} \times \mathbf{B}]_2$$

$$F_{3\nu}u_{\nu} = \frac{c}{\sqrt{1 - v^2/c^2}}\,[\mathbf{E} + \mathbf{v} \times \mathbf{B}]_3$$

Let us now evaluate the spatial components of $dp_{\mu}/d\tau$

$$\frac{d\mathbf{p}}{d\tau} = \frac{1}{\sqrt{1 - v^2/c^2}}\,\frac{d\mathbf{p}}{dt}$$

where we have used Eqs. (8.53) and (8.57).

A combination of these results reproduces Newton's second law from the covariant equations of motion

$$\frac{d\mathbf{p}}{dt} = q\,(\mathbf{E} + \mathbf{v} \times \mathbf{B})$$

(c) Let us now evaluate the fourth component of $F_{\mu\nu}u_{\nu}$

$$F_{4\nu}u_{\nu} = F_{41}u_1 + F_{42}u_2 + F_{43}u_3 = \frac{i\,\mathbf{v} \cdot \mathbf{E}}{\sqrt{1 - v^2/c^2}}$$

Similarly we have

$$\frac{dp_4}{d\tau} = \frac{i/c}{\sqrt{1 - v^2/c^2}}\,\frac{dE_p}{dt}$$

Substitution of these expressions into the covariant equation of motion gives

$$\frac{dE_p}{dt} = q\mathbf{v} \cdot \mathbf{E}$$

We can now derive this power equation from the time derivative of the particle's energy $E_p = (c^2\mathbf{p}^2 + m_0^2 c^4)^{1/2}$

$$\frac{dE_p}{dt} = \frac{c^2\mathbf{p}}{E_p} \cdot \frac{d\mathbf{p}}{dt} = \mathbf{v} \cdot \frac{d\mathbf{p}}{dt}$$

where $\mathbf{v}/c = \mathbf{p}c/E_p$ [see Eqs. (8.58)]. Now use Newton's second law

$$\frac{dE_p}{dt} = q\mathbf{v} \cdot (\mathbf{E} + \mathbf{v} \times \mathbf{B}) = q\mathbf{v} \cdot \mathbf{E}$$

Problem 8.11 The weak vector boson Z^0 has the mass $m_Z c^2 = 91.19\,\text{GeV}$. It can be produced in the reaction $e^+ + e^- \to Z^0$. What is the threshold energy for production of this particle in the C-M system? What is the threshold energy for production of this particle with a positron beam incident on electrons at rest? Discuss.

Solution to Problem 8.11

In the C-M system, the electron and positron four-momenta are

$$p_\mu^- = \left(\mathbf{p}, i\frac{E_p}{c}\right) \qquad ; \ p_\mu^+ = \left(-\mathbf{p}, i\frac{E_p}{c}\right)$$

where $E_p = (p^2c^2 + m_e^2c^4)^{1/2}$. The threshold energy for the production of a Z^0 boson in the C-M system can be obtained from the conservation of the energy

$$2E_p = m_Z c^2 \qquad ; \ E_p = 45.6\,\text{GeV}$$

In the lab frame where the electron is at rest, we have for the same process

$$p_\mu^- = (0, \, im_e c) \qquad ; \ p_\mu^+ = \left(\mathbf{p}_L, \, i\sqrt{p_L^2 + m_e^2 c^2}\right)$$

$$p_\mu^Z = \left(\mathbf{p}_L, \, i\sqrt{p_L^2 + m_Z^2 c^2}\right)$$

We can calculate the Lorentz-invariant quantity $s = -(p^+ + p^-)^2$ in the lab frame

$$\begin{aligned} s &= -(p^+ + p^-)^2 \\ &= -p_L^2 + \left(m_e c + \sqrt{p_L^2 + m_e^2 c^2}\right)^2 \\ &= 2(m_e c)^2 + 2m_e E_L \qquad ; \ \text{lab frame} \end{aligned}$$

and in the C-M system, with $p^+ + p^- = p^Z$,

$$s = -(p^Z)^2 = (m_Z c)^2 \qquad ; \ \text{C-M system}$$

If these two expressions are equated, one obtains the threshold laboratory energy for creation of the Z^0

$$E_L^{\text{thresh}} = \frac{(m_Z c^2)^2}{2m_e c^2} - m_e c^2 \approx 8.14 \times 10^6\,\text{GeV}$$

While in the C-M the Z^0 boson is produced at rest, in the case of a positron beam incident on electrons at rest, the Z^0 boson is produced with non-zero momentum, since the conservation of momentum requires that it carries the same momentum as the incident positron.

Problem 8.12 Start from Eqs. (8.83) and (8.87). Use Eqs. (8.72) and (8.73) to determine the momentum $\mathbf{p}^2(p_L)$ in the C-M system. Assume elastic scattering of equal mass particles, with $m_1 = m_2 \equiv m$.

(a) Take the limit $p_L/mc \ll 1$ and derive the non-relativistic results for $\mathbf{p}^2, p'_L$, and $\cos\theta_{CM}$ as a function of $(p_L, \cos\theta_L)$ as given in Eqs. (8.88);

(b) Take the limit $p_L/mc \gg 1$ and derive the extreme-relativistic results for $\mathbf{p}^2, p'_L$, and $\cos\theta_{CM}$ as a function of $(p_L, \cos\theta_L)$ as given in Eqs. (8.89).

Solution to Problem 8.12

For $m_1 = m_2 = m$, Eq. (8.73) for the total energy in the C-M system reduces to

$$W = 2(\mathbf{p}^2c^2 + m^2c^4)^{1/2}$$

For this case, Eq. (8.72) reads

$$E_L = \frac{1}{2mc^2}[W^2 - 2(mc^2)^2]$$

Substitution of the above provides a relation between p and p_L

$$2p^2 = m(E_L - mc^2)$$

Here $E_L = \sqrt{p_L^2 c^2 + m^2c^4}$ is the initial energy of the incident particle in the laboratory.

(a) If we take the non-relativistic limit $p_L/mc \ll 1$, we may approximate $E_L \approx mc^2 + p_L^2/2m + \cdots$, and from the previous equation we thus obtain

$$4p^2 \approx p_L^2$$

From Eqs. (8.83) and (8.87), we then have for the kinematics of elastic scattering

$$mc^2(E_L - E'_L) = E_L E'_L - (mc^2)^2 - p_L p'_L c^2 \cos\theta_L$$

In this non-relativistic limit, the condition $p_L/mc \ll 1$ also implies $p'_L/mc \ll 1$. If we use $E'_L \approx mc^2 + p_L'^2/2m + \cdots$ in the above equation, we obtain the relation

$$\frac{1}{2}(p_L^2 - p_L'^2) \approx \frac{1}{2}(p_L^2 + p_L'^2) - p_L p'_L \cos\theta_L$$

which implies

$$p'_L = p_L \cos \theta_L$$

From Eq. (8.87), we then have

$$m(E_L - E'_L) = \mathbf{p}^2(1 - \cos \theta_{CM})$$

Substitution of the above gives

$$\frac{1}{2}(p_L^2 - p'^2_L) = \frac{p_L^2}{4}(1 - \cos \theta_{CM})$$

$$1 - \cos^2 \theta_L = \frac{1}{2}(1 - \cos \theta_{CM})$$

$$\cos \theta_{CM} = 2 \cos^2 \theta_L - 1$$

which implies

$$\cos \theta_{CM} = \cos(2\theta_L)$$

(b) If we take the limit $p_L/mc \gg 1$, we may approximate $E_L \approx p_L c + \cdots$, and from the relation $2p^2 = m(E_L - mc^2)$ in part (a)

$$2p^2 \approx (mc)p_L$$

In this limit we also have $E'_L \approx p'_L c + \cdots$. In this case, from the equation in part (a) obtained by equating the l.h.s. of Eqs. (8.83) and (8.87), we have

$$\frac{1}{p'_L} - \frac{1}{p_L} = \frac{1}{mc}(1 - \cos \theta_L) \qquad ; \ E'_L \approx p'_L c$$

which is just the Compton formula (see below).

Equation (8.87) reads

$$m(E_L - E'_L) = \mathbf{p}^2(1 - \cos \theta_{CM})$$

Substitution of the above results then gives

$$mc(p_L - p'_L) = \frac{mc}{2} p_L(1 - \cos \theta_{CM})$$

$$p'_L(1 - \cos \theta_L) = \frac{mc}{2}(1 - \cos \theta_{CM})$$

$$2p'_L \sin^2 \frac{\theta_L}{2} = (mc) \sin^2 \frac{\theta_{CM}}{2}$$

which implies

$$\frac{\sin \theta_L/2}{\sin \theta_{CM}/2} = \left(\frac{mc}{2p'_L}\right)^{1/2}$$

Problem 8.13 Assume the proper relativistic relation between energy and momentum, write the equations for conservation of energy and momentum in the laboratory system, and show that the Compton formula of Eq. (3.54) holds without approximation. Compare with the result in Prob. 8.12(b).

Solution to Problem 8.13

Let us call l_μ and l'_μ the initial and final four-momenta of the photon and p_μ and p'_μ the initial and final four-momenta of the electron

$$l_\mu = (\mathbf{l}_0, il_0) \qquad ; \; l'_\mu = (\mathbf{l}_1, il_1)$$
$$p_\mu = (\mathbf{0}, im_e c) \qquad ; \; p'_\mu = (\mathbf{p}, i\sqrt{p^2 + m_e^2 c^2})$$

where the four-vectors are written in the laboratory frame.

The conservation of energy and momentum may be expressed as

$$l'_\mu - l_\mu = p_\mu - p'_\mu$$

If we square this relation, we obtain the scalar equation

$$-2l_\mu l'_\mu = -2(m_e c)^2 - 2p_\mu p'_\mu$$

Notice that

$$l_\mu l'_\mu = \mathbf{l}_0 \cdot \mathbf{l}_1 - l_0 l_1$$
$$= -\frac{h^2 \nu_0 \nu_1}{c^2}(1 - \cos\theta)$$

where $l_0 = h\nu_0/c$ and $l_1 = h\nu_1/c$.

Similarly, with the use of conservation of energy, we have

$$p_\mu p'_\mu = -m_e E_p$$
$$= -m_e(h\nu_0 - h\nu_1 + m_e c^2)$$
$$= -m_e c\left[h\frac{(\nu_0 - \nu_1)}{c} + m_e c\right]$$

After substituting these relations into the scalar equation, and noting that the frequency of the light ν is related to its wavelength by $\nu = c/\lambda$, we obtain Eq. (3.54)

$$\lambda_1 - \lambda_0 = \frac{h}{m_e c}(1 - \cos\theta)$$

If we look at the result of Prob. 8.12(b), and take into account the fact that for a massless particle $p_L = h\nu/c = h/\lambda$, we recover the above equation.

Problem 8.14 Consider the behavior of the four-velocity u_μ under a Lorentz transformation to a new inertial frame moving with velocity V along the z-axis relative to the first (compare Fig. 8.12 in the text). Assume the spatial velocity $\mathbf{v}$ in the four-velocity u_μ lies in the z-direction. We know the four-velocity transforms as a four-vector so that $u'_\mu = a_{\mu\nu}(V)u_\nu$. The mixing of the z- and fourth-components follows from Eq. (8.42).

(a) Show the transformed result can be obtained through matrix multiplication as

$$\begin{bmatrix} u'_z \\ u'_4 \end{bmatrix} = \begin{bmatrix} \dfrac{v'}{\sqrt{1-v'^2/c^2}} \\ \dfrac{ic}{\sqrt{1-v'^2/c^2}} \end{bmatrix} = \begin{bmatrix} \dfrac{1}{\sqrt{1-V^2/c^2}} & \dfrac{iV/c}{\sqrt{1-V^2/c^2}} \\ \dfrac{-iV/c}{\sqrt{1-V^2/c^2}} & \dfrac{1}{\sqrt{1-V^2/c^2}} \end{bmatrix} \begin{bmatrix} \dfrac{v}{\sqrt{1-v^2/c^2}} \\ \dfrac{ic}{\sqrt{1-v^2/c^2}} \end{bmatrix}$$

(b) Show

$$\begin{bmatrix} \dfrac{v'}{\sqrt{1-v'^2/c^2}} \\ \dfrac{ic}{\sqrt{1-v'^2/c^2}} \end{bmatrix} = \frac{1}{\sqrt{1-V^2/c^2}}\frac{1}{\sqrt{1-v^2/c^2}} \begin{bmatrix} v - V \\ ic(1 - Vv/c^2) \end{bmatrix}$$

(c) Show that

$$v' = \frac{v - V}{1 - Vv/c^2}$$

This is the law for the addition of the velocities in special relativity.

(d) Show this reduces to the galilean transformation to the new inertial frame if terms of order $(\text{velocity}/c)^2$ are neglected.

(e) Show the component relations in part (b) are individually satisfied.

Solution to Problem 8.14

(a) We notice the particular form of the matrix $\underline{a}$ of Eq. (8.42), which refers to a Lorentz transformation along the z-axis. The components of this matrix which correspond to the transverse components (x and y components) form a 2×2 identity matrix, implying that these components are not affected by the transformation. As a result, one can work with the remaining components (z and time components). We therefore consider the relevant components of the four-velocity, $u_\mu = (0, 0, v/\sqrt{1 - v^2/c^2}, ic/\sqrt{1 - v^2/c^2})$ and write the matrix equation

$$\begin{bmatrix} u'_z \\ u'_4 \end{bmatrix} = \begin{bmatrix} \frac{v'}{\sqrt{1-v'^2/c^2}} \\ \frac{ic}{\sqrt{1-v'^2/c^2}} \end{bmatrix} = \begin{bmatrix} \frac{1}{\sqrt{1-V^2/c^2}} & \frac{iV/c}{\sqrt{1-V^2/c^2}} \\ \frac{-iV/c}{\sqrt{1-V^2/c^2}} & \frac{1}{\sqrt{1-V^2/c^2}} \end{bmatrix} \begin{bmatrix} \frac{v}{\sqrt{1-v^2/c^2}} \\ \frac{ic}{\sqrt{1-v^2/c^2}} \end{bmatrix}$$

(b) The matrix multiplication of part (a) can be carried out explicitly to obtain

$$\begin{bmatrix} \frac{v'}{\sqrt{1-v'^2/c^2}} \\ \frac{ic}{\sqrt{1-v'^2/c^2}} \end{bmatrix} = \frac{1}{\sqrt{1-V^2/c^2}} \frac{1}{\sqrt{1-v^2/c^2}} \begin{bmatrix} v - V \\ ic(1 - Vv/c^2) \end{bmatrix}$$

(c) Write the equation obtained by dividing both sides of the first component by the corresponding sides of the the second component. In this way, we obtain the the law for the addition of the velocities in special relativity

$$v' = \frac{v - V}{1 - Vv/c^2}$$

(d) If we neglect the term Vv/c^2 in the result in part (c), we obtain the Galilean law of addition of velocities:

$$v' = v - V$$

(e) The individual component relations in part (b) are

$$\frac{v'}{\sqrt{1-v'^2/c^2}} = \frac{1}{\sqrt{1-v^2/c^2}} \frac{v-V}{\sqrt{1-V^2/c^2}}$$

$$\frac{1}{\sqrt{1-v'^2/c^2}} = \frac{1}{\sqrt{1-v^2/c^2}} \frac{1-Vv/c^2}{\sqrt{1-V^2/c^2}}$$

The first equation is the Lorentz transformation law for the vector part of the four-velocity, and since the *ratio* of the two equations is guaranteed to be correct by the result in part (c), the second equation is also satisfied.

Problem 8.15 Derive the second boundary condition $f'(0) = 0$ in Eqs. (8.111) from Eq. (8.104).

Solution to Problem 8.15

We use Eqs. (8.105) and (8.109) to express the mass density ρ and pressure P in the Thomas-Fermi theory of a white dwarf star in terms of the dimensionless function f according to

$$\bar{\lambda}^3 \rho \equiv f^3 \qquad ; \eta \equiv \frac{\hbar c}{4} \left(\frac{6\pi^2}{g} \right)^{1/2} \frac{1}{(2m_p)^{4/3}} \qquad ; \bar{\lambda}^2 \equiv \frac{4\pi G R^2}{4\eta}$$

$$P = \eta \left(\frac{f}{\bar{\lambda}} \right)^4$$

Therefore, with $\xi = r/R$,

$$\frac{dP}{dr} = \frac{4\eta}{R\bar{\lambda}^4} f^3 \frac{df}{d\xi} = \frac{4\eta}{R\bar{\lambda}} \rho \frac{df}{d\xi} \qquad ; \xi \equiv \frac{r}{R}$$

Equation (8.104) tells us that $dP(0)/dr = 0$, while $\rho(0) \neq 0$. If we use these conditions in the above equation, we conclude that $f'(0) = 0$.

Problem 8.16 In direct analogy to the calculation in the text, carry out a calculation of the properties of a white dwarf star in the non-relativistic limit (NRL), which is appropriate for small mass stars.

(a) Show the equation of state in the NRL is

$$P = \zeta \rho^{5/3} \qquad ; \zeta = \frac{2}{5} \frac{\hbar^2}{2m_e} \left(\frac{6\pi^2}{g} \right)^{2/3} \frac{1}{(2m_p)^{5/3}}$$

(b) Show the dimensionless equation for the density distribution is[10]

$$\frac{1}{\xi^2} \frac{d}{d\xi} \left(\xi^2 \frac{d}{d\xi} F \right) = -F^{3/2}$$

$$F(1) = 0 \qquad ; F'(0) = 0$$

Here

$$\bar{\Lambda}^3 \rho \equiv F^{3/2} \qquad ; \bar{\Lambda} \equiv \left(\frac{8\pi G}{5\zeta} \right) R^2$$

(c) Show the mass and radius of the star are related by

$$MR^3 = 4\pi \left(\frac{5\zeta}{8\pi G} \right)^3 \int_0^1 F^{3/2} \xi^2 \, d\xi$$

(d) Interpolate between the NRL and ERL to make a sketch of $M(R)$ for all M up to the Chandrasekhar limit for a white dwarf star. Discuss.[11]

---

[10]The use of the "cold" equation of state ($T = 0$) depends on the inequality $k_B T \ll \varepsilon_F$ where ε_F is the Fermi energy of the electrons. We leave the demonstration of the precise region of applicability as an exercise for the dedicated reader.

[11]Note that the calculation in the text in the ERL, by itself, does not determine R!

Solution to Problem 8.16

(a) We work in the non-relativistic limit (NRL) where the relation between a particle's energy and momentum is

$$E = \frac{p^2}{2m} = \frac{\hbar^2 k^2}{2m}$$

This replaces the corresponding Eq. (8.90), obtained in the extreme relativistic limit (ERL). Equation (8.91), on the other hand, still holds, and it can be used to obtain the analogs of Eqs. (8.92) and (8.93) for the particle number and energy

$$N = \frac{gV}{(2\pi)^3} \int_0^{k_F} d^3k = \frac{gV}{6\pi^2} k_F^3$$

$$E = \frac{gV}{(2\pi)^3} \int_0^{k_F} \frac{\hbar^2 k^2}{2m} d^3k = \frac{gV}{10\pi^2} \frac{\hbar^2 k_F^5}{2m}$$

Hence, with $N/V \equiv n$,

$$\frac{E}{N} = \frac{3}{5} \frac{\hbar^2 k_F^2}{2m} = \frac{3}{5} \frac{\hbar^2}{2m} \left(\frac{6\pi^2 n}{g} \right)^{2/3}$$

We are now in position to calculate the pressure

$$P = -\left(\frac{\partial E}{\partial V} \right)_N = -N^{5/3} \frac{3}{5} \frac{\hbar^2}{2m} \left(\frac{6\pi^2}{g} \right)^{2/3} \frac{\partial}{\partial V} \frac{1}{V^{2/3}}$$

$$= \frac{2}{5} \frac{\hbar^2}{2m} \left(\frac{6\pi^2}{g} \right)^{2/3} n^{5/3}$$

If we now use Eq. (8.98) to express the mass density ρ in terms of the electron density n, we obtain

$$P = \zeta \rho^{5/3} \qquad ; \zeta = \frac{2}{5} \frac{\hbar^2}{2m_e} \left(\frac{6\pi^2}{g} \right)^{2/3} \frac{1}{(2m_p)^{5/3}}$$

(b) If the expression for the pressure obtained in part (a) is used in Eq. (8.102), which expresses the condition of hydrostatic equilibrium, we obtain

$$\frac{1}{\rho} \frac{dP}{dr} = \zeta \frac{5}{2} \frac{d}{dr} \rho^{2/3} = -\frac{4\pi G}{r^2} \int_0^r \rho(r') r'^2 dr'$$

With the use of the dimensionless variable $\xi \equiv r/R$, this equation can be re-cast in the form

$$\frac{5\zeta}{8\pi GR^2} \frac{d}{d\xi} \rho^{2/3} = -\frac{1}{\xi^2} \int_0^\xi \rho(\xi')\xi'^2 d\xi'$$

Multiply this equation by ξ^2, and take the derivative with respect to ξ. With the introduction of

$$\bar{\Lambda}^3 \rho \equiv F^{3/2} \qquad\qquad ; \bar{\Lambda} \equiv \left(\frac{8\pi G}{5\zeta}\right) R^2$$

we finally obtain the equation

$$\frac{1}{\xi^2} \frac{d}{d\xi} \left(\xi^2 \frac{d}{d\xi} F\right) = -F^{3/2}$$

$$F(1) = 0 \qquad\qquad ; F'(0) = 0$$

where the boundary conditions are obtained from Eqs. (8.103) and (8.104).

(c) We obtain the total mass of the star by integrating the mass density out to the star's radius, as done in Eq. (8.112),

$$
\begin{aligned}
M &= 4\pi \int_0^R \rho(r) r^2 dr = 4\pi R^3 \int_0^1 \rho(\xi)\xi^2 d\xi \\
&= \frac{4\pi R^3}{\bar{\Lambda}^3} \int_0^1 F^{3/2}(\xi)\xi^2 d\xi \\
&= 4\pi \left(\frac{5\zeta}{8\pi GR}\right)^3 \int_0^1 F^{3/2}(\xi)\xi^2 d\xi
\end{aligned}
$$

Therefore

$$MR^3 = 4\pi \left(\frac{5\zeta}{8\pi G}\right)^3 \int_0^1 F^{3/2}\xi^2 \, d\xi$$

(d) We may use the differential equation obeyed by $F(\xi)$ in part (b), in the equation for the mass obtained in part (c), to obtain

$$
\begin{aligned}
MR^3 &= -4\pi \left(\frac{5\zeta}{8\pi G}\right)^3 \int_0^1 \frac{d}{d\xi}\left(\xi^2 \frac{d}{d\xi} F\right) d\xi \\
&= -4\pi \left(\frac{5\zeta}{8\pi G}\right)^3 F'(1)
\end{aligned}
$$

Thus we see that the mass of the star is inversely proportional to its volume in the NRL, whereas it is independent of R in the ERL [see Eq. (8.115)]. For numerical results and a discussion here, see the solution to Prob. 8.18.

****

(*Aside*) In order to assess the range of validity of the NRL and ERL for the white dwarf, we need to keep in mind that we must have

$$\left.\frac{E}{N}\right|_{\text{NRL}} \ll \left.\frac{E}{N}\right|_{\text{ERL}}$$

The energy can be obtained by integrating the pressure over the volume of the star, keeping the number of particles fixed. In the ERL we have

$$E_{\text{ERL}} = -\eta \int \rho^{4/3} dV = -\eta \int \left(\frac{f}{\lambda}\right)^4 dV$$

$$= -\frac{4\eta^3}{\pi G^2 R} \int_0^1 \xi^2 f^4(\xi) d\xi$$

where $\eta^3/G^2 R$ has dimensions of energy. In the NRL we have

$$E_{\text{NRL}} = -\zeta \int \rho^{5/3} dV = -\zeta \int \left(\frac{F}{\bar{\Lambda}^2}\right)^{5/2} dV$$

$$= -\frac{5^5 \zeta^6}{2^{13} \pi^4 G^5 R^7} \int_0^1 \xi^2 F^{5/2}(\xi) d\xi$$

where $\zeta^6/G^5 R^7$ has dimensions of energy.

Clearly the integrals in these expressions are dimensionless (their evaluation requires solving the corresponding differential equations for f and F— see below), and therefore we may conclude on purely dimensional grounds that the validity of the NRL requires $R/R_0 \gg 1$, where R_0 is a characteristic radius at which the ERL is valid.

****

Problem 8.17 Write, or obtain, a program to integrate the non-linear differential Eq. (8.110) subject to the boundary conditions in Eqs. (8.111), and reproduce Fig. 8.17 in the text.

Solution to Problem 8.17

Equation (8.110) can be written explicitly as

$$f''(\xi) + \frac{2}{\xi} f'(\xi) + f^3(\xi) = 0$$

with the boundary conditions

$$f'(0) = 0 \qquad ; \ f(1) = 0$$

We may obtain the numerical solution to this equation by imposing a second boundary condition at $\xi = 1$, that is $f'(1) = \alpha$, expressed in term of a parameter α. This parameter must be varied so that the boundary condition at $\xi = 0$ is achieved once the differential equation is integrated in numerically from $\xi = 1$ to $\xi = 0$. A procedure of this kind is usually referred to as the "shooting method".

The original problem is now converted to a standard problem, and we may solve the differential equation either by discretizing the equation ourselves, or by using any of the built-in commands for the numerical solution of ordinary differential equations that most programs like Maple, Mathcad, Mathematica, Matlab or similar programs have. In both cases, we need to integrate the equation down to a small but finite value of ξ, that is $\xi = \epsilon > 0$, since at $\xi = 0$ the term containing the first derivative is singular (of course, the singularity is not present when the exact boundary condition $f'(0) = 0$ is imposed). In our code we have used $\epsilon = 10^{-5}$.

In Fig. 8.1 we plot $f'(\epsilon)$ as a function of $f'(1)$, obtained from the numerical solution of the differential equation. The correct solution corresponds to $f'(1) \approx -2.018$, in agreement with Eq. (8.116) of the book.[12]

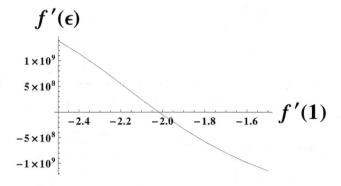

Fig. 8.1 $f'(\epsilon)$ as a function of $f'(1)$, obtained from the numerical solution of the differential equation.

In Fig. 8.2 we plot the numerical solution $f(\xi)$ corresponding to the above value $f'(1) = -2.018$.

---

[12]We quote these numbers to four significant figures, although the calculation determines them much further. The actual values depend sensitively on such things as the value of ϵ, program and machine accuracy, and round-off errors. Readers should start from these values, and then use these procedures to determine the numbers for themselves.

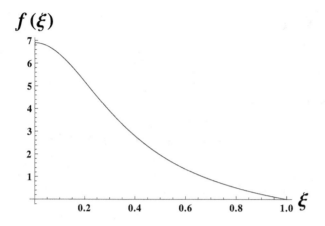

Fig. 8.2   Numerical solution $f(\xi)$ corresponding to $f'(1) = -2.018$.

With this solution we obtain the second value of Eq. (8.116), $f(0) = 6.897$. Thus for our solution

$$f'(1) = -2.018 \qquad ; \ f(0) = 6.897$$

Problem 8.18 (a) Use the program from Prob. 8.17 to integrate the nonlinear differential equation in Prob. 8.16(b) subject to the boundary conditions stated there. Make a plot of $F(\xi)$ versus ξ. Compare with Fig. 8.17 in the text.

(b) Show that the r.h.s of the expression for MR^3 in Prob. 8.16(c) has the correct dimensions. Calculate the required integral numerically, and then calculate the numerical value of MR^3 in the NRL.

Solution to Problem 8.18

(a) With the use of the code of Prob. 8.17, we may obtain the numerical solution to the nonlinear differential equation of Prob. 8.16

$$F''(\xi) + \frac{2}{\xi}F'(\xi) + F^{3/2}(\xi) = 0$$
$$F(1) = 0 \qquad ; \ F'(0) = 0$$

In Fig. 8.3 we plot $F'(\epsilon)$ as a function of $F'(1)$, obtained from the numerical solution of the differential equation: the correct solution corresponds to $F'(1) = -132.3$.

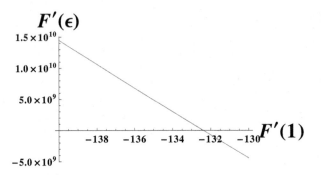

Fig. 8.3 $F'(\epsilon)$ as a function of $F'(1)$, obtained from the numerical solution of the differential equation.

The numerical solution for $F(\xi)$ obtained using the value $F'(1) = -132.3$ is shown in Fig. 8.4.

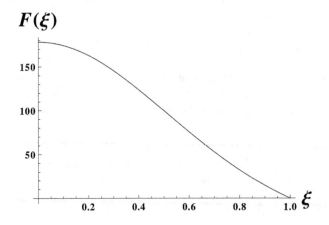

Fig. 8.4 Numerical solution $F(\xi)$ corresponding to $F'(1) = -132.3$.

With this solution we find $F(0) = 178.2$. Hence, in this case,

$$F'(1) = -132.3 \qquad ; \quad F(0) = 178.2$$

(b) It is straightforward to verify that MR^3 in Problem 8.16(c) has the correct dimensions of $[ML^3]$. If we use the expression obtained in part (d) of Prob. 8.16, together with $F'(1) = -132.3$, we obtain

$$M^{NRL}R^3 = 1.33 \times 10^{51} \text{ kg-m}^3$$

This should be compared with the Chandrasekhar limit from Eq. (8.118)

$$M^{ERL} = 2.85 \times 10^{30}\,\text{kg} \qquad ; \text{ Chandrasekhar limit}$$

In Fig. 8.5 we make a log plot of the star mass ratio $M^{\text{NRL}}/M^{\text{ERL}}$ as a function of the star radius R in meters. The straight line exhibits the R^{-3} dependence.

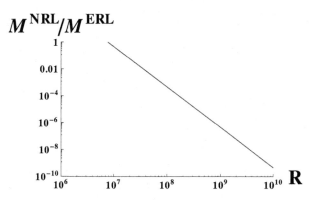

Fig. 8.5 Ratio $M^{\text{NRL}}/M^{\text{ERL}}$ as a function of R in meters. For $R_0 = 7.75 \times 10^6$ m, one has $M^{\text{NRL}}/M^{\text{ERL}} = 1$.

One finds $M^{\text{NRL}}/M^{\text{ERL}} = 1$ at a radius $R_0 = 7.75 \times 10^6$ m

$$\frac{M^{\text{NRL}}}{M^{\text{ERL}}} = 1 \qquad ; R_0 = 7.75 \times 10^6\,\text{m}$$

The NRL holds only for $R \gg R_0$.

Chapter 9

Relativistic Quantum Mechanics

Problem 9.1 The Dirac matrices $\gamma_\mu = (\gamma_1, \gamma_2, \gamma_3, \gamma_4)$ are defined in Eqs. (9.43).

(a) Show that they are hermitian with $\gamma_\mu^\dagger = \gamma_\mu$ (See Prob. B.3);

(b) Show that they satisfy the following Clifford algebra

$$\gamma_\mu \gamma_\nu + \gamma_\nu \gamma_\mu = 2\delta_{\mu\nu}$$

(c) Derive the standard representation for γ

$$\gamma = \begin{pmatrix} 0 & -i\boldsymbol{\sigma} \\ i\boldsymbol{\sigma} & 0 \end{pmatrix}$$

Solution to Problem 9.1

(a) The Dirac matrices are defined as [see Eqs. (9.43)][1]

$$\gamma \equiv i\boldsymbol{\alpha}\beta$$
$$\gamma_4 \equiv \beta$$

where the explicit forms of $\boldsymbol{\alpha}$ and β are given in Eqs. (9.27). In 2×2 block form they read

$$\boldsymbol{\alpha} = \begin{pmatrix} 0 & \boldsymbol{\sigma} \\ \boldsymbol{\sigma} & 0 \end{pmatrix} \qquad ; \; \beta = \begin{pmatrix} 1 & 0 \\ 0 & -1 \end{pmatrix}$$

Clearly γ_4 is hermitian, since β is hermitian. Thus we only need to prove that γ is hermitian

$$\gamma^\dagger = (i\boldsymbol{\alpha}\beta)^\dagger = -i\beta\boldsymbol{\alpha}^\dagger = -i\beta\boldsymbol{\alpha} = i\boldsymbol{\alpha}\beta = \gamma$$

---

[1] As with the Pauli matrices, we do not underline the 4×4 Dirac matrices. We also usually suppress the 4×4 identity matrix.

where we have used the property that the adjoint of the product of two matrices is the product of the adjoints in the reverse order [Prob. (B.3)] and the anticommutation of α and β [the first of Eqs. (9.26)]

$$\alpha\beta + \beta\alpha = 0$$

Therefore, we conclude that the gamma matrices are hermitian

$$\gamma_\mu^\dagger = \gamma_\mu \qquad ; \text{ hermitian}$$

(b) Let us first consider the case where both μ and ν correspond to spatial indices

$$\gamma_k\gamma_l + \gamma_l\gamma_k = -\alpha^k\beta\alpha^l\beta - \alpha^l\beta\alpha^k\beta = \alpha^k\alpha^l\beta^2 + \alpha^l\alpha^k\beta^2$$
$$= 2\delta_{kl}$$

where we have used Eqs. (9.26)

$$\alpha^k\alpha^l + \alpha^l\alpha^k = 2\delta_{kl}$$

Next we consider the case where μ is a spatial index, while ν is a temporal index

$$\gamma_k\gamma_4 + \gamma_4\gamma_k = i\alpha_k\beta^2 + i\beta\alpha_k\beta$$
$$= 0$$

Finally we must consider the case where both μ and ν are temporal indices:

$$2\gamma_4^2 = 2\beta^2 = 2$$

which completes the proof of the Clifford algebra

$$\gamma_\mu\gamma_\nu + \gamma_\nu\gamma_\mu = 2\delta_{\mu\nu} \qquad ; \text{ Clifford algebra}$$

(c) The explicit form of γ is obtained in 2×2 block form as

$$\gamma = i\alpha\beta = i\begin{pmatrix} 0 & \sigma \\ \sigma & 0 \end{pmatrix}\begin{pmatrix} 1 & 0 \\ 0 & -1 \end{pmatrix} = \begin{pmatrix} 0 & -i\sigma \\ i\sigma & 0 \end{pmatrix}$$

Problem 9.2 Consider the Dirac Eq. (9.51) in the presence of a static magnetic field $\mathbf{B} = \nabla \times \mathbf{A}$.

(a) Show that Eqs. (9.33) are modified to read

$$c\boldsymbol{\sigma} \cdot (\mathbf{p} - e\mathbf{A})\chi + m_0 c^2 \phi = E\phi$$
$$c\boldsymbol{\sigma} \cdot (\mathbf{p} - e\mathbf{A})\phi - m_0 c^2 \chi = E\chi$$

Show that these relations imply

$$\frac{c^2}{2m_0 c^2 + \varepsilon} [\boldsymbol{\sigma} \cdot (\mathbf{p} - e\mathbf{A})][\boldsymbol{\sigma} \cdot (\mathbf{p} - e\mathbf{A})]\phi = \varepsilon\phi \qquad ; \varepsilon \equiv E - m_0 c^2$$

(b) Show that the Pauli matrices satisfy

$$\sigma_i \sigma_j = \delta_{ij} + i\varepsilon_{ijk}\sigma_k$$

where ε_{ijk} is the completely antisymmetric tensor in three dimensions [see Prob. 8.7(b)], and use this to prove the following relation[2]

$$[\boldsymbol{\sigma} \cdot (\mathbf{p} - e\mathbf{A})][\boldsymbol{\sigma} \cdot (\mathbf{p} - e\mathbf{A})] = \mathbf{p}^2 - e(\mathbf{A} \cdot \mathbf{p} + \mathbf{p} \cdot \mathbf{A}) + e^2 \mathbf{A}^2$$
$$-e\hbar\boldsymbol{\sigma} \cdot (\boldsymbol{\nabla} \times \mathbf{A})$$

(c) Hence conclude that when $|\varepsilon|/2m_0 c^2 \ll 1$, the Dirac equation reduces to the following Schrödinger equation for the upper components ϕ

$$H\phi = \varepsilon\phi$$
$$H = \frac{\mathbf{p}^2}{2m_0} - \frac{e}{2m_0}(\mathbf{A} \cdot \mathbf{p} + \mathbf{p} \cdot \mathbf{A}) + \frac{e^2 \mathbf{A}^2}{2m_0} - \frac{e\hbar}{2m_0}\boldsymbol{\sigma} \cdot \mathbf{B}$$

(d) Use this result to obtain Eqs. (9.52).

Solution to Problem 9.2

(a) Equation (9.51) in the presence of a static magnetic field reads

$$\left[c\boldsymbol{\alpha} \cdot (\mathbf{p} - e\mathbf{A}) + \beta m_0 c^2\right] \Psi = i\hbar\frac{\partial\Psi}{\partial t}$$

Look for stationary-state solutions $\Psi = \psi e^{-iEt/\hbar}$, and write the Dirac wave function ψ in two-component form as

$$\psi = \begin{pmatrix} \phi \\ \chi \end{pmatrix}$$

---

[2]We here use the shorthand convention that repeated Latin indices are summed from 1 to 3; then $[p_i, A_j] = (\hbar/i)\nabla_i A_j$ and $\varepsilon_{ijk}\sigma_k \nabla_i A_j = \boldsymbol{\sigma} \cdot (\boldsymbol{\nabla} \times \mathbf{A})$.

With the use of the explicit form of $\boldsymbol{\alpha}$ and β given in Eqs. (9.27), the Dirac equation then takes the 2×2 form

$$\begin{bmatrix} m_0c^2 & c\boldsymbol{\sigma} \cdot (\mathbf{p} - e\mathbf{A}) \\ c\boldsymbol{\sigma} \cdot (\mathbf{p} - e\mathbf{A}) & -m_0c^2 \end{bmatrix} \begin{pmatrix} \phi \\ \chi \end{pmatrix} = E \begin{pmatrix} \phi \\ \chi \end{pmatrix}$$

Once the matrix multiplication is performed, we obtain the two equations

$$c\boldsymbol{\sigma} \cdot (\mathbf{p} - e\mathbf{A})\chi + m_0c^2\phi = E\phi$$
$$c\boldsymbol{\sigma} \cdot (\mathbf{p} - e\mathbf{A})\phi - m_0c^2\chi = E\chi$$

We may now use the second equation to express χ in terms of ϕ

$$\chi = \frac{c}{E + m_0c^2}\boldsymbol{\sigma} \cdot (\mathbf{p} - e\mathbf{A})\phi$$

Substitution of this expression in the first equation gives

$$\frac{c^2}{2m_0c^2 + \varepsilon} [\boldsymbol{\sigma} \cdot (\mathbf{p} - e\mathbf{A})] [\boldsymbol{\sigma} \cdot (\mathbf{p} - e\mathbf{A})] \phi = \varepsilon\phi$$

$$\varepsilon \equiv E - m_0c^2$$

(b) Let us explicitly verify the stated property of the Pauli matrices. It is useful to decompose their product as

$$\sigma_i\sigma_j = \frac{1}{2}\{\sigma_i, \sigma_j\} + \frac{1}{2}[\sigma_i, \sigma_j]$$

where $\{\sigma_i, \sigma_j\} \equiv \sigma_i\sigma_j + \sigma_j\sigma_i$ is the anticommutator, and $[\sigma_i, \sigma_j] \equiv \sigma_i\sigma_j - \sigma_j\sigma_i$ is the commutator. Clearly the anticommutator and the commutator are respectively symmetric and antisymmetric in the indices (i, j), and they are given in Eqs. (B.13) of the book[3]

$$\{\sigma_i, \sigma_j\} = 2\delta_{ij} \qquad ; \qquad [\sigma_i, \sigma_j] = 2i\varepsilon_{ijk}\sigma_k$$

With the use of these relations we have

$$\sigma_i\sigma_j = \delta_{ij} + i\varepsilon_{ijk}\sigma_k$$

****

(*Aside*) Alternatively, one may work with the explicit matrix representation

---

[3] Here we also suppress the 2×2 identity matrix.

of the σ_i and calculate all the relevant products. For example

$$\sigma_x \sigma_y = \begin{pmatrix} 0 & 1 \\ 1 & 0 \end{pmatrix} \begin{pmatrix} 0 & -i \\ i & 0 \end{pmatrix} = \begin{pmatrix} i & 0 \\ 0 & -i \end{pmatrix} = i\sigma_z$$

$$\sigma_x \sigma_z = \begin{pmatrix} 0 & 1 \\ 1 & 0 \end{pmatrix} \begin{pmatrix} 1 & 0 \\ 0 & -1 \end{pmatrix} = \begin{pmatrix} 0 & -1 \\ 1 & 0 \end{pmatrix} = -i\sigma_y$$

$$\sigma_y \sigma_z = \begin{pmatrix} 0 & -i \\ i & 0 \end{pmatrix} \begin{pmatrix} 1 & 0 \\ 0 & -1 \end{pmatrix} = \begin{pmatrix} 0 & i \\ i & 0 \end{pmatrix} = i\sigma_x \qquad ; \text{etc.}$$

****

We will now use the above property of the Pauli matrices in the expression $[\boldsymbol{\sigma} \cdot (\mathbf{p} - e\mathbf{A})] [\boldsymbol{\sigma} \cdot (\mathbf{p} - e\mathbf{A})]$

$$[\boldsymbol{\sigma} \cdot (\mathbf{p} - e\mathbf{A})] [\boldsymbol{\sigma} \cdot (\mathbf{p} - e\mathbf{A})] =$$
$$= \sum_{ij} \left(\delta_{ij} + i \sum_k \varepsilon_{ijk} \sigma_k \right) (p - eA)_i \, (p - eA)_j$$
$$= \mathbf{p}^2 - e(\mathbf{A} \cdot \mathbf{p} + \mathbf{p} \cdot \mathbf{A}) + e^2 \mathbf{A}^2 - ie \sum_{ijk} \varepsilon_{ijk} \, \sigma_k (p_i A_j + A_i p_j)$$

Let us concentrate on the last term

$$-ie \sum_{ijk} \varepsilon_{ijk} \, \sigma_k (p_i A_j + A_i p_j) = -ie \sum_{ijk} \varepsilon_{ijk} \, \sigma_k p_i A_j - ie \sum_{ijk} \varepsilon_{jik} \, \sigma_k A_j p_i$$
$$= -ie \sum_{ijk} \varepsilon_{ijk} \, \sigma_k \, [p_i, A_j]$$
$$= -e\hbar \sum_{ijk} \varepsilon_{ijk} \, \sigma_k \, [\nabla_i, A_j]$$
$$= -e\hbar \, \boldsymbol{\sigma} \cdot (\boldsymbol{\nabla} \times \mathbf{A})$$

Here we have re-labeled the indices $i \rightleftharpoons j$ in the last term in the first line, and then used the antisymmetry of ε_{ijk} to obtain the commutator $[p_i, A_j]$.

(c) Now assume $|\varepsilon|/2m_0 c^2 \ll 1$, and use $\mathbf{B} = \boldsymbol{\nabla} \times \mathbf{A}$. The last equation of part (a) becomes

$$H\phi = \varepsilon\phi$$
$$H = \frac{\mathbf{p}^2}{2m_0} - \frac{e}{2m_0} (\mathbf{A} \cdot \mathbf{p} + \mathbf{p} \cdot \mathbf{A}) + \frac{e^2 \mathbf{A}^2}{2m_0} - \frac{e\hbar}{2m_0} \boldsymbol{\sigma} \cdot \mathbf{B}$$

(d) We can just read off the magnetic moment of the electron from this equation

$$\mu_{\mathrm{el}} = \frac{e\hbar}{2m_e} 2\mathbf{S} \qquad ; \mathbf{S} = \frac{1}{2}\boldsymbol{\sigma}$$

It is a striking success of Dirac's equation that it yields the correct magnetic moment for the electron.

Problem 9.3 Consider the Dirac Eq. (9.51) for a particle in a static, central, electric field given by $\mathbf{E} = -\boldsymbol{\nabla}\Phi(r)$, and repeat Prob. 9.2.

(a) Show that in this case the Dirac equation can be reduced to

$$\left[c\boldsymbol{\sigma}\cdot\mathbf{p}\,\frac{1}{2m_0c^2 + \varepsilon - e\Phi(r)}\,c\boldsymbol{\sigma}\cdot\mathbf{p} + e\Phi(r) \right]\phi = \varepsilon\phi \qquad ; \varepsilon \equiv E - m_0c^2$$

(b) Show that[4]

$$\boldsymbol{\sigma}\cdot\mathbf{p}\,\Phi(r)\,\boldsymbol{\sigma}\cdot\mathbf{p} = \Phi\mathbf{p}^2 + (\mathbf{p}\Phi)\cdot\mathbf{p} + \frac{\hbar}{r}\left(\frac{d\Phi}{dr}\right)\boldsymbol{\sigma}\cdot(\mathbf{r}\times\mathbf{p})$$

(c) Hence conclude that if $|\varepsilon|/2m_0c^2 \ll 1$, and $e\Phi\left\{1 + O[(p/m_0c)^2]\right\} \approx e\Phi$, then the Dirac equation reduces to the following Schrödinger equation for the upper components ϕ

$$H\phi = \varepsilon\phi$$
$$H = \frac{\mathbf{p}^2}{2m_0} + e\Phi(r) + \frac{e\hbar^2}{(2m_0c)^2}\frac{1}{r}\left(\frac{d\Phi}{dr}\right)\boldsymbol{\sigma}\cdot\mathbf{l}$$

Here the spin-dependent contribution has been retained, and the angular momentum identified as $\mathbf{r}\times\mathbf{p} = \hbar\mathbf{l}$.

(d) Use this result to obtain Eq. (9.53).

Solution to Problem 9.3

(a) We proceed as in Prob. 9.2 and write Eq. (9.51) in the presence of a static, central electric field

$$\left[c\boldsymbol{\alpha}\cdot\mathbf{p} + \beta m_0c^2 + e\Phi(r) \right]\Psi = i\hbar\frac{\partial\Psi}{\partial t}$$

As before, we write the stationary-state solutions to this equation in 2×2 form using the explicit matrix representation of $\boldsymbol{\alpha}$ and β

$$\begin{bmatrix} m_0c^2 + e\Phi(r) & c\boldsymbol{\sigma}\cdot\mathbf{p} \\ c\boldsymbol{\sigma}\cdot\mathbf{p} & -m_0c^2 + e\Phi(r) \end{bmatrix}\begin{pmatrix}\phi\\\chi\end{pmatrix} = E\begin{pmatrix}\phi\\\chi\end{pmatrix}$$

---

[4]Note that $\boldsymbol{\nabla}\Phi(r) = (\mathbf{r}/r)(d\Phi/dr)$.

The components ϕ and χ of the Dirac wave function therefore obey the two equations

$$[m_0c^2 + e\Phi(r)]\phi + c\boldsymbol{\sigma} \cdot \mathbf{p}\chi = E\phi$$
$$[-m_0c^2 + e\Phi(r)]\chi + c\boldsymbol{\sigma} \cdot \mathbf{p}\phi = E\chi$$

We use the second equation to express χ in terms of ϕ:

$$\chi = \frac{1}{E + m_0c^2 - e\Phi(r)}c\boldsymbol{\sigma} \cdot \mathbf{p}\phi$$

and then substitute this in the first equation, obtaining

$$\left[c\boldsymbol{\sigma} \cdot \mathbf{p}\, \frac{1}{2m_0c^2 + \varepsilon - e\Phi(r)}\, c\boldsymbol{\sigma} \cdot \mathbf{p} + e\Phi(r)\right]\phi = \varepsilon\phi$$

$$\varepsilon \equiv E - m_0c^2$$

(b) Some care is necessary in working out the first term. We have

$$\boldsymbol{\sigma} \cdot \mathbf{p}\Phi(r)\boldsymbol{\sigma} \cdot \mathbf{p} = \sum_i p_i\Phi(r)p_i + i\sum_{ijk}\varepsilon_{ijk}\sigma_k p_i\Phi(r)p_j$$

$$= [\mathbf{p},\, \Phi(r)] \cdot \mathbf{p} + \Phi(r)\mathbf{p}^2 + \hbar\sum_{ijk}\varepsilon_{ijk}\sigma_k\left[\nabla_i, \Phi(r)\right]p_j$$

$$+ i\Phi(r)\sum_{ijk}\varepsilon_{ijk}\sigma_k p_i p_j$$

Notice that the last term in the above expression clearly vanishes since it contains the contraction of a symmetric and antisymmetric tensor. Moreover, as explained in the footnote, we have $[\boldsymbol{\nabla}, \Phi(r)] = \boldsymbol{\nabla}\Phi(r) = (\mathbf{r}/r)d\Phi(r)/dr$. Therefore

$$\boldsymbol{\sigma} \cdot \mathbf{p}\, \Phi(r)\, \boldsymbol{\sigma} \cdot \mathbf{p} = \Phi\mathbf{p}^2 + (\mathbf{p}\Phi) \cdot \mathbf{p} + \frac{\hbar}{r}\left(\frac{d\Phi}{dr}\right)\boldsymbol{\sigma} \cdot (\mathbf{r} \times \mathbf{p})$$

(c) Assume $|\varepsilon|/2m_0c^2 \ll 1$ and $e\Phi\left\{1 + O[(p/m_0c)^2]\right\} \approx e\Phi$. The denominator in the last expression in part (a) can then be expanded, and the result in part (b) employed, to give

$$c\boldsymbol{\sigma} \cdot \mathbf{p}\frac{1}{2m_0c^2 + \varepsilon - e\Phi(r)}c\boldsymbol{\sigma} \cdot \mathbf{p} \approx \boldsymbol{\sigma} \cdot \mathbf{p}\frac{1}{2m_0}\left[1 + \frac{e\Phi(r)}{2m_0c^2} + \cdots\right]\boldsymbol{\sigma} \cdot \mathbf{p}$$

$$\approx \frac{\mathbf{p}^2}{2m_0} + \frac{e\hbar}{(2m_0c)^2}\frac{1}{r}\frac{d\Phi}{dr}\,\boldsymbol{\sigma} \cdot (\mathbf{r} \times \mathbf{p})$$

Here the spin-dependent contribution has been retained. Upon substitution in the eigenvalue equation of part (a), and the identification of the angular momentum $\mathbf{r} \times \mathbf{p} = \hbar \mathbf{l}$, we finally obtain a Schrödinger equation for the upper components ϕ

$$H\phi = \varepsilon\phi$$

$$H = \frac{\mathbf{p}^2}{2m_0} + e\Phi(r) + \frac{e\hbar^2}{(2m_0 c)^2} \frac{1}{r} \left(\frac{d\Phi}{dr} \right) \boldsymbol{\sigma} \cdot \mathbf{l}$$

(d) With the further identification of the spin $\mathbf{S} = \boldsymbol{\sigma}/2$, we may write the spin-orbit interaction term in the above equation as in Eq. (9.53)

$$V_{SO} = e \left(\frac{\hbar}{2m_0 c} \right)^2 \frac{1}{r} \left(\frac{d\Phi}{dr} \right) 2\mathbf{S} \cdot \mathbf{l}$$

It is a further striking success of Dirac's equation that it yields the correct spin-orbit interaction for an electron, where $m_0 = m_e$.

Problem 9.4 (a) Consider the covariant form of the Dirac Eq. (9.48). Make the following *gauge-invariant replacement* [see part (b)]

$$\frac{\partial}{\partial x_\mu} \rightarrow \frac{\partial}{\partial x_\mu} - \frac{ie}{\hbar} A_\mu \qquad ; \; A_\mu = (\mathbf{A}, \frac{i}{c}\Phi)$$

Equation (9.48) then becomes

$$\left[\gamma_\mu \left(\frac{\partial}{\partial x_\mu} - \frac{ie}{\hbar} A_\mu \right) + M \right] \Psi = 0 \qquad ; \; \text{Dirac eqn in E-M field}$$

Verify that this is equivalent to the Dirac equation written as

$$\left[c\boldsymbol{\alpha} \cdot (\mathbf{p} - e\mathbf{A}) + \beta m_0 c^2 + e\Phi \right] \Psi = i\hbar \frac{\partial\Psi}{\partial t}$$

This is Eq. (9.51).

(b) The gauge transformation of Prob. 8.9 leaves the electromagnetic fields $(\mathbf{E}, \mathbf{B})$ unchanged, and hence it should leave physical results unchanged. It is also true that since physics is bilinear in the wave function, physical results should be independent of the overall phase of the wave function. Show that the covariant form of the Dirac equation in part (a) is unchanged under the following replacements

$$A_\mu \rightarrow A_\mu + \frac{\partial\Lambda(x)}{\partial x_\mu} \qquad ; \; \text{gauge invariance}$$

$$\Psi \rightarrow \exp\left\{ \frac{ie}{\hbar} \Lambda(x) \right\} \Psi$$

This is the statement of the *gauge invariance* of the theory.

Solution to Problem 9.4

(a) We use $\gamma \equiv -i\beta\boldsymbol{\alpha}$ and $\gamma_4 = \beta$, and write

$$\left[\gamma_\mu \left(\frac{\partial}{\partial x_\mu} - \frac{ie}{\hbar}A_\mu\right) + \frac{m_0 c}{\hbar}\right]\Psi$$

$$= \left[-i\beta \sum_{j=1}^{3} \alpha_j \left(\frac{\partial}{\partial x_j} - \frac{ie}{\hbar}A_j\right) - \beta\frac{i}{c}\frac{\partial}{\partial t} + \beta\frac{e\Phi}{\hbar c} + \frac{m_0 c}{\hbar}\right]\Psi$$

$$= \frac{1}{\hbar c}\beta \left[c\boldsymbol{\alpha} \cdot (\mathbf{p} - e\mathbf{A}) - i\hbar\frac{\partial}{\partial t} + e\Phi + \beta m_0 c^2\right]\Psi = 0$$

If we multiply both sides of the equation on the left by $\hbar c\beta$, and use $\beta^2 = 1$, we obtain Eq. (9.51)

$$\left[c\boldsymbol{\alpha} \cdot (\mathbf{p} - e\mathbf{A}) + e\Phi + \beta m_0 c^2\right]\Psi = i\hbar\frac{\partial\Psi}{\partial t}$$

(b) A gauge transformation is defined by

$$A_\mu(x) \to A_\mu(x) + \frac{\partial\Lambda(x)}{\partial x_\mu} \qquad ; \text{ gauge transformation}$$

$$\Psi(x) \to e^{ie\Lambda(x)/\hbar}\Psi(x)$$

Under this transformation, the derivative of Ψ transforms as

$$\frac{\partial}{\partial x_\mu}\Psi(x) \to e^{ie\Lambda/\hbar}\left(\frac{\partial}{\partial x_\mu} + \frac{ie}{\hbar}\frac{\partial\Lambda}{\partial x_\mu}\right)\Psi(x)$$

while the term $(\partial/\partial x_\mu - ieA_\mu/\hbar)\Psi$ transforms as

$$\left(\frac{\partial}{\partial x_\mu} - \frac{ie}{\hbar}A_\mu\right)\Psi \to e^{ie\Lambda/\hbar}\left(\frac{\partial}{\partial x_\mu} + \frac{ie}{\hbar}\frac{\partial\Lambda}{\partial x_\mu}\right)\Psi - e^{ie\Lambda/\hbar}\frac{ie}{\hbar}\left(A_\mu + \frac{\partial\Lambda}{\partial x_\mu}\right)\Psi$$

$$= e^{ie\Lambda/\hbar}\left(\frac{\partial}{\partial x_\mu} - \frac{ie}{\hbar}A_\mu\right)\Psi$$

Moreover, the mass term in the Dirac equation, which only involves Ψ, acquires a phase $e^{ie\Lambda/\hbar}$ under the gauge transformation. As a result, the Dirac equation transforms into

$$e^{ie\Lambda/\hbar}\left[\gamma_\mu\left(\frac{\partial}{\partial x_\mu} - \frac{ie}{\hbar}A_\mu\right) + M\right]\Psi = 0$$

which implies

$$\left[\gamma_\mu \left(\frac{\partial}{\partial x_\mu} - \frac{ie}{\hbar} A_\mu\right) + M\right] \Psi = 0$$

Therefore the Dirac equation is *invariant* under the gauge transformation.

Problem 9.5 A picture that has proven useful for understanding much of nuclear structure is that of Dirac nucleons moving in effective, strong, Lorentz vector and scalar fields $(V_\mu, \mathcal{S})$. Here the Dirac equation reads

$$\left[\gamma_\mu \left(\frac{\partial}{\partial x_\mu} - \frac{ig_v}{\hbar c} V_\mu\right) + \left(M_N - \frac{g_s}{\hbar c}\mathcal{S}\right)\right] \Psi = 0 \qquad ; \text{ Dirac nucleons}$$

Here (g_v, g_s) are coupling constants, analogous to the electric charge e. In the static case, in isotropic or spherically symmetric systems, the vector field has only a fourth component

$$V_\mu = i\delta_{\mu 4} V_0$$

(a) Show that in such systems, the above is equivalent to a Dirac equation of the form

$$(H_0 + H_1)\Psi = i\hbar \frac{\partial \Psi}{\partial t}$$
$$H_0 = c\boldsymbol{\alpha} \cdot \mathbf{p} + \beta m_N c^2$$
$$H_1 = g_v V_0 - \beta g_s \mathcal{S}$$

(b) Show that the binding energy in the Schrödinger equation in Prob. 9.3(c) now arises from the *difference* of the strong potentials $g_v V_0 - g_s \mathcal{S}$.

(c) Show that with central potentials $[g_v V_0(r), g_s \mathcal{S}(r)]$, the spin-orbit interaction in Prob. 9.3(c) is given by

$$V_{SO} = \frac{\hbar^2}{(2m_N c)^2} \frac{1}{r} \left[\frac{d}{dr}(g_v V_0 + g_s \mathcal{S})\right] \boldsymbol{\sigma} \cdot \mathbf{l}$$

It is the *sum* of the strong potentials $g_v V_0 + g_s \mathcal{S}$ that contributes to the spin-orbit interaction. In this way, one is able to understand the strong spin-orbit force of the nuclear shell model.[5]

---

[5]Relativistic Hartree calculations in this picture were first carried out by [Horowitz and Serot (1981)]; this approach is discussed further in [Walecka (2004)].

Solution to Problem 9.5

(a) We write the Dirac equation explicitly, using $\gamma = -i\beta\alpha$ and $\gamma_4 = \beta$,

$$\left[-i\beta\alpha \cdot \nabla + \beta\frac{1}{ic}\frac{\partial}{\partial t} + \beta\frac{g_v}{\hbar c}V_0 + \left(M_N - \frac{g_s}{\hbar c}S \right) \right] \Psi = 0$$

Multiplication on the left by $\hbar c\beta$ then gives

$$\left[-i\hbar c\alpha \cdot \nabla - i\hbar\frac{\partial}{\partial t} + g_v V_0 + \beta(m_N c^2 - g_s S) \right] \Psi = 0$$

which can be now easily be re-cast in the form

$$(H_0 + H_1)\Psi = i\hbar\frac{\partial\Psi}{\partial t}$$
$$H_0 = c\alpha \cdot \mathbf{p} + \beta m_N c^2$$
$$H_1 = g_v V_0 - \beta g_s S$$

(b) We may write this Dirac equation explicitly in 2×2 form, as done previously in Prob. 9.3,

$$(m_N c^2 - g_s S + g_v V_0)\phi + c\sigma \cdot \mathbf{p}\chi = E\phi$$
$$(-m_N c^2 + g_s S + g_v V_0)\chi + c\sigma \cdot \mathbf{p}\phi = E\chi$$

The second equation can again be used to express χ in terms of ϕ

$$\chi = \frac{1}{E + m_N c^2 - g_s S - g_v V_0} c\sigma \cdot \mathbf{p}\phi$$

If we substitute this relation into the first equation, we obtain

$$\left[c\sigma \cdot \mathbf{p}\frac{1}{2m_N c^2 + \varepsilon - g_s S - g_v V_0} c\sigma \cdot \mathbf{p} + (-g_s S + g_v V_0) \right] \phi = \varepsilon\phi$$

$$\varepsilon \equiv E - m_N c^2$$

The last term in the l.h.s. of the first equation provides the binding potential $(-g_s S + g_v V_0)$. This involves the *difference* of the contributions of the vector and scalar potentials, which themselves may be very large in the nuclear case.

(c) The spin-orbit interaction for this problem can be obtained exactly as in Prob. 9.3. It is evident from the above relation that we only need to make the substitution $e\Phi(r) \to g_s S(r) + g_v V_0(r)$ in the denominator of the first term. Hence

$$V_{SO} = \left(\frac{\hbar}{2m_N c} \right)^2 \frac{1}{r} \left[\frac{d(g_s S + g_v V_0)}{dr} \right] \sigma \cdot \mathbf{l}$$

The spin-orbit interaction arises from the *sum* of the contributions of the large vector and scalar potentials $(g_s S + g_v V_0)$. In this manner, one can understand the strong spin-orbit interaction of the nuclear shell model.

Problem 9.6 (a) The positive-energy spinors $u(\mathbf{k})$ in Eq. (9.29) satisfy the Dirac equation

$$(\hbar c\, \boldsymbol{\alpha} \cdot \mathbf{k} + \beta m_0 c^2)u(\mathbf{k}) = E_k u(\mathbf{k})$$
$$E_k = \sqrt{(\hbar k c)^2 + (m_0 c^2)^2}$$

Use the arguments in Eqs. (9.31)–(9.38) to solve this equation. Show the solutions, normalized to $u_\lambda^\dagger u_{\lambda'} = \delta_{\lambda\lambda'}$, can be written as

$$u_\lambda(\mathbf{k}) = \left(\frac{E_k + m_0 c^2}{2E_k}\right)^{1/2} \left(\begin{array}{c} \eta_\lambda \\ \left[\hbar c\, \boldsymbol{\sigma} \cdot \mathbf{k}/(E_k + m_0 c^2)\right] \eta_\lambda \end{array}\right)$$

Here the η_λ, with $\lambda = (\uparrow, \downarrow)$, are the two-component spinors in Eqs. (4.160).

(b) Consider the negative-energy solutions $v(-\mathbf{k})$ with energy eigenvalue $E = -E_k$ and momentum eigenvalue $\mathbf{p} = -\hbar \mathbf{k}$, again normalized to $v_\lambda^\dagger v_{\lambda'} = \delta_{\lambda\lambda'}$. Show they can be written

$$v_\lambda(-\mathbf{k}) = \left(\frac{E_k + m_0 c^2}{2E_k}\right)^{1/2} \left(\begin{array}{c} \left[\hbar c\, \boldsymbol{\sigma} \cdot \mathbf{k}/(E_k + m_0 c^2)\right] \eta_\lambda \\ \eta_\lambda \end{array}\right)$$

Solution to Problem 9.6

(a) We seek stationary-state solutions to the Dirac equation of the form $\Psi(\mathbf{x}, t) = \psi e^{i\mathbf{p}\cdot\mathbf{x}/\hbar} e^{-iEt/\hbar}$, where $\mathbf{p} = \hbar \mathbf{k}$ is the momentum eigenvalue. If we write the four-component $\psi = \begin{pmatrix} \phi \\ \chi \end{pmatrix}$, and use the standard representation of the Dirac matrices, the Dirac equation can be re-cast in the 2×2 form of Eq. (9.32), which implies that the two Eqs. (9.33) must be simultaneously satisfied

$$c\boldsymbol{\sigma} \cdot \mathbf{p}\chi + m_0 c^2 \phi = E\phi$$
$$c\boldsymbol{\sigma} \cdot \mathbf{p}\phi - m_0 c^2 \chi = E\chi$$

The lower component χ can be related to the upper component ϕ of the Dirac wave function using the second of these relations

$$\chi = \frac{c\boldsymbol{\sigma} \cdot \mathbf{p}}{E + m_0 c^2}\phi$$

Substitution into the first relation, and the use of Eq. (9.38), then gives

$$\frac{(c\boldsymbol{\sigma}\cdot\mathbf{p})^2}{E+m_0c^2}\phi = \frac{(c\mathbf{p})^2}{E+m_0c^2}\phi = (E-m_0c^2)\phi$$

The Dirac equation is thus satisfied, provided the energy eigenvalues obey the relativistic dispersion relation

$$E^2 = (c\mathbf{p})^2 + (m_0c^2)^2$$

For the positive root $E = E_k$ with $E_k \equiv \sqrt{(\hbar kc)^2 + (m_0c^2)^2}$, we can use the above to relate χ to ϕ through $\chi = [\hbar c\boldsymbol{\sigma}\cdot\mathbf{k}/(E_k+m_0c^2)]\phi$, and the solutions to the Dirac equation can then be written in the form

$$u_\lambda(\mathbf{k}) = N_k \begin{pmatrix} \eta_\lambda \\ [\hbar c\boldsymbol{\sigma}\cdot\mathbf{k}/(E_k+m_0c^2)]\,\eta_\lambda \end{pmatrix}$$

where η_λ represents one of the complete set of two-component spinors in Eqs. (4.160), and N_k is a normalization constant. The latter quantity can be obtained by imposing the normalization condition

$$
\begin{aligned}
u_\lambda^\dagger u_{\lambda'} &= N_k^2 \left(\eta_\lambda^\dagger , \; \eta_\lambda^\dagger \frac{\hbar c\boldsymbol{\sigma}\cdot\mathbf{k}}{E_k+m_0c^2} \right) \left(\begin{matrix} \eta_{\lambda'} \\ [\hbar c\boldsymbol{\sigma}\cdot\mathbf{k}/(E_k+m_0c^2)]\,\eta_{\lambda'} \end{matrix} \right) \\
&= N_k^2 \left[\eta_\lambda^\dagger \eta_{\lambda'} + \eta_\lambda^\dagger \frac{\hbar c\boldsymbol{\sigma}\cdot\mathbf{k}}{E_k+m_0c^2}\frac{\hbar c\boldsymbol{\sigma}\cdot\mathbf{k}}{E_k+m_0c^2}\eta_{\lambda'} \right] \\
&= N_k^2\,\eta_\lambda^\dagger \eta_{\lambda'}\left[1 + \frac{\hbar^2c^2k^2}{(E_k+m_0c^2)^2} \right] \\
&= N_k^2\,\delta_{\lambda\lambda'}\frac{2E_k}{E_k+m_0c^2}
\end{aligned}
$$

Since $u_\lambda^\dagger u_{\lambda'} = \delta_{\lambda\lambda'}$, we have

$$N_k = \left(\frac{E_k+m_0c^2}{2E_k} \right)^{1/2}$$

(b) For the negative root $E = -E_k$, We may repeat the analysis of part (a), this time solving for ϕ in terms of χ. We obtain[6]

$$v_\lambda(-\mathbf{k}) = N_k' \begin{pmatrix} [-\hbar c\boldsymbol{\sigma}\cdot\mathbf{k}/(-E_k-m_0c^2)]\eta_\lambda \\ \eta_\lambda \end{pmatrix}$$

$$= N_k' \begin{pmatrix} [\hbar c\boldsymbol{\sigma}\cdot\mathbf{k}/(E_k+m_0c^2)]\eta_\lambda \\ \eta_\lambda \end{pmatrix}$$

---

[6]Note the $-\mathbf{k}$ in the argument of this negative-energy spinor.

where N_k' is again a normalization factor. Exactly as in part (a), the normalization condition $v_\lambda^\dagger v_{\lambda'} = \delta_{\lambda\lambda'}$ allows one to obtain

$$N_k' = \left(\frac{E_k + m_0 c^2}{2E_k} \right)^{1/2}$$

Problem 9.7 (a) Suppose the η_λ is Prob. 9.6 are two-component eigenstates of helicity so that $\boldsymbol{\sigma} \cdot (\mathbf{k}/k)\eta_\lambda = \lambda\eta_\lambda$.[7] Show that the Dirac spinors in Prob. 9.6 are then corresponding eigenstates of the Dirac helicity $\boldsymbol{\Sigma} \cdot (\mathbf{k}/k)$.

(b) Show that the Dirac spinors in Prob. 9.6 satisfy the following orthonormality relations

$$u_{\lambda'}^\dagger(\mathbf{k})u_\lambda(\mathbf{k}) = v_{\lambda'}^\dagger(-\mathbf{k})v_\lambda^\dagger(-\mathbf{k}) = \delta_{\lambda'\lambda}$$
$$u_{\lambda'}^\dagger(\mathbf{k})v_\lambda(\mathbf{k}) = v_{\lambda'}^\dagger(\mathbf{k})u_\lambda(\mathbf{k}) = 0$$

Solution to Problem 9.7

(a) If η_λ are the two-component eigenstates of the helicity, then from Prob. 9.6 we have

$$u_\lambda(\mathbf{k}) = \left(\frac{E_k + m_0 c^2}{2E_k} \right)^{1/2} \left(\begin{array}{c} \eta_\lambda \\ [\hbar c\boldsymbol{\sigma} \cdot \mathbf{k}/(E_k + m_0 c^2)]\,\eta_\lambda \end{array} \right)$$

$$v_\lambda(-\mathbf{k}) = \left(\frac{E_k + m_0 c^2}{2E_k} \right)^{1/2} \left(\begin{array}{c} [\hbar c\boldsymbol{\sigma} \cdot \mathbf{k}/(E_k + m_0 c^2)]\eta_\lambda \\ \eta_\lambda \end{array} \right)$$

From Eq. (9.69), the spin operator is given by the 4×4 matrix

$$\boldsymbol{\Sigma} = -\gamma_5\boldsymbol{\alpha} = \left(\begin{array}{cc} \boldsymbol{\sigma} & 0 \\ 0 & \boldsymbol{\sigma} \end{array} \right)$$

Therefore

$$\boldsymbol{\Sigma} \cdot \frac{\mathbf{k}}{k} u_\lambda(\mathbf{k}) = \left(\frac{E_k + m_0 c^2}{2E_k} \right)^{1/2} \lambda \left(\begin{array}{c} \eta_\lambda \\ [\hbar c\boldsymbol{\sigma} \cdot \mathbf{k}/(E_k + m_0 c^2)]\,\eta_\lambda \end{array} \right)$$
$$= \lambda\, u_\lambda(\mathbf{k})$$

$$\boldsymbol{\Sigma} \cdot \frac{\mathbf{k}}{k} v_\lambda(-\mathbf{k}) = \left(\frac{E_k + m_0 c^2}{2E_k} \right)^{1/2} \lambda \left(\begin{array}{c} [\hbar c\boldsymbol{\sigma} \cdot \mathbf{k}/(E_k + m_0 c^2)]\eta_\lambda \\ \eta_\lambda \end{array} \right)$$
$$= \lambda\, v_\lambda(-\mathbf{k})$$

---

[7]The easiest way to ensure this is to let $\mathbf{k}$ define the z-axis; see also the *Aside* in the solution to Prob. 4.33(b).

(b) The normalization of the Dirac spinors, $u_{\lambda'}^{\dagger}(\mathbf{k})u_\lambda(\mathbf{k}) = v_{\lambda'}^{\dagger}(-\mathbf{k})v_\lambda^{\dagger}(-\mathbf{k}) = \delta_{\lambda'\lambda}$, has been already imposed in Prob. 9.6, so we will concentrate on the orthogonality relations. We have[8]

$$u_{\lambda'}^{\dagger}(\mathbf{k})v_\lambda(\mathbf{k}) = \left(\frac{E_k + m_0c^2}{2E_k} \right) \times$$

$$\left(\eta_{\lambda'}^{\dagger} \,,\, \eta_{\lambda'}^{\dagger} \frac{\hbar c \boldsymbol{\sigma} \cdot \mathbf{k}}{(E_k + m_0c^2)} \right) \left(\begin{matrix} [-\hbar c \boldsymbol{\sigma} \cdot \mathbf{k}/(E_\mathbf{k} + m_0c^2)]\eta_\lambda \\ \eta_\lambda \end{matrix} \right)$$

$$= 0$$

Similarly we have

$$v_{\lambda'}^{\dagger}(\mathbf{k})u_\lambda(\mathbf{k}) = \left[u_\lambda^{\dagger}(\mathbf{k})v_{\lambda'}(\mathbf{k}) \right]^{\dagger} = 0$$

Problem 9.8 The matrix γ_5 is defined by $\gamma_5 \equiv \gamma_1\gamma_2\gamma_3\gamma_4$. Show the following:

(a) $\gamma_5^{\dagger} = \gamma_5$;

(b) $\gamma_5^2 = 1$;

(c) $\gamma_5\gamma_\mu + \gamma_\mu\gamma_5 = 0$;

(d) Verify its standard representation in Eq. (9.68).

Solution to Problem 9.8

(a) We will use the hermiticity of the gamma matrices, and the fact that they anticommute, $\{\gamma_\mu, \gamma_\nu\} = 2\delta_{\mu\nu}$, to obtain

$$\gamma_5^{\dagger} = (\gamma_1\gamma_2\gamma_3\gamma_4)^{\dagger} = \gamma_4\gamma_3\gamma_2\gamma_1 = \gamma_1\gamma_2\gamma_3\gamma_4 = \gamma_5$$

(b) With the aid of part (a), we have

$$\gamma_5^2 = \gamma_5^{\dagger}\gamma_5 = \gamma_4\gamma_3\gamma_2\gamma_1\gamma_1\gamma_2\gamma_3\gamma_4$$

$$= \gamma_4\gamma_3\gamma_2\gamma_2\gamma_3\gamma_4 = \gamma_4\gamma_3\gamma_3\gamma_4 = \gamma_4\gamma_4 = 1$$

(c) Let us start with $\mu = 4$

$$\gamma_5\gamma_4 + \gamma_4\gamma_5 = \gamma_1\gamma_2\gamma_3 + \gamma_4\gamma_1\gamma_2\gamma_3\gamma_4$$

$$= \gamma_1\gamma_2\gamma_3 - \gamma_1\gamma_2\gamma_3 = 0$$

Consider now the case $\mu = k$, with $k = 1, 2, 3$

$$\gamma_5\gamma_k + \gamma_k\gamma_5 = \gamma_1\gamma_2\gamma_3\gamma_4\gamma_k + \gamma_k\gamma_1\gamma_2\gamma_3\gamma_4 = (-\gamma_1\gamma_2\gamma_3\gamma_k + \gamma_k\gamma_1\gamma_2\gamma_3)\,\gamma_4$$

---

[8]Note this is now $v_\lambda(\mathbf{k})$.

With use of the anticommutation relations between γ matrices, it is easy to see that the commutator $(-\gamma_1\gamma_2\gamma_3\gamma_k + \gamma_k\gamma_1\gamma_2\gamma_3) \equiv [\gamma_k, \gamma_1\gamma_2\gamma_3]$ vanishes for all values of $k = (1, 2, 3)$. Therefore $\gamma_5\gamma_k + \gamma_k\gamma_5 = 0$, and, in summary,

$$\gamma_\mu\gamma_5 + \gamma_5\gamma_\mu = 0$$

(d) We have

$$\gamma_5 = \gamma_1\gamma_2\gamma_3\gamma_4 = (i\alpha_1\beta)(i\alpha_2\beta)(i\alpha_3\beta)\beta = i\alpha_1\alpha_2\alpha_3$$

$$= i\begin{pmatrix} 0 & \sigma_x \\ \sigma_x & 0 \end{pmatrix}\begin{pmatrix} 0 & \sigma_y \\ \sigma_y & 0 \end{pmatrix}\begin{pmatrix} 0 & \sigma_z \\ \sigma_z & 0 \end{pmatrix}$$

$$= i\begin{pmatrix} 0 & \sigma_x\sigma_y\sigma_z \\ \sigma_x\sigma_y\sigma_z & 0 \end{pmatrix} = \begin{pmatrix} 0 & -1 \\ -1 & 0 \end{pmatrix}$$

Problem 9.9 (a) The wave function ϕ is defined in terms of the Dirac wave function ψ by $\phi \equiv (1/2)(1 + \gamma_5)\psi$. Use the results in Prob. 9.8, and prove the following relation satisfied by the weak-interaction vertex [recall Eq. (9.44)]

$$\bar{\psi}\gamma_\mu(1 + \gamma_5)\psi = 2\bar{\phi}\gamma_\mu\phi$$

(b) There are 16 linearly independent 4×4 matrices, and they can be taken as $1, \gamma_5, \gamma_\mu, \gamma_\mu\gamma_5$, and $\sigma_{\mu\nu} \equiv (1/2i)[\gamma_\mu, \gamma_\nu]$. Show that when written in terms of the ϕ, the weak vertex is unique. Show[9]

$$\bar{\phi}\gamma_\mu\gamma_5\phi = \bar{\phi}\gamma_\mu\phi$$

$$\bar{\phi}\phi = \bar{\phi}\gamma_5\phi = \bar{\phi}\sigma_{\mu\nu}\phi = 0$$

Solution to Problem 9.9

(a) We will first start by using property (b) of Prob. 9.8 to establish

$$\frac{1}{2}(1 + \gamma_5)\frac{1}{2}(1 + \gamma_5) = \frac{1}{4}(1 + 2\gamma_5 + \gamma_5^2) = \frac{1}{2}(1 + \gamma_5)$$

If we use this result, together with property (c) of Prob. 9.8, we also have

$$\gamma_\mu\frac{1}{2}(1 + \gamma_5) = \gamma_\mu\frac{1}{2}(1 + \gamma_5)\frac{1}{2}(1 + \gamma_5) = \frac{1}{2}(1 - \gamma_5)\gamma_\mu\frac{1}{2}(1 + \gamma_5)$$

---

[9]This is basis of the V-A theory [Feynman and Gell-Mann (1958)].

We now use the definition of this new four-component ϕ to obtain its adjoint, according to Eq. (9.44)

$$\bar{\phi} = \left[\frac{1}{2}(1+\gamma_5)\psi\right]^\dagger \gamma_4 = \psi^\dagger \frac{1}{2}(1+\gamma_5)\gamma_4 = \bar{\psi}\frac{1}{2}(1-\gamma_5)$$

With this result, the relation that we want to prove follows

$$2\bar{\phi}\gamma_\mu\phi = 2\bar{\psi}\frac{1}{2}(1-\gamma_5)\gamma_\mu\frac{1}{2}(1+\gamma_5)\psi$$
$$= \bar{\psi}\gamma_\mu(1+\gamma_5)\psi$$

(b) We first notice the property:

$$\gamma_5(1+\gamma_5) = (1+\gamma_5)$$

and use it to obtain

$$\frac{1}{2}(1-\gamma_5)\gamma_\mu\gamma_5\frac{1}{2}(1+\gamma_5) = \frac{1}{2}(1-\gamma_5)\gamma_\mu\frac{1}{2}(1+\gamma_5)$$

From this relation it clearly follows

$$\bar{\phi}\gamma_\mu\gamma_5\phi = \bar{\phi}\gamma_\mu\phi$$

We now notice that

$$\frac{1}{2}(1-\gamma_5)\frac{1}{2}(1+\gamma_5) = 0$$

and therefore

$$\frac{1}{2}(1-\gamma_5)\gamma_5\frac{1}{2}(1+\gamma_5) = \frac{1}{2}(1-\gamma_5)\frac{1}{2}(1+\gamma_5) = 0$$

Next we observe that $\sigma_{\mu\nu}$ commutes with γ_5, since it is a product of two gamma matrices. Therefore

$$\frac{1}{2}(1-\gamma_5)\sigma_{\mu\nu}\frac{1}{2}(1+\gamma_5) = \frac{1}{2}(1-\gamma_5)\frac{1}{2}(1+\gamma_5)\sigma_{\mu\nu} = 0$$

The second set of relations follows from these results

$$\bar{\phi}\phi = \bar{\phi}\gamma_5\phi = \bar{\phi}\sigma_{\mu\nu}\phi = 0$$

Problem 9.10 (a) Make use of the Dirac spinor in Prob. 9.6(a) and the standard representation for the gamma matrices. Explicitly evaluate $\bar{u}(\mathbf{k}')u(\mathbf{k})$ and $\bar{u}(\mathbf{k}')\gamma_5 u(\mathbf{k})$;

(b) Note that under a spatial reflection: $(\mathbf{r},\mathbf{p}) \to (-\mathbf{r},-\mathbf{p})$,[10] and

---

[10] Or $\mathbf{k} \to -\mathbf{k}$.

$(1, \boldsymbol{\sigma}) \to (1, \boldsymbol{\sigma})$. Hence show that $\bar{u}u$ is a scalar under the parity transformation, and $\bar{u}\gamma_5 u$ is a *pseudoscalar*;

(c) Repeat for $\bar{u}\gamma_\mu u$ and $\bar{u}\gamma_\mu\gamma_5 u$. Show the first represents an ordinary polar vector, and the second an *axial vector*.

Solution to Problem 9.10

(a) Let us work out the inner product using the explicit forms from Prob. 9.6(a)

$$
\bar{u}_{\lambda'}(\mathbf{k}')u_\lambda(\mathbf{k}) = \left(\frac{E_{k'} + m_0c^2}{2E_{k'}}\right)^{1/2}\left(\frac{E_k + m_0c^2}{2E_k}\right)^{1/2} \times
$$
$$
\left(\eta_{\lambda'}^\dagger, \ -\eta_{\lambda'}^\dagger\frac{\hbar c\,\boldsymbol{\sigma}\cdot\mathbf{k}'}{(E_{k'} + m_0c^2)}\right)\left(\begin{array}{c}\eta_\lambda \\ \left[\hbar c\,\boldsymbol{\sigma}\cdot\mathbf{k}/(E_k + m_0c^2)\right]\eta_\lambda\end{array}\right)
$$
$$
= \left(\frac{E_{k'} + m_0c^2}{2E_{k'}}\right)^{1/2}\left(\frac{E_k + m_0c^2}{2E_k}\right)^{1/2} \times
$$
$$
\left(\eta_{\lambda'}^\dagger\eta_\lambda - \eta_{\lambda'}^\dagger\frac{\hbar c\,\boldsymbol{\sigma}\cdot\mathbf{k}'}{(E_{k'} + m_0c^2)}\frac{\hbar c\,\boldsymbol{\sigma}\cdot\mathbf{k}}{(E_k + m_0c^2)}\eta_\lambda\right)
$$

We may use the properties of the Pauli matrices to write[11]

$$
\boldsymbol{\sigma}\cdot\mathbf{k}'\,\boldsymbol{\sigma}\cdot\mathbf{k} = k'_i k_j\left(\delta_{ij} + i\epsilon_{ijl}\sigma_l\right) = \mathbf{k}'\cdot\mathbf{k} + i\boldsymbol{\sigma}\cdot(\mathbf{k}'\times\mathbf{k})
$$

In a similar fashion, we find

$$
\bar{u}_{\lambda'}(\mathbf{k}')\gamma_5 u_\lambda(\mathbf{k}) = \left(\frac{E_{k'} + m_0c^2}{2E_{k'}}\right)^{1/2}\left(\frac{E_k + m_0c^2}{2E_k}\right)^{1/2} \times
$$
$$
\left(\eta_{\lambda'}^\dagger, \ -\eta_{\lambda'}^\dagger\frac{\hbar c\,\boldsymbol{\sigma}\cdot\mathbf{k}'}{(E_{k'} + m_0c^2)}\right)\left(\begin{array}{c}-\left[\hbar c\,\boldsymbol{\sigma}\cdot\mathbf{k}/(E_k + m_0c^2)\right]\eta_\lambda \\ -\eta_\lambda\end{array}\right)
$$
$$
= \left(\frac{E_{k'} + m_0c^2}{2E_{k'}}\right)^{1/2}\left(\frac{E_k + m_0c^2}{2E_k}\right)^{1/2} \times
$$
$$
\left(-\eta_{\lambda'}^\dagger\frac{\hbar c\,\boldsymbol{\sigma}\cdot\mathbf{k}}{(E_k + m_0c^2)}\eta_\lambda + \eta_{\lambda'}^\dagger\frac{\hbar c\,\boldsymbol{\sigma}\cdot\mathbf{k}'}{(E_{k'} + m_0c^2)}\eta_\lambda\right)
$$

(b) Under a spatial reflection one has $\mathbf{k} \to -\mathbf{k}$ and $\mathbf{k}' \to -\mathbf{k}'$, while $\boldsymbol{\sigma} \to \boldsymbol{\sigma}$. We then easily see that $\bar{u}_{\lambda'}(\mathbf{k}')u_\lambda(\mathbf{k})$ is unchanged, since $\boldsymbol{\sigma}\cdot\mathbf{k}'\,\boldsymbol{\sigma}\cdot\mathbf{k}$ suffers a double change of sign and is left unchanged by the reflection. On the other hand, $\bar{u}_{\lambda'}(\mathbf{k}')\gamma_5 u_\lambda(\mathbf{k})$ changes sign under a spatial reflection since $\boldsymbol{\sigma}\cdot\mathbf{k}$ does change sign, and so does $\boldsymbol{\sigma}\cdot\mathbf{k}'$.

---

[11]Here repeated Latin indices are summed from 1 to 3.

We thus conclude that $\bar{u}u$ is a scalar under the parity transformation, and $\bar{u}\gamma_5 u$ is a *pseudoscalar*.

(c) We work out $\bar{u}_{\lambda'}(\mathbf{k}')\gamma_\mu u_\lambda(\mathbf{k})$ individually. For $\mu = 4$, one has

$$\bar{u}_{\lambda'}(\mathbf{k}')\gamma_4 u_\lambda(\mathbf{k}) = u_{\lambda'}^\dagger(\mathbf{k}')u_\lambda(\mathbf{k}) = \left(\frac{E_{k'} + m_0 c^2}{2E_{k'}}\right)^{1/2}\left(\frac{E_k + m_0 c^2}{2E_k}\right)^{1/2} \times$$

$$\left(\eta_{\lambda'}^\dagger, \ \eta_{\lambda'}^\dagger \frac{\hbar c\, \boldsymbol{\sigma} \cdot \mathbf{k}'}{(E_{k'} + m_0 c^2)}\right)\left(\begin{array}{c} \eta_\lambda \\ [\hbar c\, \boldsymbol{\sigma} \cdot \mathbf{k}/(E_k + m_0 c^2)]\,\eta_\lambda \end{array}\right)$$

$$= \left(\frac{E_{k'} + m_0 c^2}{2E_{k'}}\right)^{1/2}\left(\frac{E_k + m_0 c^2}{2E_k}\right)^{1/2} \times$$

$$\left(\eta_{\lambda'}^\dagger \eta_\lambda + \eta_{\lambda'}^\dagger \frac{\hbar c\, \boldsymbol{\sigma} \cdot \mathbf{k}'}{(E_{k'} + m_0 c^2)}\frac{\hbar c\, \boldsymbol{\sigma} \cdot \mathbf{k}}{(E_k + m_0 c^2)}\eta_\lambda\right)$$

For a spatial index $\mu = j = (1, 2, 3)$

$$\bar{u}_{\lambda'}(\mathbf{k}')\gamma_j u_\lambda(\mathbf{k}) = i\left(\frac{E_{k'} + m_0 c^2}{2E_{k'}}\right)^{1/2}\left(\frac{E_k + m_0 c^2}{2E_k}\right)^{1/2} \times$$

$$\left(\eta_{\lambda'}^\dagger, \ -\eta_{\lambda'}^\dagger \frac{\hbar c\, \boldsymbol{\sigma} \cdot \mathbf{k}'}{(E_{k'} + m_0 c^2)}\right)\left(\begin{array}{cc} 0 & -\sigma_j \\ \sigma_j & 0 \end{array}\right)\left(\begin{array}{c} \eta_\lambda \\ [\hbar c\, \boldsymbol{\sigma} \cdot \mathbf{k}/(E_k + m_0 c^2)]\,\eta_\lambda \end{array}\right)$$

$$= i\left(\frac{E_{k'} + m_0 c^2}{2E_{k'}}\right)^{1/2}\left(\frac{E_k + m_0 c^2}{2E_k}\right)^{1/2} \times$$

$$\left(-\eta_{\lambda'}^\dagger \sigma_j \frac{\hbar c\, \boldsymbol{\sigma} \cdot \mathbf{k}}{(E_k + m_0 c^2)}\eta_\lambda - \eta_{\lambda'}^\dagger \frac{\hbar c\, \boldsymbol{\sigma} \cdot \mathbf{k}'}{(E_{k'} + m_0 c^2)}\sigma_j \eta_\lambda\right)$$

Thus we see that under a spatial reflection the time component of $\bar{u}\gamma_\mu u$ is invariant, whereas the spatial component changes sign.

A similar calculation for $\bar{u}_{\lambda'}(\mathbf{k}')\gamma_\mu\gamma_5 u_\lambda(\mathbf{k})$ gives for $\mu = 4$

$$\bar{u}_{\lambda'}(\mathbf{k}')\gamma_4\gamma_5 u_\lambda(\mathbf{k}) = \left(\frac{E_{k'} + m_0 c^2}{2E_{k'}}\right)^{1/2}\left(\frac{E_k + m_0 c^2}{2E_k}\right)^{1/2} \times$$

$$\left(\eta_{\lambda'}^\dagger, \ \eta_{\lambda'}^\dagger \frac{\hbar c\, \boldsymbol{\sigma} \cdot \mathbf{k}'}{(E_{k'} + m_0 c^2)}\right)\left(\begin{array}{cc} 0 & -1 \\ -1 & 0 \end{array}\right)\left(\begin{array}{c} \eta_\lambda \\ [\hbar c\, \boldsymbol{\sigma} \cdot \mathbf{k}/(E_k + m_0 c^2)]\,\eta_\lambda \end{array}\right)$$

$$= \left(\frac{E_{k'} + m_0 c^2}{2E_{k'}}\right)^{1/2}\left(\frac{E_k + m_0 c^2}{2E_k}\right)^{1/2} \times$$

$$\left(-\eta_{\lambda'}^\dagger \frac{\hbar c\, \boldsymbol{\sigma} \cdot \mathbf{k}}{(E_k + m_0 c^2)}\eta_\lambda - \eta_{\lambda'}^\dagger \frac{\hbar c\, \boldsymbol{\sigma} \cdot \mathbf{k}'}{(E_{k'} + m_0 c^2)}\eta_\lambda\right)$$

For a spatial index $\mu = j = (1, 2, 3)$

$$\bar{u}_{\lambda'}(\mathbf{k}')\gamma_j\gamma_5 u_\lambda(\mathbf{k}) = i \left(\frac{E_{k'} + m_0c^2}{2E_{k'}}\right)^{1/2} \left(\frac{E_k + m_0c^2}{2E_k}\right)^{1/2} \times$$

$$\left(\eta_{\lambda'}^\dagger, \; \eta_{\lambda'}^\dagger \frac{\hbar c\,\boldsymbol{\sigma}\cdot\mathbf{k}'}{(E_{k'} + m_0c^2)}\right) \begin{pmatrix} \sigma_j & 0 \\ 0 & \sigma_j \end{pmatrix} \left(\begin{matrix} \eta_\lambda \\ [\hbar c\,\boldsymbol{\sigma}\cdot\mathbf{k}/(E_k + m_0c^2)]\,\eta_\lambda \end{matrix}\right)$$

$$= i \left(\frac{E_{k'} + m_0c^2}{2E_{k'}}\right)^{1/2} \left(\frac{E_k + m_0c^2}{2E_k}\right)^{1/2} \times$$

$$\left(\eta_{\lambda'}^\dagger \sigma_j \eta_\lambda + \eta_{\lambda'}^\dagger \frac{\hbar c\,\boldsymbol{\sigma}\cdot\mathbf{k}'}{(E_{k'} + m_0c^2)} \sigma_j \frac{\hbar c\,\boldsymbol{\sigma}\cdot\mathbf{k}}{(E_k + m_0c^2)}\eta_\lambda\right)$$

Thus we see that under a spatial reflection the time component of $\bar{u}\gamma_\mu\gamma_5 u$ changes sign, whereas the spatial component is invariant.

We thus conclude that $\bar{u}\gamma_\mu u$ is a polar four-vector, whereas $\bar{u}\gamma_\mu\gamma_5 u$ is an axial four-vector.

Problem 9.11 (a) Start from the stationary-state Dirac Eq. (9.21). Multiply on the left by $\psi^\dagger \beta$, multiply the adjoint of that equation on the right by $\beta\psi$, and add. Show

$$m_0c^2\,\psi^\dagger\psi = E\,\bar{\psi}\psi$$

(b) Since $u^\dagger u$ represents a probability *density* (probability/volume), it is not invariant under a Lorentz transformation. The quantity $\bar{u}u$ *is* Lorentz invariant. Why is this true? Show that the Dirac spinors $U(\mathbf{k})$ with invariant norm are given by

$$U(\mathbf{k}) = \left(\frac{E_k}{m_0c^2}\right)^{1/2} u(\mathbf{k}) \qquad ; \; \bar{U}U = 1$$

Here $E_k = \hbar c\sqrt{\mathbf{k}^2 + M^2}$.

Solution to Problem 9.11

(a) Upon multiplying Eq. (9.21) on the left by $\psi^\dagger \beta$, we obtain

$$\psi^\dagger \beta \left(c\boldsymbol{\alpha}\cdot\mathbf{p} + \beta m_0c^2\right)\psi = E\psi^\dagger\beta\psi$$

We now multiply the adjoint of Eq. (9.21) on the right by $\beta\psi$, obtaining

$$\psi^\dagger \left(c\boldsymbol{\alpha}\cdot\mathbf{p} + \beta m_0c^2\right)\beta\psi = E\psi^\dagger\beta\psi$$

If we recall that β and $\boldsymbol{\alpha}$ anticommute with $\beta\boldsymbol{\alpha} + \boldsymbol{\alpha}\beta = 0$, we may sum the two equations to obtain

$$m_0 c^2 \psi^\dagger \psi = E \bar{\psi}\psi$$

(b) $\psi^\dagger \psi$ is the time component of the four-vector $\bar{\psi}\gamma_\mu\psi$, while E is the time component of the four-momentum p_μ. Therefore, if $\bar{\psi}\psi$ is a Lorentz scalar, the two sides of the above equation transform in the same way under a Lorentz transformation.

The Dirac spinors $U(\mathbf{k})$ with invariant norm are defined by

$$U(\mathbf{k}) \equiv \left(\frac{E_k}{m_0 c^2}\right)^{1/2} u(\mathbf{k}) \qquad ; E_k = \hbar c \sqrt{\mathbf{k}^2 + M^2}$$

From the calculation in part (a), the Lorentz-invariant quantity $\bar{U}U$ is then given by

$$\bar{U}(\mathbf{k})U(\mathbf{k}) = \frac{E_k}{m_0 c^2} \bar{u}(\mathbf{k})u(\mathbf{k}) = u(\mathbf{k})^\dagger u(\mathbf{k}) = 1$$

Here the last equality follows from Prob. (9.6).

Problem 9.12 For a free particle, the combination $k_\mu^2 = \mathbf{k}^2 - k_0^2 = -M^2$ is unchanged under a Lorentz transformation. Show that this represents a two-sheeted hyperboloid in four-dimensional k-space (the "mass hyperboloid"). Proper Lorentz transformations preserve the sign of k_0, and are thus confined to one sheet of the mass hyperboloid. Now define $k_\mu^2 \equiv k^2$, and

(a) Demonstrate the following relation for the Dirac delta function

$$\delta(k^2 + M^2) = \frac{1}{2\sqrt{\mathbf{k}^2 + M^2}} \left[\delta\left(k_0 - \sqrt{\mathbf{k}^2 + M^2}\right) + \delta\left(k_0 + \sqrt{\mathbf{k}^2 + M^2}\right)\right]$$

(b) Show that $d^4 k \equiv d^3 k \, dk_0$ is a Lorentz invariant.[12]
(c) Show the Lorentz-invariant phase space volume is

$$\frac{d^3 k}{2E_k} = \frac{1}{\hbar c}\delta(k^2 + M^2)\theta(k_0) \, d^4 k \qquad ; \text{ invariant phase space}$$

where the θ-function is $\theta(k_0) = 1$ if $k_0 > 0$, and $\theta(k_0) = 0$ if $k_0 < 0$.

---

[12]Hint: The transformation between volume elements is $d^4 k' = \frac{\partial(k_1' \cdots k_0')}{\partial(k_1 \cdots k_0)} d^4 k$, where $\frac{\partial(k_1' \cdots k_0')}{\partial(k_1 \cdots k_0)}$ is the jacobian determinant. If this is unfamiliar, just accept part (b).

Solution to Problem 9.12

The equation $k_0^2 = \mathbf{k}^2 + M^2$ represents a hyperboloid in the four-dimensional momentum space where k_0 is the z-axis, and the other axes are the spatial components of $\mathbf{k}$.[13] There are two sheets to the hyperboloid corresponding to $k_0 > 0$ and $k_0 < 0$.

(a) We can factor

$$\mathbf{k}^2 + M^2 - k_0^2 = \left[\sqrt{\mathbf{k}^2 + M^2} + k_0\right]\left[\sqrt{\mathbf{k}^2 + M^2} - k_0\right]$$

Thus there are two places where the argument of the delta function vanishes. Use the following properties of the delta function in the vicinity of each zero

$$\delta(-x) = \delta(x) \qquad ; \; \delta(ax) = \frac{1}{|a|}\delta(x)$$

It follows that

$$\delta(k^2 + M^2) = \frac{1}{2\sqrt{\mathbf{k}^2 + M^2}}\left[\delta\left(k_0 - \sqrt{\mathbf{k}^2 + M^2}\right) + \delta\left(k_0 + \sqrt{\mathbf{k}^2 + M^2}\right)\right]$$

****

(*Aside*) Consider the general case $\delta[f(x)]$, where $f(x)$ is a real function with N simple zeros, x_i, $i = 1, \cdots, N$. The case of question (a) corresponds to a specific form of the function $f(x)$, with 2 zeroes. As the delta function only picks the contributions at the points where its argument vanishes we may write

$$\delta[f(x)] = \sum_{i=1}^{N} \delta[f'(x_i)(x - x_i)]$$

after Taylor expanding $f(x)$ to first order around each of the zeros.

Let us now consider the integral

$$\int \delta[f(x)]g(x)dx = \sum_{i=1}^{N}\int \delta[f'(x_i)(x - x_i)]g(x)dx$$

$$= \sum_{i=1}^{N}\frac{1}{|f'(x_i)|}g(x_i)$$

where the absolute value is needed to preserve the correct sign of the expression.

---

[13] Just make a sketch of it for yourselves with two spatial dimensions.

Since we could have obtained this result directly from the integral

$$\sum_{i=1}^{N} \frac{1}{|f'(x_i)|} \int \delta(x - x_i)g(x)dx$$

we conclude that

$$\delta[f(x)] = \sum_{i=1}^{N} \frac{1}{|f'(x_i)|} \delta(x - x_i)$$

In this above case we have $f(k_0) = -k_0^2 + \mathbf{k}^2 + M^2$ and

$$\delta(k^2 + M^2) = \frac{1}{2\sqrt{\mathbf{k}^2 + M^2}} \left[\delta\left(k_0 - \sqrt{\mathbf{k}^2 + M^2} \right) + \delta\left(k_0 + \sqrt{\mathbf{k}^2 + M^2} \right) \right]$$

****

(b) As suggested in the footnote, the transformation between volume elements is $d^4k' = \frac{\partial(k_1' \cdots k_0')}{\partial(k_1 \cdots k_0)} d^4k$, where $\frac{\partial(k_1' \cdots k_0')}{\partial(k_1 \cdots k_0)}$ is the jacobian determinant. The momenta k and k' are related by a Lorentz transformation, and therefore the jacobian is the determinant of that Lorentz transformation. As we see from Eq. (8.38) of the book, this determinant can be either 1 or -1, the first value being taken for *proper* Lorentz transformations. The invariance of d^4k under proper Lorentz transformations then follows.

(c) We first recall the definition of Eq. (9.47), $M \equiv m_0c/\hbar$, and then use the result of part (a) to write

$$\frac{1}{\hbar c} \delta(k^2 + M^2)\theta(k_0)\, d^4k = \delta\left(k_0 - \sqrt{\mathbf{k}^2 + (m_0c/\hbar)^2} \right) \frac{d^4k}{2\hbar c\sqrt{\mathbf{k}^2 + (m_0c/\hbar)^2}}$$

$$= \frac{d^3k}{2E_k}$$

Here we have used the step function to eliminate the contribution corresponding to $k_0 < 0$, and then used $\int dk_0\, \delta(k_0 - \sqrt{\mathbf{k}^2 + (m_0c/\hbar)^2}) = 1$.

Problem 9.13 Consider a scattering process in a Lorentz frame where the incident momentum $\hbar\mathbf{k}_1$ and the target momentum $\hbar\mathbf{k}_2$ are parallel to each other (lab frame, C-M frame, *etc.*). Show the incident flux in such a frame can be written in terms of the four-vectors (k_1, k_2) as [recall Prob. 7.3]

$$I_{\text{inc}} = \frac{c}{\Omega} \frac{(\hbar c)^2}{E_1 E_2} \left[(k_1 \cdot k_2)^2 - M_1^2 M_2^2 \right]^{1/2}$$

Solution to Problem 9.13

Let $p_{1\mu} = \hbar k_{1\mu}$ and $p_{2\mu} = \hbar k_{2\mu}$ be the four-momenta of the incident particle and of the target. Since these momenta are parallel, $\mathbf{p}_1 \cdot \mathbf{p}_2 = \pm p_1 p_2$. First of all, we see from Eq. (8.58) of the book that $\mathbf{v} = c^2 \mathbf{p}/E_p$ and use this result to express the incident flux of Prob. (7.3) as

$$I_{\text{inc}} = \frac{1}{\Omega} |\mathbf{v}(1) - \mathbf{v}(2)| = \frac{c^2}{\Omega} \left| \frac{\mathbf{p}_1}{E_1} - \frac{\mathbf{p}_2}{E_2} \right|$$

However

$$\left(\frac{\mathbf{p}_1}{E_1} - \frac{\mathbf{p}_2}{E_2} \right)^2 = \frac{\mathbf{p}_1^2}{E_1^2} + \frac{\mathbf{p}_2^2}{E_2^2} - 2 \frac{\mathbf{p}_1 \cdot \mathbf{p}_2}{E_1 E_2}$$

$$= \frac{1}{E_1^2 E_2^2} \left(2\mathbf{p}_1^2 \mathbf{p}_2^2 c^2 + m_2^2 c^4 \mathbf{p}_1^2 + m_1^2 c^4 \mathbf{p}_2^2 - 2E_1 E_2 \, \mathbf{p}_1 \cdot \mathbf{p}_2 \right)$$

Similarly we have

$$(p_1 \cdot p_2)^2 - m_1^2 m_2^2 c^4$$

$$= (\mathbf{p}_1 \cdot \mathbf{p}_2)^2 + \mathbf{p}_1^2 \, \mathbf{p}_2^2 + m_1^2 c^2 \mathbf{p}_2^2 + m_2^2 c^2 \mathbf{p}_1^2 - 2 \frac{E_1 E_2}{c^2} \, \mathbf{p}_1 \cdot \mathbf{p}_2$$

$$= 2\mathbf{p}_1^2 \, \mathbf{p}_2^2 + m_1^2 c^2 \mathbf{p}_2^2 + m_2^2 c^2 \mathbf{p}_1^2 - 2 \frac{E_1 E_2}{c^2} \, \mathbf{p}_1 \cdot \mathbf{p}_2$$

where the last line has been obtained taking into account the fact that the momenta are parallel.

We may use this second expression to express the first one as

$$\left(\frac{\mathbf{p}_1}{E_1} - \frac{\mathbf{p}_2}{E_2} \right)^2 = \frac{c^2}{E_1^2 E_2^2} \left[(p_1 \cdot p_2)^2 - m_1^2 m_2^2 c^4 \right]$$

$$= \frac{\hbar^4 c^2}{E_1^2 E_2^2} \left[(k_1 \cdot k_2)^2 - M_1^2 M_2^2 \right]$$

Upon inserting this expression in the equation for the incident flux, we finally have

$$I_{\text{inc}} = \frac{c}{\Omega} \frac{(\hbar c)^2}{E_1 E_2} \left[(k_1 \cdot k_2)^2 - M_1^2 M_2^2 \right]^{1/2}$$

Problem 9.14 Consider inelastic electron scattering from a proton where the initial and final electron four-momenta are $(k_1, k_2) = (k_1, k_1 - q)$ (compare Fig. 7.10 in the book). The response surfaces $W_{1,2}$ can be written as functions of the two Lorentz scalars q^2 and $\nu = -q \cdot p/mc$ where $m = m_p$.

(a) Show that in the laboratory frame, in the ERL,

$$q^2 = 4\tilde{k}_1 \tilde{k}_2 \sin^2 \frac{\theta}{2} \qquad ; \nu = \tilde{k}_1 - \tilde{k}_2 \qquad ; \text{ lab frame}$$

Here $\tilde{k} \equiv |\mathbf{k}|$.

(b) The functions $W_{1,2}(q^2, \nu)$ can be calculated under any kinematic conditions. Consider the "$\mathbf{p} \to \infty$ frame" where the electron and proton approach each other with $v \to c$. In this frame the proton is flattened by Lorentz contraction, and any internal motion is slowed down by time dilation. Suppose a quark in the proton now carries a fraction x of the proton's four-momentum p, and that the scattering takes place incoherently from the free, point-like quark. Show from the kinematics that in the *deep-inelastic* regime, where $(q^2, \nu) \to \infty$ and all masses are negligible, the quantity x is given by

$$x = \frac{q^2}{2M\nu} \qquad ; \text{ Bjorken-x}$$

where $M = mc/\hbar$. In this regime, the response surfaces become functions of the single variable x. This is the basis of the quark-parton model [Bjorken and Paschos (1969)].[14]

Solution to Problem 9.14

(a) Let p and $p' = p + q$ be the initial and final four-momenta of the proton.[15] In particular, in the lab frame we have $p_\mu = (\mathbf{0}, iM_p)$. Since for electron scattering the momentum transfer is $q = k_1 - k_2$ where the electron four-momenta are $k_\mu = (\mathbf{k}, iE_k/\hbar c)$, one has

$$q^2 = (k_1 - k_2)^2 = -2M_e^2 - 2\left(-\frac{E_1 E_2}{(\hbar c)^2} + \mathbf{k}_1 \cdot \mathbf{k}_2\right)$$

$$\approx 4\tilde{k}_1 \tilde{k}_2 \sin^2 \frac{\theta}{2}$$

where we work in the ERL and neglect the rest mass of the electron, so that $E_k \approx \hbar c \tilde{k}$.

Similarly, in the lab frame

$$\nu = -\frac{q \cdot p}{M_p} = \frac{1}{\hbar c}(E_1 - E_2) \approx \tilde{k}_1 - \tilde{k}_2$$

---

[14] For an introduction to electron scattering, see [Walecka (2001)].

[15] We here speak of (p, p', q) as four-momenta, but they are all really just inverse lengths; the actual four-momenta are $\hbar(p, p', q)$. The statement of the problem is inconsistent on this point.

where we have once again used the ERL to obtain the last result.

(b) In the *deep-inelastic* regime where $(q^2, \nu) \to \infty$, all masses are negligible. Let $\mathcal{P}_\mu$ be the four-momentum of a quark inside the proton, carrying a fraction x of the total four-momentum of the proton in the "$\mathbf{p} \to \infty$ frame", as described in the statement of the problem. In this frame

$$\mathcal{P}_\mu \approx x p_\mu$$

Let $\mathcal{P}'_\mu$ be the final four-momentum of the quark. We can use the conservation of energy and momentum to write

$$\mathcal{P}'^2 = (\mathcal{P} + q)^2 \approx 2xq \cdot p + q^2 \approx 0$$

where we have neglected the quark mass so that $\mathcal{P}'^2 = \mathcal{P}^2 \approx 0$. This relation can be solved for x, to obtain

$$x = -\frac{q^2}{2q \cdot p} = \frac{q^2}{2\nu M_p}$$

where we have used the definition of ν in the last result. Recall that here all momenta are defined as inverse lengths. The quantity x is Bjorken's celebrated dimensionless scaling variable for deep-inelastic scattering.[16]

---

[16]For the experimental deep-inelastic electron scattering results, see [Friedman and Kendall (1972)]. For the interpretation of these results in terms of the quark-parton model of Feynman and Bjorken and Paschos, see [Bjorken and Paschos (1969)].

Chapter 10

General Relativity

Problem 10.1 This problem concerns the Doppler shift in special relativity. A light signal is sent out from a source at $\bar{O}$ moving with velocity V in the z-direction. In its rest frame the source undergoes dN oscillations in a proper time $d\tau$. All observers can agree on this number dN. The proper frequency in the rest frame of the source is $\bar{\nu} = dN/d\tau$ or $dN = \bar{\nu}\,d\tau$. Now Lorentz transform to the laboratory frame where the source is moving with velocity V.

(a) Show the corresponding time interval dt in the lab is given by the relation

$$d\tau = \left(1 - \frac{V^2}{c^2}\right)^{1/2} dt$$

(b) During the time dt, the light front has traveled a distance $dl = cdt$ in the laboratory frame and hence arrives at an origin O a distance $dl = cdt$ away as shown in Fig. 10.1(a).

In the time dt, the source has moved to a new position a distance $dl' = cdt + Vdt$ away from O. The wavelength of the radiation received at O is thus increased as indicated in Fig. 10.1(b). Since λ is the actual distance the signal has traveled divided by the number of oscillations, show that

$$\lambda = \frac{(c+V)dt}{dN} = \bar{\lambda}\frac{(1+V/c)}{(1-V^2/c^2)^{1/2}}$$

Here $\bar{\lambda} = c/\bar{\nu}$ is the proper wavelength of the radiation for a source at rest.

(c) Show that in the limit $V/c \to 0$, this reduces to the familiar freshman physics result. Note that $V = \pm|V|$ can have either sign in these arguments.

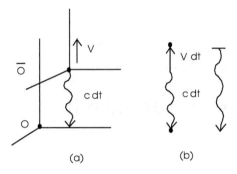

Fig. 10.1 Configuration for calculation of the Doppler shift in special relativity.

Solution to Problem 10.1

(a) Let us call O' and O the origins of two distinct frames: in the first frame the source is not moving, whereas in the second frame the source is moving with a constant velocity V in the z-direction. The two origins initially coincide.[1] After a given time the source has not moved in the first frame and therefore

$$z' = 0$$
$$t' = \tau$$

where τ is the proper time. In the second frame the coordinate of the source will be

$$z = Vt$$

where t is the time which has passed according to the observer's watch after the instant when the two origins coincided.

The coordinates of the two frames are related by a Lorentz transformation

$$z = \frac{(z' + Vt')}{\sqrt{1 - V^2/c}}$$
$$t = \frac{(t' + Vz'/c^2)}{\sqrt{1 - V^2/c}}$$

With the use of the above equations, we obtain the relation

$$t = \frac{\tau}{\sqrt{1 - V^2/c^2}}$$

---

[1] See Fig. 8.12 in the text.

In the present case, after making the identifications $\tau \to d\tau$ and $t \to dt$, we obtain the anticipated relation between the proper time in the rest frame, and the time interval in the laboratory frame where the source is moving with velocity V

$$d\tau = \sqrt{1 - V^2/c^2}\, dt$$

(b) The configuration for the Doppler shift is illustrated in Fig. 10.1. λ is the wavelength of the signal received at O, and it is obtained as the actual distance the signal has traveled divided by the number of oscillations

$$\lambda = \frac{(c+V)dt}{dN} = \frac{(c+V)dt}{\bar{\nu}d\tau}$$

where $\bar{\nu}$ is the proper frequency of the signal, that is, the frequency in the frame where the source is at rest.

With the use of the proper wavelength of the signal $\bar{\lambda} = c/\bar{\nu}$, and the use of the result from part (a) that $d\tau = \sqrt{1 - V^2/c^2}\, dt$, we obtain the Doppler-shifted wavelength

$$\lambda = \bar{\lambda}\,\frac{(1+V/c)}{(1-V^2/c^2)^{1/2}} \qquad ; \text{Doppler-shift}$$

(c) In the non-relativistic limit $V/c \to 0$ the above formula reduces to the freshman-physics result for the Doppler shift of a wave

$$\lambda \approx \bar{\lambda}\,(1+V/c) \qquad ; |V|/c \ll 1$$

The corresponding fractional shift in frequency will then be

$$\frac{\Delta\nu}{\bar{\nu}} \equiv \frac{1}{\bar{\nu}}(\nu - \bar{\nu}) = \frac{1}{\bar{\nu}}\left(\frac{c}{\lambda} - \frac{c}{\bar{\lambda}}\right) \approx -\frac{V}{c}$$

Note that in these arguments, V can have either sign.

Problem 10.2 (a) Consider a curve $q(x)$. Show that a little element of length along the curve is given by $ds = [1+(q')^2]^{1/2}dx$ where $q' \equiv dq(x)/dx$. Hence show that the total length of the curve S is given by

$$S = \int_{x_1}^{x_2} [1 + (q')^2]^{1/2}\, dx$$

(b) Use the result in Prob. 10.3 to prove that the shortest distance between two points is a straight line.

Solution to Problem 10.2

(a) We consider a small element of the curve $y = q(x)$ whose extrema are the points $[x, q(x)]$ and $[x + dx, q(x + dx)]$. This element is represented in Fig. 10.2.

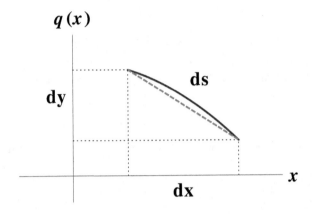

Fig. 10.2 Small element of a curve $y = q(x)$ between the points $[x, q(x)]$ and $[x + dx, q(x + dx)]$. Note that here $q' \equiv dq(x)/dx < 0$.

With the assumption that dx is infinitesimal, we have that $dy \approx (dq/dx)dx$ and we can approximate the length of the element of curve (the continuous line in the figure) with the hypotenuse of the triangle of sides dx and dy

$$ds \approx \sqrt{dx^2 + dy^2} = dx\sqrt{1 + (q')^2}$$

Here $q' \equiv dq(x)/dx$, and this expression is exact to first order in differentials.

The length of an element of curve will then be obtained by integrating ds between the initial and final points of the element

$$S = \int_{x_1}^{x_2} \left[1 + (q')^2\right]^{1/2} dx$$

(b) We apply the Euler-Lagrange equation proven in Prob. 10.3 to the functional $S = \int_{x_1}^{x_2} F(q, q'; x)dx$ that we have just obtained

$$S = \int_{x_1}^{x_2} F(q, q'; x)dx$$

$$F = \left[1 + (q')^2\right]^{1/2}$$

Notice that F here only depends on q'. The Euler-Lagrange equation therefore reads

$$\frac{d}{dx}\left(\frac{\partial F}{\partial q'}\right) = 0$$

(10.1)

This is integrated to give

$$\frac{\partial F}{\partial q'} = \frac{q'}{\sqrt{1+(q')^2}} = \text{constant}$$

Since a constant slope, $q' = \text{constant}$, is a solution to this equation, we conclude that the shortest distance between the two points (x_1, x_2) is a straight line.

Problem 10.3 You are given an integral of the following form

$$S = \int_{x_1}^{x_2} F(q, q'; x)\,dx$$

where $q' \equiv dq(x)/dx$ and $F(q, q'; x)$ is a specified function of the indicated variables (the last one indicates any additional explicit x dependence). This is known as a *functional* of $q(x)$. It assigns to every function $q(x)$ a number S. Suppose you want to find that curve $q(x)$ that *minimizes S*.

(a) Let $q(x)$ be the solution to the problem. Consider small *variations* about that curve of the form $q(x) \to q(x) + \lambda \eta(x)$ where λ is an infinitesimal and $\eta(x)$ is arbitrary except for the condition of fixed endpoints $\eta(x_1) = \eta(x_2) = 0$ (see Fig. 10.3).

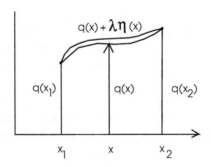

Fig. 10.3   Curve $q(x)$ that minimizes the functional S, and small variation about that curve $q(x) + \lambda \eta(x)$ where λ is an infinitesimal and $\eta(x)$ is arbitrary except for the fixed endpoints $\eta(x_1) = \eta(x_2) = 0$.

It follows that $q'(x) \to q'(x) + \lambda \eta'(x)$. Now make a Taylor series expansion in λ, and show that S takes the form

$$S(\lambda) = S_0 + \lambda \int_{x_1}^{x_2} \left[\eta(x) \left(\frac{\partial F}{\partial q} \right) + \frac{d\eta(x)}{dx} \left(\frac{\partial F}{\partial q'} \right) \right] dx + O(\lambda^2)$$

Here the partial derivatives indicate that all the other variables in $F(q, q'; x)$ are held fixed in carrying out the differentiation;

(b) If $q(x)$ is the solution to the problem, then any other curve must increase the value of S; hence $S(\lambda)$ must have a minimum at $\lambda = 0$. Show this condition becomes

$$\int_{x_1}^{x_2} \left[\eta(x) \left(\frac{\partial F}{\partial q} \right) + \frac{d\eta(x)}{dx} \left(\frac{\partial F}{\partial q'} \right) \right] dx = 0$$

Note that this condition only ensures that $S(\lambda)$ has an extremum at $\lambda = 0$ — minimum, maximum, or point of inflection; which it is can usually be argued from the physics of the problem;

(c) Carry out a partial integration on the second term using the condition of fixed endpoints, and show that the condition in part (b) becomes

$$\int_{x_1}^{x_2} \left[\left(\frac{\partial F}{\partial q} \right) - \frac{d}{dx} \left(\frac{\partial F}{\partial q'} \right) \right] \eta(x) \, dx = 0$$

(d) Since $\eta(x)$ is arbitrary, conclude that the curve $q(x)$ that minimizes the functional S is determined by the following second-order differential equation

$$\frac{d}{dx} \left(\frac{\partial F}{\partial q'} \right) - \left(\frac{\partial F}{\partial q} \right) = 0 \qquad ; \text{ Euler-Lagrange equation}$$

Since the solution is required to pass through the two endpoints, it is determined. This differential equation is known as the Euler-Lagrange equation for this variational problem.[2] Note that the d/dx here is a *total* derivative.

Solution to Problem 10.3

(a) Let $q(x)$ be a solution to the problem, and consider small variations around the curve of the form $q(x) \to q(x) + \lambda \eta(x)$. The functional S now reads

$$S[q + \lambda \eta, q' + \lambda \eta'] = \int_{x_1}^{x_2} F(q + \lambda \eta, q' + \lambda \eta'; x) \, dx$$

---

[2]See the section on "calculus of variations" in [Fetter and Walecka (2003)].

If we make a Taylor expansion to order λ, we obtain

$$S\left[q + \lambda\eta, q' + \lambda\eta'\right] = S\left[q, q'\right] + \lambda \int_{x_1}^{x_2} \left[\left(\frac{\partial F}{\partial q}\right)\eta(x) + \left(\frac{\partial F}{\partial q'}\right)\eta'(x)\right] dx$$
$$+ O\left(\lambda^2\right)$$

(b) Consider the functional S, expressed as a simple function of the single parameter λ. A Taylor expansion then gives

$$S(\lambda) = S(0) + \lambda S'(0) + \frac{1}{2}\lambda^2 S''(0) + \cdots$$

The condition that S have a minimum at $\lambda = 0$ requires that the coefficient of the term of order λ vanish

$$\left.\frac{dS}{d\lambda}\right|_{\lambda=0} = 0$$

More generally, this condition only implies that the function $S(\lambda)$ is *stationary* at the origin, having a minimum, maximum, or point of inflection there.[3] Which it is, can usually be argued from the physics of the problem.

Now apply this argument to the $S(\lambda)$ in part (a). The condition for a minimum (extremum) is

$$\int_{x_1}^{x_2} \left[\left(\frac{\partial F}{\partial q}\right)\eta(x) + \left(\frac{\partial F}{\partial q'}\right)\eta'(x)\right] dx = 0$$

(c) Let us use the condition obtained in part (b), and integrate by parts to obtain

$$\int_{x_1}^{x_2} \left\{\left(\frac{\partial F}{\partial q}\right)\eta(x) - \left[\frac{d}{dx}\left(\frac{\partial F}{\partial q'}\right)\right]\eta(x) + \frac{d}{dx}\left[\left(\frac{\partial F}{\partial q'}\right)\eta(x)\right]\right\} dx = 0$$

The last term in this expression clearly vanishes

$$\int_{x_1}^{x_2} \frac{d}{dx}\left[\left(\frac{\partial F}{\partial q'}\right)\eta(x)\right] dx = \left[\left(\frac{\partial F}{\partial q'}\right)\eta(x)\right]_{x_2} - \left[\left(\frac{\partial F}{\partial q'}\right)\eta(x)\right]_{x_1} = 0$$

since the endpoints are held fixed

$$\eta(x_2) = \eta(x_1) = 0$$

---

[3] We say that $S(\lambda)$ has an *extremum* at the origin.

We therefore obtain the following condition for finding an extremum of the functional S

$$\int_{x_1}^{x_2} \left[\left(\frac{\partial F}{\partial q} \right) - \frac{d}{dx} \left(\frac{\partial F}{\partial q'} \right) \right] \eta(x)\, dx = 0$$

(d) The integral condition obtained in part (c) must be fulfilled for any $\eta(x)$; however, the function $\eta(x)$, apart from vanishing at the endpoints, is completely arbitrary. The only possibility then is that the term multiplying $\eta(x)$ in the integrand must vanish identically

$$\frac{d}{dx} \left(\frac{\partial F}{\partial q'} \right) - \left(\frac{\partial F}{\partial q} \right) = 0 \qquad ;\ \text{Euler-Lagrange equation}$$

This is known as the *Euler-Lagrange equation* for this variational problem. In order to minimize the functional S (more generally, make it an extremum), the integrand $F(q, q'; x)$ must satisfy this differential equation along the path. This yields a second-order differential equation for $q(x)$, and since the solution is required to pass through the two endpoints, it is determined.

Problem 10.4 You are given a lagrangian $L(q, \dot{q}; t) = T - V$ for a particle with generalized coordinate $q(t)$ and kinetic and potential energies (T, V) respectively. Here $\dot{q} \equiv dq(t)/dt$. The lagrangian is a given function of the indicated variables (the last denotes any additional explicit t dependence). The *action* is defined by

$$S \equiv \int_{t_1}^{t_2} L(q, \dot{q}; t)\, dt \qquad ;\ \text{action}$$

Hamilton's principle states that the path the particle takes between the points $q(t_1)$ and $q(t_2)$ is the one that minimizes the action (more generally, makes it *stationary*).

(a) Use the results in Prob. 10.3 to derive *Lagrange's equation* for the subsequent motion

$$\frac{d}{dt} \left(\frac{\partial L}{\partial \dot{q}} \right) - \frac{\partial L}{\partial q} = 0 \qquad ;\ \text{Lagrange's equation}$$

(b) Show that if one has a problem with n dimensions, there will be one Lagrange equation for each linearly independent coordinate q_i with $i = 1, 2, \cdots, n$.

Solution to Problem 10.4

(a) *Hamilton's principle* in mechanics states that the path a particle takes will be one that minimizes the action (more generally, makes it stationary)

$$S \equiv \int_{t_1}^{t_2} L(q, \dot{q}; t)\, dt \qquad ; \text{ action}$$

Here t is the time, and the *lagrangian* is given by $L(q, \dot{q}; t) = T - V$ with $\dot{q} \equiv dq(t)/dt$. The Euler-Lagrange equation for this variational problem follows directly from the result of part (d) of Prob. 10.3, after making the substitutions $F \to L$ and $x \to t$

$$\frac{d}{dt}\left(\frac{\partial L}{\partial \dot{q}}\right) - \frac{\partial L}{\partial q} = 0 \qquad ; \text{ Lagrange's equation}$$

In mechanics, this second-order differential equation derived from the lagrangian is known as *Lagrange's equation* for the motion.

(b) Let us now consider the case in which there are n linearly independent coordinates q_i with $i = 1, 2, \cdots, n$. The action functional then reads

$$S = \int_{t_1}^{t_2} L(q_1, \ldots, q_n, \dot{q}_1, \ldots, \dot{q}_n; t)\, dt \qquad ; \text{ action}$$

Now consider small variations of each of the coordinates

$$q_i(t) \to q_i(t) + \lambda \eta_i(t) \qquad ; i = 1, \cdots, n$$

keeping the endpoints fixed

$$\eta_i(t_1) = \eta_i(t_2) = 0$$

Under these variations the functional changes to

$$S\left[q_i + \lambda \eta_i, \dot{q}_i + \lambda \dot{\eta}_i\right] = S\left[q_i, \dot{q}_i\right]$$
$$+ \lambda \int_{t_1}^{t_2} \sum_{i=1}^{n} \left[\left(\frac{\partial L}{\partial q_i}\right)\eta_i(t) + \left(\frac{\partial L}{\partial \dot{q}_i}\right)\dot{\eta}_i(t)\right] dt$$
$$+ O\left(\lambda^2\right)$$

After repeating the steps of parts (b) and (c) of Prob. 10.3, we finally obtain the condition

$$\int_{t_1}^{t_2} \sum_{i=1}^{n} \left[\left(\frac{\partial L}{\partial q_i}\right) - \frac{d}{dt}\left(\frac{\partial L}{\partial \dot{q}_i}\right)\right]\eta_i(t)\, dt = 0$$

Once again, this integral condition must be fulfilled for arbitrary, independent $\eta_i(t)$ with $i = 1, \cdots, n$, and therefore the only possibility is that the coefficient of each term $\eta_i(t)$ must vanish identically

$$\frac{\partial L}{\partial q_i} - \frac{d}{dt}\frac{\partial L}{\partial \dot{q}_i} = 0 \qquad ; i = 1, 2, \cdots, n$$

There is one Lagrange equation for each independent generalized coordinate $q_i(t)$ with $i = 1, \cdots, n$. This leads, quite generally, to a set of n coupled second-order differential equations for the motion.

Problem 10.5 You are given the lagrangian in cartesian coordinates for a non-relativistic particle moving in a potential $V(\mathbf{x})$

$$L = \frac{1}{2}m(\dot{x}^2 + \dot{y}^2 + \dot{z}^2) - V(\mathbf{x})$$

Use Lagrange's equations in Prob. 10.4 to derive the equations of motion. Show they are exactly those given by Newton's second law.

Solution to Problem 10.5

There are three coordinates $(q_1, q_2, q_3) = (x, y, z)$ in the problem, and Lagrange's equations are

$$\frac{d}{dt}\frac{\partial L}{\partial \dot{q}_i} = \frac{\partial L}{\partial q_i} \qquad ; i = 1, 2, 3$$

We observe from the structure of L that

$$\frac{\partial L}{\partial q_i} = -\frac{\partial V}{\partial q_i} \qquad ; i = 1, 2, 3$$

$$\frac{\partial L}{\partial \dot{q}_i} = m\dot{q}_i$$

Therefore Lagrange's equations here take the form

$$m\ddot{q}_i = -\frac{\partial V}{\partial q_i} \qquad ; i = 1, 2, 3$$

In vector notation, with $\mathbf{x} = (x, y, z)$, this reads

$$m\frac{d^2\mathbf{x}}{dt^2} = -\boldsymbol{\nabla}V$$

If we recall that $\mathbf{F} \equiv -\boldsymbol{\nabla}V$ is the force acting on the particle, then this equation is precisely Newton's second law of motion.

Hence Hamilton's principle and Lagrange's equations are seen to be fully equivalent to Newton's law, at least in this case.

Problem 10.6 The lifetime of a free muon is $\tau = 2.197 \times 10^{-6}$ sec. How close to the Schwarzschild radius, $r/R_S = 1 + \epsilon$, must a muon be to have its laboratory lifetime extended to 1 sec?

Solution to Problem 10.6

According to Eq. (10.32) of the book, the time dilation in the global laboratory frame at a radial distance r with the Schwarzschild metric is

$$dt = \frac{d\bar{t}}{\sqrt{1 - R_S/r}}$$

where $R_S = 2MG/c^2$ is the Schwarzschild radius. In this problem, $d\bar{t}$ is the lifetime of the free muon while dt is the desired laboratory lifetime

$$d\bar{t} = \tau = 2.197 \times 10^{-6} \text{ sec}$$
$$dt = 1 \text{ sec}$$

If we solve the above equation for r/R_S, we obtain

$$\frac{r}{R_S} = \frac{1}{1 - (d\bar{t}/dt)^2} \approx 1 + \left(\frac{d\bar{t}}{dt}\right)^2$$

The last approximation follows from the observation that $d\bar{t}/dt = 2.197 \times 10^{-6} \ll 1$. We thus conclude that

$$\frac{r}{R_S} = 1 + \epsilon \approx 1 + (2.197 \times 10^{-6})^2$$

Hence

$$\epsilon \approx 4.83 \times 10^{-12}$$

This is how close to the Schwarzschild radius one must be, $r/R_S = 1 + \epsilon$, in order to have the laboratory lifetime of the muon extended to one second.

Problem 10.7 Prove that in the Robertson-Walker metric with $k = 0$, the horizon recedes with a constant velocity $(v/c)_H = 2/3$.

Solution to Problem 10.7

From Eq. (10.55) of the book, the relative velocity of the horizon at the time t in the Robertson-Walker metric with $k = 0$ is given by

$$\frac{v_r}{c} = \frac{d\Lambda}{dt} \frac{l}{\Lambda} \frac{1}{c} \qquad ; \text{ velocity of horizon}$$

where l is the distance to the horizon

$$l = c(t - t_0) \qquad ; \text{ distance to horizon}$$

From the first of Eqs. (10.48), the Einstein field equations imply that the scale parameter in the metric satisfies

$$h(t) = \frac{1}{\Lambda(t)} \frac{d\Lambda(t)}{dt} = \frac{2}{3(t - t_0)} \qquad ; \text{ scale parameter}$$

A combination of these results gives

$$\left(\frac{v}{c}\right)_{\text{H}} = \frac{2}{3}$$

where this relation holds for all $t - t_0$.

Chapter 11

Quantum Fluids

Problem 11.1 Make use of Fig. 11.7, which gives the low-temperature excitation spectrum $\varepsilon_k(k)$ in He-II, to answer the following:

(a) What is the velocity of sound c_{sound} in m/s?

(b) What is the critical velocity v_{critical} in m/s?

Solution to Problem 11.1 The first thing to do here is to deal with the units, since the ordinate in Fig. 11.7 is ε_k/k_B in units of °K and the abscissa is k in units of Å^{-1}. With the aid of appendix L in the text, one has

$$1\text{Å} = 10^{-10}\,\text{m}$$

$$\frac{k_B}{\hbar} = 1.31 \times 10^{11}\,\frac{1}{\text{s-°K}} \times \left[\frac{10^{-10}\,\text{m}}{1\text{Å}}\right] = 13.1\,\frac{\text{m}}{\text{s}}\left[\frac{\text{Å}^{-1}}{\text{°K}}\right]$$

(a) The speed of sound in He-II is determined from the slope at the origin in Fig. 11.7

$$c_{\text{sound}} = \frac{\varepsilon}{\hbar k}$$

From the figure

$$\frac{\varepsilon/k_B}{k} \approx \frac{12}{0.8}\left[\frac{\text{°K}}{\text{Å}^{-1}}\right]$$

Hence, after multiplication by $k_B/\hbar$,

$$c_{\text{sound}} \approx 197\,\frac{\text{m}}{\text{s}}$$

(b) The critical velocity in He-II is determined by the minimum value

227

of this quantity

$$v_{\text{critical}} = \left(\frac{\varepsilon}{\hbar k}\right)_{\min}$$

This occurs at the roton minimum, and from the figure

$$\left(\frac{\varepsilon}{\hbar k}\right)_{\min} \approx \frac{10}{2} \left[\frac{°\text{K}}{\text{Å}^{-1}}\right]$$

Therefore

$$v_{\text{critical}} \approx 66 \, \frac{\text{m}}{\text{s}}$$

Problem 11.2 It is shown in [Fetter and Walecka (2003a)] that with the interaction in Eq. (11.15), and with the same set of approximations as used in the text, the excitation spectrum in the weakly interacting Bose gas is that of a sound wave with a speed of sound given by $c_{\text{sound}} = (n_0 g/m)^{1/2}$. Use the results in Prob. 11.1(a), and Eqs. (11.30) and (11.26), to estimate the vortex core size in He-II. Compare with the observed value given in the text.

Solution to Problem 11.2

The expression for ξ in Eq. (11.26) is

$$\xi = \left(\frac{\hbar^2}{2mgn_0}\right)^{1/2}$$

With the use of $c_{\text{sound}} = (n_0 g/m)^{1/2}$ as stated in the problem, this is rewritten as

$$\xi = \frac{1}{\sqrt{2}} \frac{\hbar}{m} \frac{1}{c_{\text{sound}}} \qquad ; \; m \approx 4m_p$$

Recall that here the mass is $m_{^4\text{He}} \approx 4m_p$. With the use of

$$\frac{\hbar}{m_p c} = 2.103 \times 10^{-16} \, \text{m} \qquad ; \; c = 2.998 \times 10^8 \, \frac{\text{m}}{\text{s}}$$

and the value of c_{sound} from Prob. 11.1(a)

$$c_{\text{sound}} \approx 197 \, \frac{\text{m}}{\text{s}}$$

one finds

$$\xi \approx 5.66 \times 10^{-11} \, \text{m} \approx 0.57 \, \text{Å}$$

As quoted in the text, the observed value of the vortex core size in He-II is given by

$$\xi_{\text{obs}} \approx 1 \text{ Å}$$

Problem 11.3 Write or obtain a program to integrate the dimensionless, non-linear Gross-Pitaevskii equation for the vortex in Eqs. (11.27) and (11.28), and reproduce Fig. 11.6.

Solution to Problem 11.3

We consider the Gross-Pitaevskii equation in its reduced form given in Eq. (11.27) of the book

$$\frac{d^2 f}{d\zeta^2} + \frac{1}{\zeta}\frac{df}{d\zeta} - \frac{1}{\zeta^2}f + f - f^3 = 0$$

and obeying the boundary conditions

$$f(\zeta) = C\zeta \qquad\qquad ; \zeta \to 0$$

and

$$f(\zeta) = 1 - \frac{1}{2\zeta^2} + \cdots \qquad ; \zeta \to \infty$$

As done before, we may solve this nonlinear differential equation numerically using any of a variety of existing programs. We start at a point ζ_0 very close to the origin, and integrate out to a large value of ζ.[1]

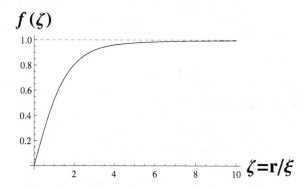

Fig. 11.1 Numerical integration of the Gross-Pitaevskii Eqs. (11.27) and (11.28) for a vortex with unit circulation.

---

[1] Although the combination $f'/\zeta - f/\zeta^2$ is finite as $\zeta \to 0$, each term is singular there.

Since the boundary conditions are imposed at different points, it is convenient to apply the shooting method to the problem. By suitably choosing the slope of $f(\zeta)$ at $\zeta = \zeta_0$, one can fulfill the second boundary condition at a second point $\zeta = \zeta_1$, with $\zeta_1 \gg 1$.

In our program we have used $\zeta_0 = 10^{-10}$ and we have found that $f'(\zeta_0) = C = 1.166 \cdots$ allows one to obtain a numerical solution up to $\zeta \approx 20$. The solution is plotted in Fig. 11.1. It reproduces Fig. 11.6 in the book.

Problem 11.4 Consider a slab of bulk superconductor with a surface in the $x = 0$ plane and occupying the half-space $x \geq 0$. Assume there is a static magnetic field which for $x \ll 0$ is $\mathbf{B} = B_0 \hat{\mathbf{e}}_z$. In this case, London augmented Maxwell's equations with an additional relation inside the superconductor which allows one to trace the magnetic field through the surface[2]

$$\boldsymbol{\nabla} \times \mathbf{j} = -\frac{1}{\mu_0 \lambda^2} \mathbf{B} \qquad ; \lambda^2 \equiv \frac{mc^2 \varepsilon_0}{n_s e^2}$$

Hence, when combined with Maxwell's equations, one has for static fields

$$\boldsymbol{\nabla} \times \mathbf{B} = \mu_0 \mathbf{j} \qquad ; \text{ Maxwell}$$

$$\boldsymbol{\nabla} \cdot \mathbf{B} = 0$$

$$\boldsymbol{\nabla} \times \mathbf{j} = -\frac{1}{\mu_0 \lambda^2} \mathbf{B} \qquad ; \text{ London}$$

Here $\mathbf{j} = 0$ for $x < 0$, and $\mathbf{E} = 0$ everywhere.

(a) Look for a solution to this set of equations of the form $\mathbf{B} = B(x)\hat{\mathbf{e}}_z$. Take the curl of the first equation and show that (*Hint:* see Prob. F.4)

$$\frac{d^2 B(x)}{dx^2} = 0 \qquad ; x < 0$$

$$\frac{d^2 B(x)}{dx^2} = \frac{1}{\lambda^2} B(x) \qquad ; x \geq 0$$

(b) Show the acceptable solution in (a) is

$$B(x) = B_0 \qquad ; x < 0$$

$$= B_0 e^{-x/\lambda} \qquad ; x \geq 0$$

Here λ is the *penetration depth*, with a typical value of $\lambda_{\text{exp}} = 510$ Å in superconducting tin.

---

[2]See [Fetter and Walecka (2003a)]; here n_s is the density of superconducting electrons. In the London gauge, this equation is written as $\mathbf{j} = -(1/\mu_0 \lambda^2)\mathbf{A}$ where $\mathbf{B} = \boldsymbol{\nabla} \times \mathbf{A}$.

(c) Show that for $x \geq 0$

$$\mathbf{j} = j(x)\hat{\mathbf{e}}_y \qquad\qquad ; j(x) = \frac{B_0}{\mu_0 \lambda} e^{-x/\lambda}$$

(d) Show that Maxwell's equations given above imply

$$\nabla \cdot \mathbf{B} = \nabla \cdot \mathbf{j} = 0$$

and that the obtained solutions satisfy these relations.

Solution to Problem 11.4

(a) We take the ith component of the curl of the l.h.s. of the first of Maxwell's equations

$$[\nabla \times (\nabla \times \mathbf{B})]_i = \sum_{j,k,l,m} \epsilon_{ijk}\epsilon_{klm} \frac{\partial}{\partial x_j} \frac{\partial}{\partial x_l} B_m$$

$$= \sum_{j,l,m} (\delta_{il}\delta_{jm} - \delta_{im}\delta_{jl}) \frac{\partial}{\partial x_j} \frac{\partial}{\partial x_l} B_m$$

$$= \frac{\partial}{\partial x_i}(\nabla \cdot \mathbf{B}) - \nabla^2 B_i = -\nabla^2 B_i$$

where we have used the property

$$\sum_{k=1}^{3} \epsilon_{ijk}\epsilon_{klm} = (\delta_{il}\delta_{jm} - \delta_{im}\delta_{jl})$$

The curl of the r.h.s. of the first of Maxwell's equations is, with the aid of London's equation,

$$\mu_0 \left(\nabla \times \mathbf{j}\right)_i = 0 \qquad\qquad ; x < 0$$

$$= -\frac{1}{\lambda^2} B_i \qquad ; x \geq 0$$

From these, one obtains the equations

$$\nabla^2 B_i = 0 \qquad\qquad ; x < 0$$

$$= \frac{1}{\lambda^2} B_i \qquad\qquad ; x \geq 0$$

If we look for a solution of the form $\mathbf{B} = B(x)\mathbf{e}_z$, we have[3]

$$\frac{d^2 B(x)}{dx^2} = 0 \qquad\qquad ; x < 0$$

$$\frac{d^2 B(x)}{dx^2} = \frac{1}{\lambda^2} B(x) \qquad\qquad ; x \geq 0$$

(b) It is readily established that the following $B(x)$ provides a solution that is continuous at $x = 0$

$$B(x) = B_0 \qquad\qquad ; x < 0$$
$$= B_0 e^{-x/\lambda} \qquad\qquad ; x \geq 0$$

Hence, when Maxwell's equations are augmented with London's equation, the magnetic field $\mathbf{B} = B(x)\mathbf{e}_z$ dies off exponentially over a distance λ, known as the *penetration depth*, as one moves into the superconductor. The fact that the magnetic field vanishes inside the bulk superconductor is known as the *Meissner effect*.

(c) We write the third component of London's equation as

$$(\boldsymbol{\nabla} \times \mathbf{j})_z = \sum_{k,l=1}^3 \epsilon_{3kl} \frac{\partial}{\partial x_k} j_l = \frac{\partial}{\partial x} j_y - \frac{\partial}{\partial y} j_x = -\frac{1}{\mu_0 \lambda^2} B_z$$

With the use of the previous solution for $\mathbf{B} = B(x)\mathbf{e}_z$, this becomes

$$\frac{\partial}{\partial x} j_y - \frac{\partial}{\partial y} j_x = -\frac{B_0}{\mu_0 \lambda^2} e^{-x/\lambda} \qquad ; x \geq 0$$

Since $\mathbf{j} = 0$ for $x < 0$, we continue to take $j_x = 0$ for $x \geq 0$, and this equation reduces to[4]

$$\frac{\partial}{\partial x} j_y = -\frac{B_0}{\mu_0 \lambda^2} e^{-x/\lambda}$$

The corresponding solution for the current $\mathbf{j}(x) = j(x)\mathbf{e}_y$ is therefore

$$j(x) = \frac{B_0}{\mu_0 \lambda} e^{-x/\lambda}$$

Thus, when Maxwell's equations are augmented with London's equation, there is a current sheet $\mathbf{j} = j(x)\mathbf{e}_y$ which also dies off exponentially over a distance λ as one moves into the bulk superconductor.

---

[3] We delete the hats on the unit vectors (see footnote on p. 289).

[4] Note from the first of Maxwell's equations $\boldsymbol{\nabla} \times [B(x)\mathbf{e}_z] = -[dB(x)/dx]\mathbf{e}_y = \mu_0 \mathbf{j}$, and, indeed, the current here has only a y-component.

(d) The divergence of the first of Maxwell's equations yields

$$\mu_0 \mathbf{\nabla} \cdot \mathbf{j} = [\mathbf{\nabla} \cdot (\mathbf{\nabla} \times \mathbf{B})] = 0$$

where we have used the property $\mathbf{\nabla} \cdot (\mathbf{\nabla} \times \mathbf{B}) = 0$. The second Maxwell's equation on the other hand states that $\mathbf{\nabla} \cdot \mathbf{B} = 0$, and therefore

$$\mathbf{\nabla} \cdot \mathbf{B} = \mathbf{\nabla} \cdot \mathbf{j} = 0$$

The solutions $\mathbf{B}$ and $\mathbf{j}$ calculated in parts (b) and (c) satisfy these equations since both $\mathbf{B}$ and $\mathbf{j}$ depend only on x, but do not have a component along x.

Problem 11.5 Carry out a partial integration on the expression for the gap Δ in Eq. (11.62) to isolate the singular part of the integral in Eq. (11.63). Make use of the fact that $\ln(t^2 - k_F^2)$ is integrable. Assume that $|\tilde{u}(t)|^2$ is well-behaved and that it vanishes sufficiently fast as $t \to \infty$.

Solution to Problem 11.5

First, carry out the integration over angles in Eq. (11.62)

$$\frac{1}{2\pi^2} \int_{k_F}^{\infty} t^2 dt \frac{|\tilde{u}(t)|^2}{\kappa_B^2 - t^2} = \frac{1}{\lambda}$$

The integral is re-written as

$$\frac{1}{2\pi^2} \int_{k_F}^{\infty} t^2 dt \frac{|\tilde{u}(t)|^2}{\kappa_B^2 - t^2} = -\frac{1}{4\pi^2} \int_{k_F}^{\infty} t dt \, |\tilde{u}(t)|^2 \frac{d}{dt} \log(t^2 - \kappa_B^2)$$

As suggested, this expression can be integrated by parts to obtain

$$\frac{1}{2\pi^2} \int_{k_F}^{\infty} t^2 dt \frac{|\tilde{u}(t)|^2}{\kappa_B^2 - t^2} = -\frac{1}{4\pi^2} \int_{k_F}^{\infty} dt \frac{d}{dt} \left\{ t|\tilde{u}(t)|^2 \log(t^2 - \kappa_B^2) \right\}$$

$$+ \frac{1}{4\pi^2} \int_{k_F}^{\infty} dt \, \log(t^2 - \kappa_B^2) \frac{d}{dt} \left[t|\tilde{u}(t)|^2 \right]$$

With the assumption that $|\tilde{u}(k)|^2$ vanishes sufficiently fast at large k, this becomes

$$\frac{1}{2\pi^2} \int_{k_F}^{\infty} t^2 dt \frac{|\tilde{u}(t)|^2}{\kappa_B^2 - t^2} = \frac{1}{4\pi^2} k_F |\tilde{u}(k_F)|^2 \log(k_F^2 - \kappa_B^2)$$

$$+ \frac{1}{4\pi^2} \int_{k_F}^{\infty} dt \, \log(t^2 - \kappa_B^2) \frac{d}{dt} \left[t|\tilde{u}(t)|^2 \right]$$

The last term is well-behaved for $\kappa_B^2 \to k_F^2$, since $\ln(t^2 - k_F^2)$ is integrable. Thus in this limit, only the singular term on the r.h.s. contributes, and Eq. (11.62) reduces to

$$\frac{k_F}{4\pi^2}|\tilde{u}(k_F)|^2 \log(k_F^2 - \kappa_B^2) = \frac{1}{\lambda} \qquad ; \ \kappa_B^2 \to k_F^2$$

Since $\lambda < 0$ in this discussion, and the addition of a constant term to the l.h.s. leaves the singularity unaffected, this expression can equally well be written

$$\frac{k_F}{4\pi^2}|\tilde{u}(k_F)|^2 \log\frac{k_F^2}{k_F^2 - \kappa_B^2} = \frac{1}{|\lambda|} \qquad ; \ \kappa_B^2 \to k_F^2$$

which is just Eq. (11.63).

Problem 11.6 Consider the ordinary solution to the eigenvalue Eq. (11.59) given in Eq. (11.61).[5] Show that it can be written as the expectation value of the potential in Eqs. (11.45) taken with the wave function in Eq. (11.46)

$$\Delta E(k) = \frac{\hbar^2}{2\mu}(\kappa^2 - k^2) = \int d^3x \, \phi_{\mathbf{k}}^\star(\mathbf{x})V(x)\phi_{\mathbf{k}}(\mathbf{x}) \Big/ \int d^3x \, \phi_{\mathbf{k}}^\star(\mathbf{x})\phi_{\mathbf{k}}(\mathbf{x})$$

where the potential is generalized to be non-local and separable as in Eqs. (11.53)–(11.55).

Solution to Problem 11.6

We want to show that for the ordinary eigenvalue in Eq. (11.61)

$$\kappa^2 - k^2 = \frac{2\mu}{\hbar^2}\Delta E(k) = \frac{2\mu}{\hbar^2}\langle \mathbf{k}|V|\mathbf{k}\rangle$$

where

$$\frac{2\mu}{\hbar^2}\langle \mathbf{k}|V|\mathbf{k}\rangle \equiv \int d^3x \, \phi_{\mathbf{k}}^\star(\mathbf{x})\lambda v(x)\phi_{\mathbf{k}}(\mathbf{x}) \Big/ \int d^3x \, \phi_{\mathbf{k}}^\star(\mathbf{x})\phi_{\mathbf{k}}(\mathbf{x})$$

The wave functions in Eqs. (11.46) are normalized, so the denominator in this expression is unity

$$\int d^3x \, \phi_{\mathbf{k}}^\star(\mathbf{x})\phi_{\mathbf{k}}(\mathbf{x}) = 1$$

---

[5] We now use an equality in Eq. (11.61); note that here λ can have either sign.

The potential in the numerator is generalized to be non-local and separable as in Eqs. (11.53)–(11.55). Therefore

$$\int d^3x\, \phi_{\mathbf{k}}^\star(\mathbf{x})\lambda v(x)\phi_{\mathbf{k}}(\mathbf{x}) \rightarrow \int d^3x \int d^3y\, \phi_{\mathbf{k}}^\star(\mathbf{x})\lambda v(\mathbf{x},\mathbf{y})\phi_{\mathbf{k}}(\mathbf{y})$$

$$\rightarrow \int d^3x \int d^3y\, \phi_{\mathbf{k}}^\star(\mathbf{x})\lambda u(x)u(y)\phi_{\mathbf{k}}(\mathbf{y})$$

$$= \frac{\lambda}{\Omega}|\tilde{u}(k)|^2$$

where from Eqs. (11.55)

$$\tilde{u}(k) \equiv \int d^3x\, e^{-i\mathbf{k}\cdot\mathbf{x}}u(x)$$

A combination of these results gives

$$\kappa^2 - k^2 = \frac{2\mu}{\hbar^2}\langle \mathbf{k}|V|\mathbf{k}\rangle = \frac{\lambda}{\Omega}|\tilde{u}(k)|^2$$

which is just Eq. (11.61).

Problem 11.7 The Schrödinger equation for a particle in a magnetic field follows from the hamiltonian in Eqs. (11.68) and the gauge-invariant replacement in Eq. (11.69)

$$i\hbar\frac{\partial\Psi}{\partial t} = \frac{1}{2m}(\mathbf{p} - e\mathbf{A})^2\Psi$$

Here the canonical momentum is given by $\mathbf{p} = (\hbar/i)\nabla$ and the vector potential $\mathbf{A}(\mathbf{x})$ is real. The probability density and current then become

$$\rho = |\Psi|^2$$

$$\mathbf{s} = \frac{1}{2m}\left\{\Psi^\star(\mathbf{p} - e\mathbf{A})\Psi + [(\mathbf{p} - e\mathbf{A})\Psi]^\star\Psi\right\}$$

(a) Show that the continuity equation given in Eqs. (4.111) is satisfied.

(b) Make use of Eqs. (9.50), and generalize this result to the case of an arbitrary electromagnetic field.

Solution to Problem 11.7

(a) We use the Schrödinger equation together with its complex conjugate

$$i\hbar\frac{\partial\Psi}{\partial t} = \frac{1}{2m}(\mathbf{p} - e\mathbf{A})^2\Psi$$

$$-i\hbar\frac{\partial\Psi^\star}{\partial t} = \frac{1}{2m}\left[(\mathbf{p} - e\mathbf{A})^2\Psi\right]^\star$$

If we multiply the first equation by $\Psi^\star$, the second equation by Ψ, and take the difference, we obtain

$$i\hbar\frac{\partial}{\partial t}|\Psi|^2 = \frac{1}{2m}\left\{\Psi^\star(\mathbf{p}-e\mathbf{A})^2\Psi - [(\mathbf{p}-e\mathbf{A})^2\Psi]^\star\,\Psi\right\}$$

We now need to check that the r.h.s. can be written a divergence. We have

$$\begin{aligned}
\text{r.h.s.} &= \frac{1}{2m}\left\{\Psi^\star(\mathbf{p}-e\mathbf{A})^2\Psi - [(\mathbf{p}-e\mathbf{A})^2\Psi]^\star\,\Psi\right\} \\
&= \frac{\hbar}{2mi}\left\{\Psi^\star\boldsymbol{\nabla}\cdot(\mathbf{p}-e\mathbf{A})\Psi + [\boldsymbol{\nabla}\cdot(\mathbf{p}-e\mathbf{A})\Psi]^\star\,\Psi\right\} \\
&\quad -\frac{e}{2m}\left\{\Psi^\star\mathbf{A}\cdot(\mathbf{p}-e\mathbf{A})\Psi - [\mathbf{A}\cdot(\mathbf{p}-e\mathbf{A})\Psi]^\star\,\Psi\right\} \\
&\equiv (A) + (B)
\end{aligned}$$

Let us first separately simplify the two parts of r.h.s.

$$\begin{aligned}
(A) &= \frac{\hbar}{2mi}\left\{\Psi^\star\boldsymbol{\nabla}\cdot(\mathbf{p}-e\mathbf{A})\Psi + [\boldsymbol{\nabla}\cdot(\mathbf{p}-e\mathbf{A})\Psi]^\star\,\Psi\right\} \\
&= \frac{\hbar}{2mi}\boldsymbol{\nabla}\cdot\left\{\Psi^\star(\mathbf{p}-e\mathbf{A})\Psi + [(\mathbf{p}-e\mathbf{A})\Psi]^\star\,\Psi\right\} \\
&\quad -\frac{\hbar}{2mi}\left\{\boldsymbol{\nabla}\Psi^\star\cdot(\mathbf{p}-e\mathbf{A})\Psi + [(\mathbf{p}-e\mathbf{A})\Psi]^\star\cdot\boldsymbol{\nabla}\Psi\right\} \\
&= \frac{\hbar}{2mi}\boldsymbol{\nabla}\cdot\left\{\Psi^\star(\mathbf{p}-e\mathbf{A})\Psi + [(\mathbf{p}-e\mathbf{A})\Psi]^\star\,\Psi\right\} \\
&\quad +\frac{e\hbar}{2mi}\left(\boldsymbol{\nabla}\Psi^\star\cdot\mathbf{A}\Psi + \Psi^\star\mathbf{A}\cdot\boldsymbol{\nabla}\Psi\right)
\end{aligned}$$

Furthermore

$$\begin{aligned}
(B) &= -\frac{e}{2m}\left\{\Psi^\star\mathbf{A}\cdot(\mathbf{p}-e\mathbf{A})\Psi - [\mathbf{A}\cdot(\mathbf{p}-e\mathbf{A})\Psi]^\star\,\Psi\right\} \\
&= -\frac{e\hbar}{2mi}\left[\Psi^\star\mathbf{A}\cdot\boldsymbol{\nabla}\Psi + \boldsymbol{\nabla}\Psi^\star\cdot\mathbf{A}\Psi\right]
\end{aligned}$$

As a result we have

$$\begin{aligned}
\text{r.h.s.} &= (A) + (B) \\
&= \frac{\hbar}{2mi}\boldsymbol{\nabla}\cdot\left\{\Psi^\star(\mathbf{p}-e\mathbf{A})\Psi + [(\mathbf{p}-e\mathbf{A})\Psi]^\star\,\Psi\right\}
\end{aligned}$$

Thus we obtain the continuity equation for a particle in a magnetic field

$$\frac{\partial\rho}{\partial t} + \boldsymbol{\nabla}\cdot\mathbf{s} = 0$$

where

$$\rho = |\Psi|^2$$

$$\mathbf{s} = \frac{1}{2m} \left\{ \Psi^*(\mathbf{p} - e\mathbf{A})\Psi + [(\mathbf{p} - e\mathbf{A})\Psi]^*\Psi \right\}$$

(b) In the case of an arbitrary electromagnetic field, the hamiltonian gets modified according to Eqs. (9.50), and therefore the Schrödinger equation reads in this case

$$i\hbar \frac{\partial \Psi}{\partial t} = \left[\frac{1}{2m}(\mathbf{p} - e\mathbf{A})^2 + e\Phi \right] \Psi$$

If we repeat the procedure followed in part (a), we see that the extra term does not modify the continuity equation, since the term $e\Psi^*\Phi\Psi$ obtained by multiplying the Schrödinger equation on the left by Ψ^* is real, and therefore exactly cancels when the complex conjugate equation is multiplied by Ψ and subtracted.

Chapter 12

Quantum Fields

Problem 12.1 Retain the next term in the expansion in Eq. (12.1), and treat the resulting change in V as a perturbation. Discuss the processes it describes when the quantum field expression in Eq. (12.30) is substituted into it.

Solution to Problem 12.1

We write the length ds of the stretched element, retaining the next term in the expansion in Eq. (12.1) which serves as a non-linear correction to the leading result

$$ds \approx dx \left\{ 1 + \frac{1}{2} \left[\frac{\partial q(x,t)}{\partial x} \right]^2 - \frac{1}{8} \left[\frac{\partial q(x,t)}{\partial x} \right]^4 + \cdots \right\}$$

The potential energy stored in an element dx of the string is then given by

$$dW \approx \frac{\tau}{2} \left[\frac{\partial q(x,t)}{\partial x} \right]^2 dx - \frac{\tau}{8} \left[\frac{\partial q(x,t)}{\partial x} \right]^4 dx$$

The total potential energy is obtained by integrating over the extension of the string

$$V \approx \frac{\tau}{2} \int_0^L \left[\frac{\partial q(x,t)}{\partial x} \right]^2 dx - \frac{\tau}{8} \int_0^L \left[\frac{\partial q(x,t)}{\partial x} \right]^4 dx$$

We may treat the last term in this expression as a perturbation

$$H' = -\frac{\tau}{8} \int_0^L \left[\frac{\partial q(x,t)}{\partial x} \right]^4 dx$$

239

Since this is a perturbation, it can be evaluated using the unperturbed field expression of Eq. (12.30)

$$q(x, t) = \sum_k \left(\frac{\hbar}{2\omega_k \sigma L}\right)^{1/2} \left[a_k e^{i(kx - \omega_k t)} + a_k^\star e^{-i(kx - \omega_k t)}\right]$$

When quantized, the normal-mode amplitudes become creation and destruction operators[1]

$$a_k \to \hat{a}_k \qquad \text{; destruction operator}$$
$$a_k^\star \to \hat{a}_k^\dagger \qquad \text{; creation operator}$$

We see that there are $2^4 = 16$ terms which originate from the product of four fields. We may group these terms into different categories:

- terms which contain four destruction operators (one term);
- terms which contain three destruction operators and one creation operator (four terms);
- terms which contain two destruction operators and two creation operators (six terms);
- terms which contain one destruction operator and three creation operator (four terms);
- terms which contain four creation operators (one term).

At each vertex the integration over x leads to a delta function in the momentum, and therefore to the conservation of momentum. For example

$$\frac{1}{L} \int_0^L dx \, e^{i(k_1 + k_2 - k_3 - k_4)x} = \delta_{k_1 + k_2, \, k_3 + k_4}$$

To be explicit, we write out the last term above

$$H'(\hat{a}^\dagger \hat{a}^\dagger \hat{a}^\dagger \hat{a}^\dagger) = -\frac{\tau}{8} \int_0^L dx \sum_{k_1, k_2, k_3, k_4} \left(\frac{\hbar}{2\sigma L}\right)^2 \frac{(-i)^4 k_1 k_2 k_3 k_4}{(\omega_{k_1} \omega_{k_2} \omega_{k_3} \omega_{k_4})^{1/2}} \times$$

$$\hat{a}_{k_1}^\dagger e^{-i(k_1 x - \omega_{k_1} t)} \hat{a}_{k_2}^\dagger e^{-i(k_2 x - \omega_{k_2} t)} \hat{a}_{k_3}^\dagger e^{-i(k_3 x - \omega_{k_3} t)} \hat{a}_{k_4}^\dagger e^{-i(k_4 x - \omega_{k_4} t)}$$

Physically, these terms allow the transition from a given state to another given state which can be connected by the perturbation. For instance, the term of the final category is responsible for the creation of a state containing four phonons from the vacuum (that is, the state with no phonons).

---

[1] More generally, the kinetic momentum density, and hence the canonical commutation relations, are unchanged in the presence of H', and hence this quantum field expansion is still valid. We here use a caret to distinguish the operators.

Problem 12.2 (a) Show

$$\int_\Omega d^3x\, q_{\mathbf{k}}(\mathbf{x},t) q_{\mathbf{k}'}(\mathbf{x},t) = \delta_{\mathbf{k}',-\mathbf{k}}\, e^{-2i\omega_k t}$$

$$[\mathbf{k}\times\mathbf{e}_{\mathbf{k}s}]\cdot[\mathbf{k}\times\mathbf{e}_{\mathbf{k}s'}] = \mathbf{k}^2\,\mathbf{e}_{\mathbf{k}s}\cdot\mathbf{e}_{\mathbf{k}s'}$$

where $q_{\mathbf{k}}(\mathbf{x},t)$ is the normal mode in Eq. (12.39);

(b) Now substitute the expression for the vector potential in Eq. (12.42) into the electromagnetic field energy in Eqs. (12.36) and (12.38) to obtain the normal-mode expansion in Eq. (12.46).

Solution to Problem 12.2

(a) We have for the first relation (remember $\omega_k = |\mathbf{k}|c$)

$$\int_\Omega d^3x\, q_{\mathbf{k}}(\mathbf{x},t) q_{\mathbf{k}'}(\mathbf{x},t) = \frac{e^{-i(\omega_k+\omega_{k'})t}}{\Omega}\int_\Omega d^3x\, e^{i(\mathbf{k}+\mathbf{k}')\cdot x} = \delta_{\mathbf{k}',-\mathbf{k}}\, e^{-2i\omega_k t}$$

Let us now prove the second relation

$$[\mathbf{k}\times\mathbf{e}_{\mathbf{k}s}]\cdot[\mathbf{k}\times\mathbf{e}_{\mathbf{k}s'}] = \sum_j\left[\sum_{lm}\varepsilon_{jlm}\,k_l\,(\mathbf{e}_{\mathbf{k}s})_m\right]\left[\sum_{pq}\varepsilon_{jpq}\,k_p\,(\mathbf{e}_{\mathbf{k}s'})_q\right]$$

$$= \sum_{lmpq}(\delta_{lp}\delta_{mq}-\delta_{lq}\delta_{mp})\,k_l\,(\mathbf{e}_{\mathbf{k}s})_m\,k_p\,(\mathbf{e}_{\mathbf{k}s'})_q$$

where we have again used the identity

$$\sum_j\varepsilon_{jlm}\varepsilon_{jpq} = (\delta_{lp}\delta_{mq}-\delta_{lq}\delta_{mp})$$

It follows that[2]

$$[\mathbf{k}\times\mathbf{e}_{\mathbf{k}s}]\cdot[\mathbf{k}\times\mathbf{e}_{\mathbf{k}s'}] = \mathbf{k}^2(\mathbf{e}_{\mathbf{k}s}\cdot\mathbf{e}_{\mathbf{k}s'}) - (\mathbf{k}\cdot\mathbf{e}_{\mathbf{k}s})(\mathbf{k}\cdot\mathbf{e}_{\mathbf{k}s'}) = \mathbf{k}^2\,\delta_{ss'}$$

(b) With the use of Eqs. (12.38) and (12.42), the fields $(\mathbf{E},\mathbf{B})$ are expressed in terms of the normal-mode expansion of the vector potential $\mathbf{A}(\mathbf{x},t)$ as

$$\mathbf{E} = i\sum_{\mathbf{k}}\sum_{s=1}^2\left(\frac{\hbar\omega_k}{2\varepsilon_0\Omega}\right)^{1/2}\mathbf{e}_{\mathbf{k}s}\left[a_{\mathbf{k}s}e^{i(\mathbf{k}\cdot\mathbf{x}-\omega_k t)} - a_{\mathbf{k}s}^*e^{-i(\mathbf{k}\cdot\mathbf{x}-\omega_k t)}\right]$$

$$\mathbf{B} = i\sum_{\mathbf{k}}\sum_{s=1}^2\left(\frac{\hbar}{2\omega_k\varepsilon_0\Omega}\right)^{1/2}(\mathbf{k}\times\mathbf{e}_{\mathbf{k}s})\left[a_{\mathbf{k}s}e^{i(\mathbf{k}\cdot\mathbf{x}-\omega_k t)} - a_{\mathbf{k}s}^*e^{-i(\mathbf{k}\cdot\mathbf{x}-\omega_k t)}\right]$$

---

[2]Recall that here $(\mathbf{k}\cdot\mathbf{e}_{\mathbf{k}s}) = 0$ for $s = (1,2)$.

We will now substitute these expressions in the relation for the field energy in Eq. (12.36)

$$E = \frac{1}{2} \int_\Omega d^3x \left(\varepsilon_0 \mathbf{E}^2 + \frac{1}{\mu_0} \mathbf{B}^2 \right)$$

and use the orthonormality of the plane waves

$$\frac{1}{\Omega} \int_\Omega d^3x \, e^{i(\mathbf{k}-\mathbf{k}')\cdot\mathbf{x}} = \delta_{\mathbf{k},\mathbf{k}'}$$

Consider first the term containing the contribution of the electric field

$$\frac{\varepsilon_0}{2} \int_\Omega d^3x \, \mathbf{E}^2 = \frac{1}{4} \sum_{\mathbf{k},s} \sum_{\mathbf{k}',s'} \hbar\omega_k \left[\delta_{\mathbf{k},\mathbf{k}'}\delta_{ss'}(a^*_{\mathbf{k}s}a_{\mathbf{k}s} + a_{\mathbf{k}s}a^*_{\mathbf{k}s}) \right.$$
$$\left. - \delta_{\mathbf{k},-\mathbf{k}'}(\mathbf{e}_{\mathbf{k}s}\cdot\mathbf{e}_{-\mathbf{k},s'})\left(a_{\mathbf{k}s}a_{-\mathbf{k},s'}\, e^{-2i\omega_k t} + a^*_{\mathbf{k}s}a^*_{-\mathbf{k},s'}\, e^{2i\omega_k t}\right) \right]$$

Now consider the term containing the contribution of the magnetic field

$$\frac{1}{2\mu_0} \int_\Omega d^3x \, \mathbf{B}^2 = \frac{1}{2\mu_0} \sum_{\mathbf{k},s} \sum_{\mathbf{k}',s'} \frac{\hbar}{2\omega_k\varepsilon_0} \left[\delta_{\mathbf{k},\mathbf{k}'}\mathbf{k}^2\delta_{ss'}(a^*_{\mathbf{k}s}a_{\mathbf{k}s} + a_{\mathbf{k}s}a^*_{\mathbf{k}s}) \right.$$
$$\left. - \delta_{\mathbf{k},-\mathbf{k}'}(\mathbf{k}\times\mathbf{e}_{\mathbf{k}s})\cdot(\mathbf{k}'\times\mathbf{e}_{\mathbf{k}'s'})\left(a_{\mathbf{k}s}a_{-\mathbf{k},s'}\, e^{-2i\omega_k t} + a^*_{\mathbf{k}s}a^*_{-\mathbf{k},s'}\, e^{2i\omega_{k'} t}\right) \right]$$

Use the previous vector identity again in the form[3]

$$(\mathbf{k}\times\mathbf{e}_{\mathbf{k}s})\cdot(\mathbf{k}\times\mathbf{e}_{-\mathbf{k},s'}) = \mathbf{k}^2(\mathbf{e}_{\mathbf{k}s}\cdot\mathbf{e}_{-\mathbf{k},s'})$$

Recall $1/\mu_0\varepsilon_0 = c^2$ and $\omega_k^2 = \mathbf{k}^2 c^2$, so that

$$\frac{\mathbf{k}^2}{\mu_0\varepsilon_0\omega_k} = \frac{\mathbf{k}^2 c^2}{\omega_k} = \omega_k$$

The magnetic contribution then takes the form

$$\frac{1}{2\mu_0} \int_\Omega d^3x \, \mathbf{B}^2 = \frac{1}{4} \sum_{\mathbf{k},s} \sum_{\mathbf{k}',s'} \hbar\omega_k \left[\delta_{\mathbf{k},\mathbf{k}'}\delta_{ss'}(a^*_{\mathbf{k}s}a_{\mathbf{k}s} + a_{\mathbf{k}s}a^*_{\mathbf{k}s}) \right.$$
$$\left. + \delta_{\mathbf{k},-\mathbf{k}'}(\mathbf{e}_{\mathbf{k}s}\cdot\mathbf{e}_{-\mathbf{k},s'})\left(a_{\mathbf{k}s}a_{-\mathbf{k},s'}\, e^{-2i\omega_k t} + a^*_{\mathbf{k}s}a^*_{-\mathbf{k},s'}\, e^{2i\omega_k t}\right) \right]$$

When the electric and magnetic contributions are combined, the final time-dependent terms cancel identically, and the energy in the field is then indeed given by the normal-mode expansion in Eq. (12.46)

$$E = \frac{1}{2} \sum_{\mathbf{k}} \sum_{s=1}^{2} \hbar\omega_k \left(a^*_{\mathbf{k}s}a_{\mathbf{k}s} + a_{\mathbf{k}s}a^*_{\mathbf{k}s} \right)$$

---

[3]Note that $(\mathbf{k}'\times\mathbf{e}_{\mathbf{k}'s'}) \to -(\mathbf{k}\times\mathbf{e}_{-\mathbf{k},s'})$; note also that $(\mathbf{k}\cdot\mathbf{e}_{-\mathbf{k}s'}) = 0$ for $s' = (1,2)$.

Problem 12.3 Consider a Dirac particle bound in some potential making a transition from a state with wave function ψ_i and energy ε_i to a state with ψ_f and ε_f, while emitting a photon of polarization $\mathbf{e_{ks}}$ and energy $\hbar\omega_k$. The transition rate can be calculated from Eqs. (I.35) and (I.32) through the use of the matrix element developed in Eq. (12.50). Show this radiative transition rate is

$$d\omega_{fi} = \frac{\alpha\omega_k}{2\pi} \left| \int_\Omega d^3x\, e^{-i\mathbf{k}\cdot\mathbf{x}} \mathbf{e_{ks}} \cdot \bar{\psi}_f \gamma \psi_i \right|^2 d\Omega_k \qquad \text{; photoemission}$$

where $\varepsilon_i = \varepsilon_f + \hbar\omega$, and α is the fine-structure constant.

Solution to Problem 12.3

With the use of Eqs. (I.35) and (I.32), we may write the radiative transition rate as

$$d\omega_{fi} = \frac{2\pi}{\hbar} |\langle\psi_f|H_I|\psi_i\rangle|^2\, \delta(\varepsilon_i - \varepsilon_f - \hbar\omega_k)\frac{\Omega d^3k}{(2\pi)^3}$$

where the matrix element is given in Eq. (12.50)

$$\langle\psi_f|H_I|\psi_i\rangle = -ie\left(\frac{\hbar c^2}{2\omega_k\varepsilon_0\Omega}\right)^{1/2} \int_\Omega d^3x\, e^{-i\mathbf{k}\cdot\mathbf{x}} \mathbf{e_{ks}} \cdot \bar{\psi}_f \gamma \psi_i$$

The Dirac delta function expresses the conservation of energy in the process, and it allows one to integrate over the momentum of the photon using $d^3k = k^2 dk\, d\Omega_k$ and $\partial k/\partial\hbar\omega_k = 1/\hbar c$

$$d\omega_{fi} = \frac{2\pi}{\hbar} |\langle\psi_f|H_I|\psi_i\rangle|^2\, \frac{\Omega}{8\pi^3\hbar c}\frac{\omega_k^2}{c^2}d\Omega_k$$

Here we have used the property of the Dirac delta function

$$\int dx\, \delta(ax)f(x) = \frac{1}{|a|}f(a)$$

With the insertion of the explicit expression for the matrix element, and the identification of the fine-structure constant as $\alpha = e^2/4\pi\hbar c\varepsilon_0$, we thus obtain the stated result

$$d\omega_{fi} = \frac{\alpha\omega_k}{2\pi} \left| \int_\Omega d^3x\, e^{-i\mathbf{k}\cdot\mathbf{x}} \mathbf{e_{ks}} \cdot \bar{\psi}_f \gamma \psi_i \right|^2 d\Omega_k$$

Problem 12.4 (a) The continuity equation in non-relativistic quantum mechanics is given in Eq. (4.111). Use the argument in Eq. (9.42) to show the corresponding electromagnetic current four-vector ej_μ is obtained from

$$\frac{e}{c}\mathbf{j} = \frac{e\hbar}{2im_0c}\left[\Psi^\star\boldsymbol{\nabla}\Psi - (\boldsymbol{\nabla}\Psi)^\star\,\Psi\right] \qquad ; \text{E-M current}$$
$$e\rho = e\Psi^\star\Psi$$

(b) Show the radiative transition rate in Prob. 12.3 now becomes

$$d\omega_{fi} = \frac{\alpha\omega_k}{2\pi}\left|\frac{\hbar}{2m_0c}\int_\Omega d^3x\,e^{-i\mathbf{k}\cdot\mathbf{x}}\mathbf{e}_{ks}\cdot\left[\psi_f^\star\boldsymbol{\nabla}\psi_i - (\boldsymbol{\nabla}\psi_f)^\star\,\psi_i\right]\right|^2 d\Omega_k$$

(c) Use the result in Prob. 9.2(c) to show that this is what is obtained with a non-relativistic reduction of the Dirac equation.

Solution to Problem 12.4

(a) The non-relativistic probability current and probability density are given in Eqs. (4.111), and we can then use Eqs. (9.42) to obtain the corresponding electromagnetic current and charge density as

$$\frac{e}{c}\mathbf{j} = \frac{e\hbar}{2im_0c}\left[\Psi^\star\boldsymbol{\nabla}\Psi - (\boldsymbol{\nabla}\Psi)^\star\,\Psi\right]$$
$$e\rho = e\Psi^\star\Psi$$

(b) The radiative transition rate of Prob. 12.3 contains the relativistic current $(e/c)\mathbf{j} = ie\bar\psi_f\boldsymbol{\gamma}\psi_i$. We may substitute the current found in part (a), with the factor of e removed, to obtain the corresponding non-relativistic result

$$d\omega_{fi} = \frac{\alpha\omega_k}{2\pi}\left|\frac{\hbar}{2m_0c}\int_\Omega d^3x\,e^{-i\mathbf{k}\cdot\mathbf{x}}\mathbf{e}_{ks}\cdot\left[\psi_f^\star\boldsymbol{\nabla}\psi_i - (\boldsymbol{\nabla}\psi_f)^\star\,\psi_i\right]\right|^2 d\Omega_k$$

(c) In Prob. 9.2(c) we found the non-relativistic reduction of the Dirac equation. The second term appearing in the non-relativistic hamiltonian is

$$H_I = -\frac{e}{2m_0}\left(\mathbf{A}\cdot\mathbf{p} + \mathbf{p}\cdot\mathbf{A}\right)$$

With ψ_i and ψ_f the wave functions of the initial and final states, the matrix element of this hamiltonian is

$$\langle f|H_I|i\rangle = -\frac{e}{2m_0}\int d^3x\,\psi_f^\star\left(\mathbf{A}\cdot\mathbf{p} + \mathbf{p}\cdot\mathbf{A}\right)\psi_i$$

where $\mathbf{p} = (\hbar/i)\boldsymbol{\nabla}$. We may integrate the second term by parts, neglecting the surface term due to the vanishing of the wave functions at infinity, and we thus obtain

$$\langle f|H_I|i\rangle = -\frac{e}{2m_0} \int d^3x \; \psi_f^* \left(\mathbf{A} \cdot \mathbf{p} + \mathbf{p} \cdot \mathbf{A}\right) \psi_i$$

$$= -\frac{e\hbar}{2im_0} \int d^3x \, \mathbf{A} \cdot \left[\psi_f^* \boldsymbol{\nabla}\psi_i - (\boldsymbol{\nabla}\psi_f)^* \psi_i\right]$$

A comparison with Eqs. (12.61) then allows us to identify the non-relativistic transition electromagnetic current as

$$\frac{e}{c}\mathbf{j} = \frac{e\hbar}{2im_0 c} \left[\psi_f^* \boldsymbol{\nabla}\psi_i - (\boldsymbol{\nabla}\psi_f)^* \psi_i\right]$$

Problem 12.5 Substitute the expansion of the Dirac field in Eq. (12.57) into Eq. (12.59) to obtain the normal-mode expression for the hamiltonian of the free Dirac field in Eq. (12.60). Make use of the Dirac equation satisfied by the free Dirac spinors and the orthonormality relations in Prob. 9.7.

Solution to Problem 12.5

Let us use the expansion of the Dirac field in Eq. (12.57) to calculate

$$\left(c\boldsymbol{\alpha} \cdot \mathbf{p} + \beta m_0 c^2\right)\hat{\psi} = \frac{1}{\sqrt{\Omega}} \sum_{\mathbf{k}} \sum_{\lambda} \left[a_{\mathbf{k}\lambda} \left(\hbar c\boldsymbol{\alpha} \cdot \mathbf{k} + \beta m_0 c^2\right) u_\lambda(\mathbf{k}) e^{i(\mathbf{k}\cdot\mathbf{x}-\omega_k t)} \right.$$

$$\left. + b_{\mathbf{k}\lambda}^\dagger \left(-\hbar c\boldsymbol{\alpha} \cdot \mathbf{k} + \beta m_0 c^2\right) v_\lambda(-\mathbf{k}) e^{-i(\mathbf{k}\cdot\mathbf{x}-\omega_k t)}\right]$$

The spinors u and v satisfy the Dirac equation (see Prob. 9.6)

$$\left(\hbar c\boldsymbol{\alpha} \cdot \mathbf{k} + \beta m_0 c^2\right) u_\lambda(\mathbf{k}) = \hbar\omega_k u_\lambda(\mathbf{k})$$
$$\left(-\hbar c\boldsymbol{\alpha} \cdot \mathbf{k} + \beta m_0 c^2\right) v_\lambda(-\mathbf{k}) = -\hbar\omega_k v_\lambda(-\mathbf{k})$$

Therefore

$$\left(c\boldsymbol{\alpha} \cdot \mathbf{p} + \beta m_0 c^2\right)\hat{\psi} = \frac{1}{\sqrt{\Omega}} \sum_{\mathbf{k}} \sum_{\lambda} \hbar\omega_k \left[a_{\mathbf{k}\lambda} u_\lambda(\mathbf{k}) e^{i(\mathbf{k}\cdot\mathbf{x}-\omega_k t)}\right.$$

$$\left. - b_{\mathbf{k}\lambda}^\dagger v_\lambda(-\mathbf{k}) e^{-i(\mathbf{k}\cdot\mathbf{x}-\omega_k t)}\right]$$

We will now use this expression in the unperturbed hamiltonian in Eq. (12.59)

$$\hat{H}_0 = \int_\Omega d^3x \, \hat{\psi}^\dagger \left[c\boldsymbol{\alpha} \cdot \mathbf{p} + \beta m_0 c^2\right] \hat{\psi}$$

Insertion of the above results gives

$$\hat{H}_0 = \frac{1}{\Omega} \int_\Omega d^3x \sum_{\mathbf{k},\mathbf{k}'} \sum_{\lambda,\lambda'} \left[a^\dagger_{\mathbf{k}\lambda} u^\dagger_\lambda(\mathbf{k}) e^{-i(\mathbf{k}\cdot\mathbf{x}-\omega_k t)} + b_{\mathbf{k}\lambda} v^\dagger_\lambda(-\mathbf{k}) e^{i(\mathbf{k}\cdot\mathbf{x}-\omega_k t)} \right] \times$$

$$\hbar\omega_{k'} \left[a_{\mathbf{k}'\lambda'} u_{\lambda'}(\mathbf{k}') e^{i(\mathbf{k}'\cdot\mathbf{x}-\omega_{k'} t)} - b^\dagger_{\mathbf{k}'\lambda'} v_{\lambda'}(-\mathbf{k}') e^{-i(\mathbf{k}'\cdot\mathbf{x}-\omega_{k'} t)} \right]$$

The integration over $\mathbf{x}$ is immediately carried out to give

$$\hat{H}_0 = \sum_{\mathbf{k},\mathbf{k}'\lambda\lambda'} \hbar\omega_{k'} \left[a^\dagger_{\mathbf{k}\lambda} a_{\mathbf{k}'\lambda'} u^\dagger_\lambda(\mathbf{k}) u_{\lambda'}(\mathbf{k}') \delta_{\mathbf{k},\mathbf{k}'} - b_{\mathbf{k}\lambda} b^\dagger_{\mathbf{k}'\lambda'} v^\dagger_\lambda(-\mathbf{k}) v_{\lambda'}(-\mathbf{k}') \delta_{\mathbf{k},\mathbf{k}'} \right.$$

$$\left. - a^\dagger_{\mathbf{k}\lambda} b^\dagger_{\mathbf{k}'\lambda'} u^\dagger_\lambda(\mathbf{k}) v_{\lambda'}(-\mathbf{k}') \delta_{\mathbf{k},-\mathbf{k}'} e^{2i\omega_k t} + b_{\mathbf{k}\lambda} a_{\mathbf{k}'\lambda'} v^\dagger_\lambda(-\mathbf{k}) u_{\lambda'}(\mathbf{k}') \delta_{\mathbf{k},-\mathbf{k}'} e^{-2i\omega_k t} \right]$$

which reduces to

$$\hat{H}_0 = \sum_{\mathbf{k}} \sum_{\lambda\lambda'} \hbar\omega_k \left[a^\dagger_{\mathbf{k}\lambda} a_{\mathbf{k}\lambda'} u^\dagger_\lambda(\mathbf{k}) u_{\lambda'}(\mathbf{k}) - b_{\mathbf{k}\lambda} b^\dagger_{\mathbf{k}\lambda'} v^\dagger_\lambda(-\mathbf{k}) v_{\lambda'}(-\mathbf{k}) \right.$$

$$\left. - a^\dagger_{\mathbf{k}\lambda} b^\dagger_{-\mathbf{k},\lambda'} u^\dagger_\lambda(\mathbf{k}) v_{\lambda'}(\mathbf{k}) e^{2i\omega_k t} + b_{\mathbf{k}\lambda} a_{-\mathbf{k},\lambda'} v^\dagger_\lambda(-\mathbf{k}) u_{\lambda'}(-\mathbf{k}) e^{-2i\omega_k t} \right]$$

This expression is now simplified using the orthonormality relations obtained in Prob. 9.7

$$u^\dagger_{\lambda'}(\mathbf{k}) u_\lambda(\mathbf{k}) = v^\dagger_{\lambda'}(-\mathbf{k}) v^\dagger_\lambda(-\mathbf{k}) = \delta_{\lambda'\lambda}$$

$$u^\dagger_{\lambda'}(\mathbf{k}) v_\lambda(\mathbf{k}) = v^\dagger_{\lambda'}(\mathbf{k}) u_\lambda(\mathbf{k}) = 0$$

which yields

$$\hat{H}_0 = \sum_{\mathbf{k}} \sum_{\lambda} \hbar\omega_k \left(a^\dagger_{\mathbf{k}\lambda} a_{\mathbf{k}\lambda} - b_{\mathbf{k}\lambda} b^\dagger_{\mathbf{k}\lambda} \right)$$

The anticommutation relations of the antiparticle creation and destruction operators in Eqs. (12.58) allow this to be re-written as Eq. (12.60)

$$\hat{H}_0 = \sum_{\mathbf{k}} \sum_{\lambda} \hbar\omega_k \left(a^\dagger_{\mathbf{k}\lambda} a_{\mathbf{k}\lambda} + b^\dagger_{\mathbf{k}\lambda} b_{\mathbf{k}\lambda} - 1 \right)$$

The last term now presents a zero-point energy for this fermion system.

Problem 12.6 (a) Show that in the ERL the Dirac spinor in Prob. 9.6(a) becomes

$$u_\lambda(\mathbf{k}) = \frac{1}{\sqrt{2}} \begin{pmatrix} \eta_\lambda \\ \boldsymbol{\sigma} \cdot \mathbf{e_k} \, \eta_\lambda \end{pmatrix} \qquad ; \; \mathbf{e_k} \equiv \frac{\mathbf{k}}{k}$$

(b) Use this result to derive Eq. (12.65).

Solution to Problem 12.6

(a) In the ERL we may approximate

$$E_k \to \hbar c |\mathbf{k}|$$

In this limit the Dirac spinor of Prob. 9.6(a) then becomes

$$u_\lambda(\mathbf{k}) = \frac{1}{\sqrt{2}} \begin{pmatrix} \eta_\lambda \\ \boldsymbol{\sigma} \cdot \mathbf{e_k} \, \eta_\lambda \end{pmatrix} \qquad ; \, \mathbf{e_k} \equiv \frac{\mathbf{k}}{k}$$

(b) Let us write Eq. (12.65) as

$$\frac{1}{2} \sum_\lambda \sum_{\lambda'} |u_{\lambda'}^\dagger(\mathbf{k}') u_\lambda(\mathbf{k})|^2 =$$

$$= \frac{1}{2} \sum_{a,b} \left[\sum_\lambda u_\lambda(\mathbf{k})_a u_\lambda^*(\mathbf{k})_b \right] \left[\sum_{\lambda'} u_{\lambda'}(\mathbf{k}')_b u_{\lambda'}^*(\mathbf{k}')_a \right]$$

$$= \frac{1}{2} \mathrm{Tr} \left[\sum_\lambda u_\lambda(\mathbf{k}) u_\lambda^\dagger(\mathbf{k}) \right] \left[\sum_{\lambda'} u_{\lambda'}(\mathbf{k}') u_{\lambda'}^\dagger(\mathbf{k}') \right]$$

Here we have written the Dirac spinor indices explicitly in the second line, and then identified the trace of the 4×4 matrix product in the third.

The 4×4 matrices appearing in this expression can be written in 2×2 form with the help of the result in part (a)[4]

$$\sum_\lambda u_\lambda(\mathbf{k}) u_\lambda^\dagger(\mathbf{k}) = \frac{1}{2} \sum_\lambda \begin{pmatrix} \eta_\lambda \\ \boldsymbol{\sigma} \cdot \mathbf{e_k} \, \eta_\lambda \end{pmatrix} \begin{pmatrix} \eta_\lambda^\dagger, & \eta_\lambda^\dagger \boldsymbol{\sigma} \cdot \mathbf{e_k} \end{pmatrix}$$

$$= \frac{1}{2} \begin{pmatrix} 1 & \boldsymbol{\sigma} \cdot \mathbf{e_k} \\ \boldsymbol{\sigma} \cdot \mathbf{e_k} & 1 \end{pmatrix}$$

As a result

$$\frac{1}{2} \sum_\lambda \sum_{\lambda'} |u_{\lambda'}^\dagger(\mathbf{k}') u_\lambda(\mathbf{k})|^2 = \frac{1}{8} \mathrm{Tr} \begin{pmatrix} 1 & \boldsymbol{\sigma} \cdot \mathbf{e_k} \\ \boldsymbol{\sigma} \cdot \mathbf{e_k} & 1 \end{pmatrix} \begin{pmatrix} 1 & \boldsymbol{\sigma} \cdot \mathbf{e_{k'}} \\ \boldsymbol{\sigma} \cdot \mathbf{e_{k'}} & 1 \end{pmatrix}$$

$$= \frac{1}{8} \mathrm{Tr} \begin{bmatrix} 1 + (\boldsymbol{\sigma} \cdot \mathbf{e_k})(\boldsymbol{\sigma} \cdot \mathbf{e_{k'}}) & \boldsymbol{\sigma} \cdot (\mathbf{e_k} + \mathbf{e_{k'}}) \\ \boldsymbol{\sigma} \cdot (\mathbf{e_k} + \mathbf{e_{k'}}) & 1 + (\boldsymbol{\sigma} \cdot \mathbf{e_k})(\boldsymbol{\sigma} \cdot \mathbf{e_{k'}}) \end{bmatrix}$$

$$= \frac{1}{4} \mathrm{Tr}[1 + (\boldsymbol{\sigma} \cdot \mathbf{e_k})(\boldsymbol{\sigma} \cdot \mathbf{e_{k'}})]$$

$$= \frac{1}{2}[1 + \mathbf{e_k} \cdot \mathbf{e_{k'}}]$$

---

[4] Note $\sum_\lambda \eta_\lambda \eta_\lambda^\dagger = \begin{pmatrix} 1 & 0 \\ 0 & 1 \end{pmatrix}$ is the unit 2×2 matrix (Prob. B.4); also, $\mathrm{Tr} \, \underline{M} \equiv \sum_a M_{aa}$.

Here we have used the property $\text{Tr}(\sigma_i \sigma_j) = 2\delta_{ij}$.

Thus

$$\frac{1}{2} \sum_\lambda \sum_{\lambda'} |u^\dagger_{\lambda'}(\mathbf{k}') u_\lambda(\mathbf{k})|^2 = \frac{1}{2}(1 + \cos\theta) = \cos^2 \frac{\theta}{2} \qquad ; \text{ERL}$$

which is just Eq. (12.65).

Problem 12.7 (a) Prove the completeness relation for the Dirac spinors in Prob. 9.6[5]

$$\sum_\lambda \left[\underline{u}_\lambda(\mathbf{k})\,\underline{u}_\lambda(\mathbf{k})^\dagger + \underline{v}_\lambda(\mathbf{k})\,\underline{v}_\lambda(\mathbf{k})^\dagger \right]_{\alpha\beta} = \delta_{\alpha\beta}$$

(b) Use this result to derive the canonical equal-time anticommutation relation for the Dirac field

$$\left\{ \hat\psi_\alpha(\mathbf{x}, t), \hat\psi^\dagger_\beta(\mathbf{x}', t') \right\}_{t=t'} = \delta_{\alpha\beta}\, \delta^{(3)}(\mathbf{x} - \mathbf{x}')$$

Solution to Problem 12.7

(a) With the use of the expressions for the positive-energy spinors given in Prob. 9.6, we have

$$\sum_\lambda \underline{u}_\lambda(\mathbf{k})\,\underline{u}_\lambda(\mathbf{k})^\dagger = \sum_\lambda \left(\frac{E_k + m_0c^2}{2E_k} \right) \times$$

$$\left(\begin{array}{c} \eta_\lambda \\ [\hbar c\boldsymbol{\sigma} \cdot \mathbf{k}/(E_\mathbf{k} + m_0c^2)]\,\eta_\lambda \end{array} \right) \left(\eta^\dagger_\lambda\,,\ \eta^\dagger_\lambda\,[\hbar c\boldsymbol{\sigma} \cdot \mathbf{k}/(E_\mathbf{k} + m_0c^2)] \right)$$

$$= \sum_\lambda \left(\frac{E_k + m_0c^2}{2E_k} \right) \times$$

$$\left(\begin{array}{cc} 1 & \hbar c(\boldsymbol{\sigma} \cdot \mathbf{k})/(E_k + m_0c^2) \\ \hbar c(\boldsymbol{\sigma} \cdot \mathbf{k})/(E_k + m_0c^2) & [\hbar c/(E_k + m_0c^2)]^2(\boldsymbol{\sigma} \cdot \mathbf{k})(\boldsymbol{\sigma} \cdot \mathbf{k}) \end{array} \right)$$

where we have used the fact that $\sum_\lambda \eta_\lambda \eta^\dagger_\lambda$ is the unit 2×2 matrix. This simplifies to

$$\sum_\lambda \underline{u}_\lambda(\mathbf{k})\,\underline{u}_\lambda(\mathbf{k})^\dagger = \frac{1}{2E_k} \left(\begin{array}{cc} E_k + m_0c^2 & \hbar c\boldsymbol{\sigma} \cdot \mathbf{k} \\ \hbar c\boldsymbol{\sigma} \cdot \mathbf{k} & E_k - m_0c^2 \end{array} \right)$$

---

[5]See Prob. B.4.

Similarly, for the negative-energy spinors we have

$$\sum_\lambda \underline{v}_\lambda(\mathbf{k})\,\underline{v}_\lambda(\mathbf{k})^\dagger = \frac{1}{2E_k}\begin{pmatrix} E_k - m_0c^2 & -\hbar c\boldsymbol{\sigma}\cdot\mathbf{k} \\ -\hbar c\boldsymbol{\sigma}\cdot\mathbf{k} & E_k + m_0c^2 \end{pmatrix}$$

The sum of these two expressions then gives the completeness relation for the Dirac spinors

$$\sum_\lambda \left[\underline{u}_\lambda(\mathbf{k})\,\underline{u}_\lambda(\mathbf{k})^\dagger + \underline{v}_\lambda(\mathbf{k})\,\underline{v}_\lambda(\mathbf{k})^\dagger\right]_{\alpha\beta} = \delta_{\alpha\beta}$$

(b) The Dirac field and its adjoint are given by [see Eqs. (12.57)]

$$\hat{\psi}(\mathbf{x},t) = \frac{1}{\sqrt{\Omega}}\sum_{\mathbf{k}}\sum_\lambda \left[a_{\mathbf{k}\lambda}u_\lambda(\mathbf{k})e^{i(\mathbf{k}\cdot\mathbf{x}-\omega_k t)} + b^\dagger_{\mathbf{k}\lambda}v_\lambda(-\mathbf{k})e^{-i(\mathbf{k}\cdot\mathbf{x}-\omega_k t)}\right]$$

$$\hat{\psi}^\dagger(\mathbf{x},t) = \frac{1}{\sqrt{\Omega}}\sum_{\mathbf{k}}\sum_\lambda \left[a^\dagger_{\mathbf{k}\lambda}u^\dagger_\lambda(\mathbf{k})e^{-i(\mathbf{k}\cdot\mathbf{x}-\omega_k t)} + b_{\mathbf{k}\lambda}v^\dagger_\lambda(-\mathbf{k})e^{i(\mathbf{k}\cdot\mathbf{x}-\omega_k t)}\right]$$

where $\hbar\omega_k = \sqrt{(\hbar k c)^2 + (m_0c^2)^2}$. The creation and destruction operators obey the anticommutation relations of Eqs. (12.58), and thus

$$\left\{\hat{\psi}_\alpha(\mathbf{x},t),\hat{\psi}^\dagger_\beta(\mathbf{x}',t')\right\}_{t=t'} = \frac{1}{\Omega}\sum_{\mathbf{k},\lambda}\sum_{\mathbf{k}',\lambda'} \times$$

$$\left[u_\lambda(\mathbf{k})_\alpha u^\dagger_{\lambda'}(\mathbf{k}')_\beta \left\{a_{\mathbf{k}\lambda}, a^\dagger_{\mathbf{k}'\lambda'}\right\} e^{i[\mathbf{k}\cdot\mathbf{x}-\mathbf{k}'\cdot\mathbf{x}'-(\omega_k-\omega_{k'})t]}\right.$$

$$\left. + v_\lambda(-\mathbf{k})_\alpha v^\dagger_{\lambda'}(-\mathbf{k}')_\beta \left\{b^\dagger_{\mathbf{k}\lambda}, b_{\mathbf{k}'\lambda'}\right\} e^{-i[\mathbf{k}\cdot\mathbf{x}-\mathbf{k}'\cdot\mathbf{x}'-(\omega_k-\omega_{k'})t]}\right]$$

where we have made the Dirac spinor indices explicit. With a change of summation variable $\mathbf{k} \to -\mathbf{k}$ in the second term, this reduces to

$$\left\{\hat{\psi}_\alpha(\mathbf{x},t),\hat{\psi}^\dagger_\beta(\mathbf{x}',t')\right\}_{t=t'} = \frac{1}{\Omega}\sum_{\mathbf{k},\lambda} e^{i\mathbf{k}\cdot(\mathbf{x}-\mathbf{x}')}\left[u_\lambda(\mathbf{k})u^\dagger_\lambda(\mathbf{k}) + v_\lambda(\mathbf{k})v^\dagger_\lambda(\mathbf{k})\right]_{\alpha\beta}$$

Therefore, with the use of the completeness result for the Dirac spinors in part (a),

$$\left\{\hat{\psi}_\alpha(\mathbf{x},t),\hat{\psi}^\dagger_\beta(\mathbf{x}',t')\right\}_{t=t'} = \delta_{\alpha\beta}\,\delta^{(3)}(\mathbf{x}-\mathbf{x}')$$

Problem 12.8 (a) Prove from Eqs. (12.52)–(12.53) that $a^\dagger\psi_0 \propto \psi_1$ and $a^\dagger\psi_1 = 0$;

(b) Prove that $a\psi_1 \propto \psi_0$ and $a\psi_0 = 0$.

The phases of the states can then be chosen so that Eqs. (12.56) are satisfied.

Solution to Problem 12.8

In this problem $(a^\dagger, a)$ are fermion creation and destruction operators satisfying the anticommutation relations of Eqs. (12.52).

(a) Consider the state $a^\dagger \psi_0$, and apply the number operator $N = a^\dagger a$ to it

$$N a^\dagger \psi_0 = a^\dagger a a^\dagger \psi_0$$
$$= a^\dagger (1 - a^\dagger a) \psi_0 = a^\dagger \psi_0$$

where we have used $a^\dagger a^\dagger = 0$. We conclude that $a^\dagger \psi_0$ is an eigenstate of N with eigenvalue 1, and therefore

$$a^\dagger \psi_0 \propto \psi_1$$

We use this result to calculate

$$a^\dagger \psi_1 \propto a^\dagger a^\dagger \psi_0 = 0$$

where we have once again used the property $a^\dagger a^\dagger = 0$.

(b) Consider the state $a\psi_1$ and apply the number operator to it

$$N a\psi_1 = a^\dagger a a \psi_1 = 0$$

where we have used the property $aa = 0$. We conclude that $a\psi_1$ is an eigenstate of N with eigenvalue 0, and therefore

$$a\psi_1 \propto \psi_0$$

With the use of this result, we also have that

$$a\psi_0 \propto a a \psi_1 = 0$$

Problem 12.9 Derive Eqs. (12.33).

Solution to Problem 12.9

The equal-time commutator of the quantum field of the string $\hat{q}(x, t)$ and corresponding momentum density $\hat{\pi}(x', t')$ is evaluated from the field expansions in Eqs. (12.30)–(12.32) and the commutation relations of the creation and destruction operators in Eqs. (12.26). We have

$$
[\hat{q}(x,t), \hat{\pi}(x',t')]_{t=t'} = \frac{1}{i}\frac{\hbar}{2L} \sum_{k,k'} \Big\{ -\Big[a_k, a_{k'}^\dagger\Big] e^{i(kx-\omega_k t)} e^{-i(k'x'-\omega_{k'}t)}
$$
$$
+ \Big[a_k^\dagger, a_{k'}\Big] e^{-i(kx-\omega_k t)} e^{i(k'x'-\omega_{k'}t)} \Big\}
$$
$$
= \frac{i\hbar}{2L} \sum_k \Big[e^{ik(x-x')} + e^{-ik(x-x')} \Big]
$$

A change of dummy summation variable $k \to -k$ in the second term reduces this to

$$
[\hat{q}(x,t), \hat{\pi}(x',t')]_{t=t'} = \frac{i\hbar}{L} \sum_k e^{ik(x-x')}
$$
$$
= i\hbar\delta(x-x')
$$

We now consider the equal-time commutator of the field itself at two distinct spatial points

$$
[\hat{q}(x,t), \hat{q}(x',t')]_{t=t'} = \frac{\hbar}{2\sigma L} \sum_{k,k'} \frac{1}{\sqrt{\omega_k \omega_{k'}}} \Big\{ \Big[a_k, a_{k'}^\dagger\Big] e^{i(kx-\omega_k t)} e^{-i(k'x'-\omega_{k'}t)}
$$
$$
+ \Big[a_k^\dagger, a_{k'}\Big] e^{-i(kx-\omega_k t)} e^{i(k'x'-\omega_{k'}t)} \Big\}
$$
$$
= \frac{\hbar}{2\sigma L} \sum_k \frac{1}{\omega_k} \Big[e^{ik(x-x')} - e^{-ik(x-x')} \Big]
$$

A change of summation variable $k \to -k$ in the second term, keeping in mind that $\omega_k = |k|c$, shows that the two terms cancel, and hence this commutator vanishes

$$
[\hat{q}(x,t), \hat{q}(x',t')]_{t=t'} = 0
$$

The equal-time commutator of the momentum density at two distinct spatial points is similarly evaluated as

$$
[\hat{\pi}(x,t), \hat{\pi}(x',t')]_{t=t'} = \frac{\hbar\sigma}{2L} \sum_{k,k'} \sqrt{\omega_k \omega_{k'}} \Big\{ \Big[a_k, a_{k'}^\dagger\Big] e^{i(kx-\omega_k t)} e^{-i(k'x'-\omega_{k'}t)}
$$
$$
+ \Big[a_k^\dagger, a_{k'}\Big] e^{-i(kx-\omega_k t)} e^{i(k'x'-\omega_{k'}t)} \Big\}
$$
$$
= \frac{\hbar\sigma}{2L} \sum_k \omega_k \Big[e^{ik(x-x')} - e^{-ik(x-x')} \Big]
$$

Once again, a change of summation variable $k \to -k$ in the second term shows that this expression also vanishes

$$[\hat{\pi}(x,t),\hat{\pi}(x',t')]_{t=t'} = 0$$

Problem 12.10 (a) Insert the field expansion in Eq. (12.68) into the hamiltonian in Eq. (12.67), and reduce that expression to a sum over modes;
(b) Discuss the processes described by the resulting hamiltonian.

Solution to Problem 12.10

(a) Let us first work on the kinetic part of the many-body hamiltonian

$$\hat{T} = \int d^3x \; \hat{\psi}^\dagger(\mathbf{x})T\hat{\psi}(\mathbf{x})$$

$$= \sum_k \sum_{k'} a_k^\dagger a_{k'} \int d^3x \; \psi_k^\dagger(\mathbf{x})T\psi_{k'}(\mathbf{x})$$

$$\equiv \sum_k \sum_{k'} t_{kk'} a_k^\dagger a_{k'}$$

where

$$t_{kk'} \equiv \int d^3x \; \psi_k^\dagger(\mathbf{x})T\psi_{k'}(\mathbf{x})$$

The $\psi_k(\mathbf{x})$ are the complete set of solutions to a one-body Schrödinger equation. If we choose the three-dimensional plane waves $\psi_{\mathbf{k}}(\mathbf{x})$ in Eqs. (4.118), which are eigenfunctions of the kinetic-energy operator, we have[6]

$$t_{\mathbf{k}\mathbf{k}'} = \delta_{\mathbf{k},\mathbf{k}'} \frac{\hbar^2\mathbf{k}^2}{2m}$$

$$\hat{T} = \sum_{\mathbf{k}} \frac{\hbar^2\mathbf{k}^2}{2m} a_{\mathbf{k}}^\dagger a_{\mathbf{k}}$$

Now consider the many-body potential

$$\hat{V} = \frac{1}{2} \sum_{k_1,k_2,k_3,k_4} a_{k_1}^\dagger a_{k_2}^\dagger a_{k_3} a_{k_4} \int d^3x \int d^3y \; \psi_{k_1}^\dagger(\mathbf{x})\psi_{k_2}^\dagger(\mathbf{y})V(\mathbf{x},\mathbf{y})\psi_{k_3}(\mathbf{y})\psi_{k_4}(\mathbf{x})$$

$$\equiv \frac{1}{2} \sum_{k_1,k_2,k_3,k_4} v_{k_1,k_2,k_3,k_4} \; a_{k_1}^\dagger a_{k_2}^\dagger a_{k_3} a_{k_4}$$

---

[6]This is for spin-zero bosons. For spin-1/2 fermions, we would use the wave functions in Eqs. (4.159).

where

$$v_{k_1,k_2,k_3,k_4} \equiv \int d^3x \int d^3y \, \psi^\dagger_{k_1}(\mathbf{x})\psi^\dagger_{k_2}(\mathbf{y})V(\mathbf{x},\mathbf{y})\psi_{k_3}(\mathbf{y})\psi_{k_4}(\mathbf{x})$$

A combination of these terms gives the many-body hamiltonian as

$$\hat{H} = \sum_k \sum_{k'} t_{kk'} a^\dagger_k a_{k'} + \frac{1}{2}\sum_{k_1,k_2,k_3,k_4} v_{k_1,k_2,k_3,k_4} \, a^\dagger_{k_1} a^\dagger_{k_2} a_{k_3} a_{k_4}$$

(b) With the use of the above expression, it is easy to check that the hamiltonian commutes with the number operator $\hat{N} = \sum_k a^\dagger_k a_k$. For instance, assume bosonic operators and consider the kinetic-energy operator. We have

$$\left[\hat{T},\hat{N}\right] = \sum_{k_1,k_2,k} t_{k_1,k_2} \left[a^\dagger_{k_1} a_{k_2}, a^\dagger_k a_k\right]$$

$$= \sum_{k_1,k_2,k} t_{k_1,k_2} \left(a^\dagger_{k_1} a_k \delta_{k,k_2} - a^\dagger_k a_{k_2} \delta_{k,k_1}\right) = 0$$

In a similar fashion, with a little more algebra, one proves that $\left[\hat{V},\hat{N}\right] = 0$. Therefore $\hat{H}$ describes processes where the number of particles is conserved.[7]

If we neglect for the moment the potential part, and assume the $\psi_k(\mathbf{r})$ to be the previous plane waves, then a state with N particles,

$$|\Psi_N\rangle = \prod_{i=1}^N a^\dagger_{\mathbf{k}_i}|0\rangle$$

is an eigenstate of $\hat{T}$

$$\hat{T}|\Psi_N\rangle = \sum_k \frac{\hbar^2 \mathbf{k}^2}{2m} a^\dagger_{\mathbf{k}} a_{\mathbf{k}} \prod_{i=1}^N a^\dagger_{\mathbf{k}_i}|0\rangle = \left[\sum_{i=1}^N \frac{\hbar^2 \mathbf{k}_i^2}{2m}\right]|\Psi_N\rangle$$

and $\left[\sum_{i=1}^N (\hbar^2 \mathbf{k}_i^2/2m)\right]$ clearly amounts to the total energy of a system of N non-interacting particles. As soon as we switch on the interaction, the potential part of the hamiltonian, $\hat{V}$, allows any two particles in the original state $|\Psi_N\rangle$ to interact and thus change their quantum numbers.

---

[7]This result is evident, since for every particle destroyed by $\hat{H}$, another is created.

Problem 12.11 The lagrangian and action for the string are obtained as follows

$$L = T - V = \frac{\sigma}{2} \int_0^l dx \left\{ \left[\frac{\partial q(x,t)}{\partial t} \right]^2 - c^2 \left[\frac{\partial q(x,t)}{\partial x} \right]^2 \right\} \quad ; \text{ lagrangian}$$

$$S = \int_{t_1}^{t_2} L\, dt = \int_{t_1}^{t_2} dt \int_0^l dx\, \mathcal{L} \left(\frac{\partial q}{\partial t}, \frac{\partial q}{\partial x}, q;\, t \right) \quad ; \text{ action}$$

The quantity $\mathcal{L}$ is known as the *lagrangian density*, and to avoid confusion, we now use l for the length of the string.

(a) Use the analysis of Hamilton's principle in Prob. 10.4, together with fixed endpoints in time and periodic boundary conditions in space, to derive the continuum form of Lagrange's equation[8]

$$\frac{\partial}{\partial t} \left[\frac{\partial \mathcal{L}}{\partial(\partial q/\partial t)} \right] + \frac{\partial}{\partial x} \left[\frac{\partial \mathcal{L}}{\partial(\partial q/\partial x)} \right] - \frac{\partial \mathcal{L}}{\partial q} = 0 \quad ; \text{ Lagrange's equation}$$

Here the partial derivatives of $\mathcal{L}$ keep all the other variables in $\mathcal{L}$ fixed, while the final partial derivatives with respect to (x,t) keep the other variable in this pair fixed;

(b) Show this reproduces the wave equation for $q(x,t)$;

(c) With the introduction of the "two-vector" $x_\mu \equiv (x, ict)$, and the convention that repeated Greek indices are summed from 1 to 2, show the lagrangian density and Lagrange's equation can be rewritten in invariant form as[9]

$$\mathcal{L} = -\frac{\tau}{2} \left(\frac{\partial q}{\partial x_\mu} \right)^2$$

$$\frac{\partial}{\partial x_\mu} \left[\frac{\partial \mathcal{L}}{\partial(\partial q/\partial x_\mu)} \right] - \frac{\partial \mathcal{L}}{\partial q} = 0$$

Solution to Problem 12.11

(a) The action defined above is a functional of $q(x,t)$, $\partial q/\partial t$ and $\partial q/\partial x$. When arbitrary variations of $q(x,t)$ are considered, subject to the conditions $\delta q(x,t_1) = \delta q(x,t_2) = 0$, then according to Hamilton's principle, the corresponding variation of the action must vanish, or $\delta S = 0$. This variation

---

[8] Continuum mechanics is developed in [Fetter and Walecka (2003)].

[9] Recall that here c is the sound velocity with $c^2 = \tau/\sigma$.

may be written explicitly as

$$\delta S = \delta \int_{t_1}^{t_2} dt \int_0^l dx \, \mathcal{L} \left(\frac{\partial q}{\partial t}, \frac{\partial q}{\partial x}, q; \, t \right)$$

$$= \int_{t_1}^{t_2} dt \int_0^l dx \left[\frac{\partial \mathcal{L}}{\partial q} \delta q + \frac{\partial \mathcal{L}}{\partial (\partial q / \partial t)} \delta \left(\frac{\partial q}{\partial t} \right) + \frac{\partial \mathcal{L}}{\partial (\partial q / \partial x)} \delta \left(\frac{\partial q}{\partial x} \right) \right]$$

Here, in our previous notation,

$$\delta q(x, t) = \lambda \eta(x, t)$$

where λ is an infinitesimal and $\eta(x, t)$ is an arbitrary function of space-time. Then

$$\delta \left(\frac{\partial q}{\partial t} \right) = \lambda \left(\frac{\partial \eta}{\partial t} \right) = \frac{\partial}{\partial t} \delta q$$

$$\delta \left(\frac{\partial q}{\partial x} \right) = \lambda \left(\frac{\partial \eta}{\partial x} \right) = \frac{\partial}{\partial x} \delta q$$

Let us focus on the second term in square brackets in δS. We make use of the relation $\delta(\partial q / \partial t) = (\partial / \partial t) \delta q$, and an integration by parts on t (at fixed x) then gives

$$\int_{t_1}^{t_2} dt \int_0^l dx \, \frac{\partial \mathcal{L}}{\partial (\partial q / \partial t)} \frac{\partial \delta q}{\partial t} = \int_0^l dx \left[\frac{\partial \mathcal{L}}{\partial (\partial q / \partial t)} \delta q \right]_{t_1}^{t_2}$$

$$- \int_{t_1}^{t_2} dt \int_0^l dx \, \frac{\partial}{\partial t} \left[\frac{\partial \mathcal{L}}{\partial (\partial q / \partial t)} \right] \delta q$$

With fixed endpoints in *time* one has $\delta q(x, t_1) = \delta q(x, t_2) = 0$, and the first term on the r.h.s. vanishes. Thus we have

$$\int_{t_1}^{t_2} dt \int_0^l dx \, \frac{\partial \mathcal{L}}{\partial (\partial q / \partial t)} \frac{\partial \delta q}{\partial t} = - \int_{t_1}^{t_2} dt \int_0^l dx \, \frac{\partial}{\partial t} \left[\frac{\partial \mathcal{L}}{\partial (\partial q / \partial t)} \right] \delta q$$

Consider the third term in square brackets in δS. We use $\delta(\partial q / \partial x) = (\partial / \partial x) \delta q$, and an integration by parts on x (at fixed t) gives

$$\int_{t_1}^{t_2} dt \int_0^l dx \, \frac{\partial \mathcal{L}}{\partial (\partial q / \partial x)} \frac{\partial \delta q}{\partial x} = \int_{t_1}^{t_2} dt \left[\frac{\partial \mathcal{L}}{\partial (\partial q / \partial x)} \delta q \right]_0^l$$

$$- \int_{t_1}^{t_2} dt \int_0^l dx \, \frac{\partial}{\partial x} \left[\frac{\partial \mathcal{L}}{\partial (\partial q / \partial x)} \right] \delta q$$

With the assumption of periodic boundary conditions in *space*, the first term on the r.h.s. again vanishes, and we have

$$\int_{t_1}^{t_2} dt \int_0^l dx \, \frac{\partial \mathcal{L}}{\partial(\partial q/\partial x)} \delta \frac{\partial q}{\partial x} = -\int_{t_1}^{t_2} dt \int_0^l dx \, \frac{\partial}{\partial x}\left[\frac{\partial \mathcal{L}}{\partial(\partial q/\partial x)}\right]\delta q$$

A combination of these results allows us to express Hamilton's principle as

$$\delta S = \int_{t_1}^{t_2} dt \int_0^l dx \left[\frac{\partial \mathcal{L}}{\partial q} - \frac{\partial}{\partial t}\frac{\partial \mathcal{L}}{\partial(\partial q/\partial t)} - \frac{\partial}{\partial x}\frac{\partial \mathcal{L}}{\partial(\partial q/\partial x)}\right]\delta q = 0$$

Given that δq is arbitrary (apart from the requirements of fixed endpoints in time and periodic boundary conditions in space), this equation can be satisfied only if the Euler-Lagrange equation is fulfilled

$$\frac{\partial}{\partial t}\left[\frac{\partial \mathcal{L}}{\partial(\partial q/\partial t)}\right] + \frac{\partial}{\partial x}\left[\frac{\partial \mathcal{L}}{\partial(\partial q/\partial x)}\right] - \frac{\partial \mathcal{L}}{\partial q} = 0 \quad ; \text{ Euler-Lagrange eqn}$$

This is the Euler-Lagrange equation in continuum mechanics in one dimension, as illustrated by the string.[10]

(b) The lagrangian for the string is

$$L = T - V = \int_0^l dx\, \mathcal{L}$$

Hence from Eqs. (12.3)–(12.4), the lagrangian density for the string is

$$\mathcal{L} = \frac{\sigma}{2}\left\{\left[\frac{\partial q(x,t)}{\partial t}\right]^2 - c^2\left[\frac{\partial q(x,t)}{\partial x}\right]^2\right\}$$

It follows that

$$\frac{\partial \mathcal{L}}{\partial q} = 0$$

$$\frac{\partial \mathcal{L}}{\partial(\partial q/\partial t)} = \sigma\frac{\partial q}{\partial t}$$

$$\frac{\partial \mathcal{L}}{\partial(\partial q/\partial x)} = -\sigma c^2\frac{\partial q}{\partial x}$$

Substitution of these expressions in the above Euler-Lagrange equation leads to the one-dimensional wave equation

$$\frac{\partial^2 q}{\partial t^2} - c^2\frac{\partial^2 q}{\partial x^2} = 0$$

---

[10]This is often simply referred to as "Lagrange's equation".

(c) Introduce the "two-vector" $x_\mu = (x, ict)$ where c is the sound velocity with $c^2 = \tau/\sigma$. Introduce also the *summation convention* that repeated Greek indices are summed from 1 to 2. Then the lagrangian density of the string can be written as[11]

$$\mathcal{L} = -\frac{\tau}{2}\left(\frac{\partial q}{\partial x_\mu}\right)^2$$

$$= -\frac{\tau}{2}\left(\frac{\partial q}{\partial x}\right)^2 + \frac{\tau}{2c^2}\left(\frac{\partial q}{\partial t}\right)^2$$

$$= \frac{\sigma}{2}\left\{\left[\frac{\partial q(x,t)}{\partial t}\right]^2 - c^2\left[\frac{\partial q(x,t)}{\partial x}\right]^2\right\}$$

In a similar manner, one has

$$\frac{\partial}{\partial x_\mu}\left[\frac{\partial \mathcal{L}}{\partial(\partial q/\partial x_\mu)}\right] = \frac{\partial}{\partial x}\left[\frac{\partial \mathcal{L}}{\partial(\partial q/\partial x)}\right] + \frac{\partial}{\partial ict}\left[\frac{\partial \mathcal{L}}{\partial(\partial q/\partial ict)}\right]$$

$$= \frac{\partial}{\partial x}\left[\frac{\partial \mathcal{L}}{\partial(\partial q/\partial x)}\right] + \frac{\partial}{\partial t}\left[\frac{\partial \mathcal{L}}{\partial(\partial q/\partial t)}\right]$$

Hence the continuum Euler-Lagrange equation can be written as

$$\frac{\partial}{\partial x_\mu}\left[\frac{\partial \mathcal{L}}{\partial(\partial q/\partial x_\mu)}\right] - \frac{\partial \mathcal{L}}{\partial q} = 0$$

---

[11] Note that now $(\partial q/\partial x_\mu)^2 = (\partial q/\partial x_\mu)(\partial q/\partial x_\mu) \equiv \sum_{\mu=1}^{2}(\partial q/\partial x_\mu)(\partial q/\partial x_\mu)$.

Appendix A

Mathematica Program for Problem 2.5

There are no problems in appendix A; however, we want to somewhere give one detailed example of the use of a widely available computer program to solve one of the problems. Here we write a simple *Mathematica* program to visualize the solution to Prob. 2.5. The appropriate formatting of instructions can be found in any description of Mathematica.

The initial profile of the string corresponds to the function

```
q0[h_, L_][x_] := Piecewise[{{2*h*x/L, x >= 0 && x <= L/2},
                            {2*h (L - x)/L, x >= L/2 && x <= L}}];
```

Since Mathematica has the capability of handling symbolic expressions we do not need to specify numerical values for h, L and x at this point.

The frequencies of the string, which appear in Eq. (2.36) of the book are defined as

```
w[h_, L_, c_][n_] := Pi*c/L*n;
```

and depend on the velocity c.

The implementation of Eq. (2.43) is straightforward and reads

```
q[h_, L_, c_, nmax_][x_, t_] :=
                Sum[(8*(-1)^k*h)/(Pi+2*k*Pi)^2*
                Cos[(c*(1+2*k)*Pi*t)/L]*
                Sin[((1+2*k)*Pi*x)/L],{k,0,nmax}];
```

Notice that Eq. (2.43) contains an infinite sum, which in general it will not be possible to calculate explicitly. However, as we have seen before, the Fourier coefficients of the solution decay rather rapidly, which means that one can still obtain a good approximation using a limited number of terms in the sum. The parameter **nmax** which appears in the last expression specifies the number of terms which are effectively used in the sum.

If we want to visualize the solution, we need to give numerical values to the parameters; for example, we can use $h = 1/10$, $L = 1$, $c = 1$ and **nmax**=20. We may plot the solution at $t = 0$ using the Mathematica **Plot** command

```
Plot[{q[0.1, 1, 1, 20][x, 0]}, {x, 0, 1},
   PlotRange -> {0, 0.11}, PlotStyle -> Blue,
   AxesLabel ->
        {Style[x, Large, Bold], Style[q[x, 0], Large, Bold]}]
```

whose output is shown in Fig. A.1(a). Figure A.1(b) shows the profile of the string at a later time, $t = 0.25$.

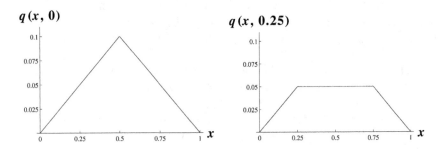

Fig. A.1   (a) Initial profile of the string at $t = 0$. (b) Profile of the string at $t = 0.25$.

The command

```
Animate[Plot[{q[0.1, 1, 1, 20][x, t]}, {x, 0, 1},
   PlotRange -> {-0.11, 0.11}, PlotStyle -> Blue,
   AxesLabel -> {Style[x, Large, Bold], Style[q, Large, Bold]}]
   ,{t, 0,2}]
```

provides a complete animation of the string motion, with the parameters specified above, between $t = 0$ and $t = 2$ (in this special case the period is $T = 2$).

With the use of the same steps that we have illustrated in this example, the reader may consider different initial profiles, including the case where the string is not at rest initially, and implement his or her own Mathematica code. As we have mentioned before, Mathematica offers the advantage of handling symbolic calculations, which, in many cases, may allow one to evaluate the Fourier coefficients A_m and the phases δ_m of Eqs. (2.49) directly on a computer.

Appendix B

Matrices

Problem B.1 Verify the statement on block multiplication of matrices made in this appendix at the start of the discussion of Dirac matrices.

Solution to Problem B.1

Let us consider two $N \times N$ matrices

$$M = \begin{pmatrix} a & b \\ c & d \end{pmatrix} \qquad ; \; \tilde{M} = \begin{pmatrix} \tilde{a} & \tilde{b} \\ \tilde{c} & \tilde{d} \end{pmatrix}$$

where we assume that a and $\tilde{a}$ are $q \times q$ matrices ($1 \leq q < N$), d and $\tilde{d}$ are $(N - q) \times (N - q)$ matrices, and $b, c, \tilde{b}$ and $\tilde{c}$ are the rectangular matrices with dimensions which are consistent with the definition of M and $\tilde{M}$.[1]

We explicitly write the matrix product $M\tilde{M}$

$$\left(M\tilde{M}\right)_{ij} = \sum_{l=1}^{N} M_{il}\tilde{M}_{lj} = \sum_{l=1}^{q} M_{il}\tilde{M}_{lj} + \sum_{l=q+1}^{N} M_{il}\tilde{M}_{lj}$$

We need to consider separately the four cases:

- $1 \leq i \leq q, \; 1 \leq j \leq q$;
  In this case we have that $M_{il} = a_{il}$ for $1 \leq l \leq q$, while $M_{il} = b_{il}$ for $q + 1 \leq l \leq N$. Similarly we have that $\tilde{M}_{lj} = \tilde{a}_{lj}$ for $1 \leq l \leq q$, while $\tilde{M}_{lj} = \tilde{c}_{lj}$ for $q + 1 \leq l \leq N$. Therefore in this case

$$\left(M\tilde{M}\right)_{ij} = \sum_{l=1}^{q} a_{il}\tilde{a}_{lj} + \sum_{l=q+1}^{N} b_{il}\tilde{c}_{lj}$$

---

[1]For simplicity, we do not underline the matrices in the solutions to Probs. B.1–B.2.

- $1 \le i \le q$, $q + 1 \le j \le N$;

  In this case we have that $M_{il} = a_{il}$ for $1 \le l \le q$, while $M_{il} = b_{il}$ for $q + 1 \le l \le N$. Similarly we have that $\tilde{M}_{lj} = \tilde{b}_{lj}$ for $1 \le l \le q$, while $\tilde{M}_{lj} = \tilde{d}_{lj}$ for $q + 1 \le l \le N$. Therefore in this case

$$\left(M\tilde{M} \right)_{ij} = \sum_{l=1}^{q} a_{il}\tilde{b}_{lj} + \sum_{l=q+1}^{N} b_{il}\tilde{d}_{lj}$$

- $q + 1 \le i \le N$, $1 \le j \le q$;

  In this case we have that $M_{il} = c_{il}$ for $1 \le l \le q$, while $M_{il} = d_{il}$ for $q + 1 \le l \le N$. Similarly we have that $\tilde{M}_{lj} = \tilde{a}_{lj}$ for $1 \le l \le q$, while $\tilde{M}_{lj} = \tilde{c}_{lj}$ for $q + 1 \le l \le N$. Therefore in this case

$$\left(M\tilde{M} \right)_{ij} = \sum_{l=1}^{q} c_{il}\tilde{a}_{lj} + \sum_{l=q+1}^{N} d_{il}\tilde{c}_{lj}$$

- $q + 1 \le i \le N$, $q + 1 \le j \le N$;

  In this case we have that $M_{il} = c_{il}$ for $1 \le l \le q$, while $M_{il} = d_{il}$ for $q + 1 \le l \le N$. Similarly we have that $\tilde{M}_{lj} = \tilde{b}_{lj}$ for $1 \le l \le q$, while $\tilde{M}_{lj} = \tilde{d}_{lj}$ for $q + 1 \le l \le N$. Therefore in this case

$$\left(M\tilde{M} \right)_{ij} = \sum_{l=q}^{N} c_{il}\tilde{b}_{lj} + \sum_{l=q+1}^{N} d_{il}\tilde{d}_{lj}$$

Thus we see that in matrix format

$$M\tilde{M} = \begin{pmatrix} a\tilde{a} + b\tilde{c} & a\tilde{b} + b\tilde{d} \\ c\tilde{a} + d\tilde{c} & c\tilde{b} + d\tilde{d} \end{pmatrix}$$

This proves the statement on block multiplication of matrices.

Problem B.2 Verify the relations in Eqs. (B.18) by direct multiplication of the Dirac matrices in Eqs. (B.17).

Solution to Problem B.2

The first two relations of Eqs. (B18) are straightforward, so we will concentrate on the remaining three equations. We use the result of Prob. B.1, and express the Dirac matrices in terms of 2×2 blocks as done in Eqs. (B.15)

$$\beta^2 = \begin{pmatrix} 1 & 0 \\ 0 & -1 \end{pmatrix} \begin{pmatrix} 1 & 0 \\ 0 & -1 \end{pmatrix} = \begin{pmatrix} 1 & 0 \\ 0 & 1 \end{pmatrix}$$

$$\alpha_i \beta + \beta \alpha_i = \begin{pmatrix} 0 & \sigma_i \\ \sigma_i & 0 \end{pmatrix} \begin{pmatrix} 1 & 0 \\ 0 & -1 \end{pmatrix} + \begin{pmatrix} 1 & 0 \\ 0 & -1 \end{pmatrix} \begin{pmatrix} 0 & \sigma_i \\ \sigma_i & 0 \end{pmatrix}$$

$$= \begin{pmatrix} 0 & -\sigma_i \\ \sigma_i & 0 \end{pmatrix} + \begin{pmatrix} 0 & \sigma_i \\ -\sigma_i & 0 \end{pmatrix} = \begin{pmatrix} 0 & 0 \\ 0 & 0 \end{pmatrix}$$

$$\alpha_i \alpha_j + \alpha_j \alpha_i = \begin{pmatrix} 0 & \sigma_i \\ \sigma_i & 0 \end{pmatrix} \begin{pmatrix} 0 & \sigma_j \\ \sigma_j & 0 \end{pmatrix} + \begin{pmatrix} 0 & \sigma_j \\ \sigma_j & 0 \end{pmatrix} \begin{pmatrix} 0 & \sigma_i \\ \sigma_i & 0 \end{pmatrix}$$

$$= \begin{pmatrix} \sigma_i \sigma_j + \sigma_j \sigma_i & 0 \\ 0 & \sigma_i \sigma_j + \sigma_j \sigma_i \end{pmatrix} = 2\delta_{ij} \begin{pmatrix} 1 & 0 \\ 0 & 1 \end{pmatrix}$$

where we have used the second of Eqs. (B.13).

Problem B.3 Prove that the adjoint of a product of matrices is the product of the adjoints in the reverse order

$$[\underline{A}\,\underline{B}]^\dagger = \underline{B}^\dagger \underline{A}^\dagger$$

Solution to Problem B.3

We first need to prove that the transpose of a product of matrices is the product of the transpose of each matrix taken in the reverse order

$$[\underline{A}\,\underline{B}]^T_{mn} = \left[\sum_r A_{mr} B_{rn} \right]^T = \sum_r A_{nr} B_{rm} = \sum_r B^T_{mr} A^T_{rn}$$

$$= [\underline{B}^T \underline{A}^T]_{mn}$$

If we then remember that $\underline{A}^\dagger \equiv \underline{A}^{T\star}$, we have

$$[\underline{A}\,\underline{B}]^\dagger = (\underline{B}^T \underline{A}^T)^\star = \underline{B}^\dagger \underline{A}^\dagger$$

Problem B.4 If $\underline{a}$ and $\underline{b}$ are n-component column vectors, then $\underline{a}\,\underline{b}^\dagger$ is an $n \times n$ matrix (just use the extended rule for matrix multiplication). Prove the completeness relation for the two-component spinors in Eqs. (4.160)

$$\sum_\lambda \underline{\eta}_\lambda \underline{\eta}^\dagger_\lambda = \underline{1} \qquad ; \text{ completeness}$$

Solution to Problem B.4

With the extended rule for matrix multiplication $[\underline{a}\,\underline{b}]_{ij} = \sum_k a_{ik}b_{kj}$, we have

$$\underline{\eta}_\downarrow\,\underline{\eta}_\downarrow^\dagger = \begin{pmatrix} 0 \\ 1 \end{pmatrix}(0\ 1) = \begin{pmatrix} 0 & 0 \\ 0 & 1 \end{pmatrix}$$

$$\underline{\eta}_\uparrow\,\underline{\eta}_\uparrow^\dagger = \begin{pmatrix} 1 \\ 0 \end{pmatrix}(1\ 0) = \begin{pmatrix} 1 & 0 \\ 0 & 0 \end{pmatrix}$$

Therefore

$$\sum_\lambda \underline{\eta}_\lambda\,\underline{\eta}_\lambda^\dagger = \begin{pmatrix} 1 & 0 \\ 0 & 1 \end{pmatrix} = \underline{1}$$

Problem B.5 Evaluate a few of the structure constants f^{abc} in Eq. (B.20). How many are there in total?

Solution to Problem B.5

The eight matrices $\underline{\lambda}^a$ are given in Eqs. (B.19). Taking into account the antisymmetry of the commutators $\left[\frac{1}{2}\underline{\lambda}^a, \frac{1}{2}\underline{\lambda}^b\right]$ with respect to the interchange of the indices, there are 28 non-trivial commutators, which explicitly provide the structure constants f^{abc} of Eqs. (B.20). Clearly, since the commutators corresponding to the same indices, $a = b$, vanish, we have that the corresponding structure constants also vanish, $f^{aac} = 0$.

The calculation of these commutators requires simple operations of matrix multiplication and can be carried out rather easily. We enumerate all the independent commutators, grouping them in sets of commutators with the same value of the first index.

For $a = 1$ we have a total of 7 commutators, which are

$$\left[\frac{1}{2}\underline{\lambda}^1, \frac{1}{2}\underline{\lambda}^2\right] = \frac{i}{2}\begin{pmatrix} 1 & 0 & 0 \\ 0 & -1 & 0 \\ 0 & 0 & 0 \end{pmatrix} = \frac{i}{2}\underline{\lambda}^3 \qquad ; f^{123} = 1$$

$$\left[\frac{1}{2}\underline{\lambda}^1, \frac{1}{2}\underline{\lambda}^3\right] = \frac{1}{2}\begin{pmatrix} 0 & -1 & 0 \\ 1 & 0 & 0 \\ 0 & 0 & 0 \end{pmatrix} = -\frac{i}{2}\underline{\lambda}^2 \qquad ; f^{132} = -1$$

$$\left[\frac{1}{2}\underline{\lambda}^1, \frac{1}{2}\underline{\lambda}^4\right] = \frac{1}{4}\begin{pmatrix} 0 & 0 & 0 \\ 0 & 0 & 1 \\ 0 & -1 & 0 \end{pmatrix} = \frac{i}{4}\underline{\lambda}^7 \qquad ; f^{147} = \frac{1}{2}$$

Furthermore

$$\left[\frac{1}{2}\lambda^1, \frac{1}{2}\lambda^5\right] = -\frac{i}{4}\begin{pmatrix} 0 & 0 & 0 \\ 0 & 0 & 1 \\ 0 & 1 & 0 \end{pmatrix} = -\frac{i}{4}\lambda^6 \qquad ; f^{156} = -\frac{1}{2}$$

$$\left[\frac{1}{2}\lambda^1, \frac{1}{2}\lambda^6\right] = \frac{1}{4}\begin{pmatrix} 0 & 0 & 1 \\ 0 & 0 & 0 \\ -1 & 0 & 0 \end{pmatrix} = \frac{i}{4}\lambda^5 \qquad ; f^{165} = \frac{1}{2}$$

$$\left[\frac{1}{2}\lambda^1, \frac{1}{2}\lambda^7\right] = -\frac{i}{4}\begin{pmatrix} 0 & 0 & 1 \\ 0 & 0 & 0 \\ 1 & 0 & 0 \end{pmatrix} = -\frac{i}{4}\lambda^4 \qquad ; f^{174} = -\frac{1}{2}$$

$$\left[\frac{1}{2}\lambda^1, \frac{1}{2}\lambda^8\right] = \begin{pmatrix} 0 & 0 & 0 \\ 0 & 0 & 0 \\ 0 & 0 & 0 \end{pmatrix} = 0$$

For $a = 2$ there are only 6 independent commutators, after eliminating the commutators that we have already calculated

$$\left[\frac{1}{2}\lambda^2, \frac{1}{2}\lambda^3\right] = \frac{i}{2}\begin{pmatrix} 0 & 1 & 0 \\ 1 & 0 & 0 \\ 0 & 0 & 0 \end{pmatrix} = \frac{i}{2}\lambda^1 \qquad ; f^{231} = 1$$

$$\left[\frac{1}{2}\lambda^2, \frac{1}{2}\lambda^4\right] = \frac{i}{4}\begin{pmatrix} 0 & 0 & 0 \\ 0 & 0 & 1 \\ 0 & 1 & 0 \end{pmatrix} = \frac{i}{4}\lambda^6 \qquad ; f^{246} = \frac{1}{2}$$

$$\left[\frac{1}{2}\lambda^2, \frac{1}{2}\lambda^5\right] = \frac{1}{4}\begin{pmatrix} 0 & 0 & 0 \\ 0 & 0 & 1 \\ 0 & -1 & 0 \end{pmatrix} = \frac{i}{4}\lambda^7 \qquad ; f^{257} = \frac{1}{2}$$

$$\left[\frac{1}{2}\lambda^2, \frac{1}{2}\lambda^6\right] = -\frac{i}{4}\begin{pmatrix} 0 & 0 & 1 \\ 0 & 0 & 0 \\ 1 & 0 & 0 \end{pmatrix} = -\frac{i}{4}\lambda^4 \qquad ; f^{264} = -\frac{1}{2}$$

$$\left[\frac{1}{2}\lambda^2, \frac{1}{2}\lambda^7\right] = -\frac{1}{4}\begin{pmatrix} 0 & 0 & 1 \\ 0 & 0 & 0 \\ -1 & 0 & 0 \end{pmatrix} = -\frac{i}{4}\lambda^5 \qquad ; f^{275} = -\frac{1}{2}$$

$$\left[\frac{1}{2}\lambda^2, \frac{1}{2}\lambda^8\right] = \begin{pmatrix} 0 & 0 & 0 \\ 0 & 0 & 0 \\ 0 & 0 & 0 \end{pmatrix} = 0$$

For $a = 3$ there are 5 independent commutators

$$\left[\frac{1}{2}\lambda^3, \frac{1}{2}\lambda^4\right] = \frac{1}{4}\begin{pmatrix} 0 & 0 & 1 \\ 0 & 0 & 0 \\ -1 & 0 & 0 \end{pmatrix} = \frac{i}{4}\lambda^5 \qquad ; f^{345} = \frac{1}{2}$$

$$\left[\frac{1}{2}\lambda^3, \frac{1}{2}\lambda^5\right] = -\frac{i}{4}\begin{pmatrix} 0 & 0 & 1 \\ 0 & 0 & 0 \\ 1 & 0 & 0 \end{pmatrix} = -\frac{i}{4}\lambda^4 \qquad ; f^{354} = -\frac{1}{2}$$

$$\left[\frac{1}{2}\lambda^3, \frac{1}{2}\lambda^6\right] = -\frac{1}{4}\begin{pmatrix} 0 & 0 & 0 \\ 0 & 0 & 1 \\ 0 & -1 & 0 \end{pmatrix} = -\frac{i}{4}\lambda^7 \qquad ; f^{367} = -\frac{1}{2}$$

$$\left[\frac{1}{2}\lambda^3, \frac{1}{2}\lambda^7\right] = \frac{i}{4}\begin{pmatrix} 0 & 0 & 0 \\ 0 & 0 & 1 \\ 0 & 1 & 0 \end{pmatrix} = \frac{i}{4}\lambda^6 \qquad ; f^{376} = \frac{1}{2}$$

$$\left[\frac{1}{2}\lambda^3, \frac{1}{2}\lambda^8\right] = \begin{pmatrix} 0 & 0 & 0 \\ 0 & 0 & 0 \\ 0 & 0 & 0 \end{pmatrix} = 0$$

For $a = 4$ there are 4 independent commutators

$$\left[\frac{1}{2}\lambda^4, \frac{1}{2}\lambda^5\right] = \frac{i}{2}\begin{pmatrix} 1 & 0 & 0 \\ 0 & 0 & 0 \\ 0 & 0 & -1 \end{pmatrix} = \frac{i}{4}\lambda^3 + \frac{i\sqrt{3}}{4}\lambda^8 \qquad ; f^{453} = \frac{1}{2}$$

$$; f^{458} = \frac{\sqrt{3}}{2}$$

$$\left[\frac{1}{2}\lambda^4, \frac{1}{2}\lambda^6\right] = \frac{1}{4}\begin{pmatrix} 0 & 1 & 0 \\ -1 & 0 & 0 \\ 0 & 0 & 0 \end{pmatrix} = \frac{i}{4}\lambda^2 \qquad ; f^{462} = \frac{1}{2}$$

$$\left[\frac{1}{2}\lambda^4, \frac{1}{2}\lambda^7\right] = \frac{i}{4}\begin{pmatrix} 0 & 1 & 0 \\ 1 & 0 & 0 \\ 0 & 0 & 0 \end{pmatrix} = \frac{i}{4}\lambda^1 \qquad ; f^{471} = \frac{1}{2}$$

$$\left[\frac{1}{2}\lambda^4, \frac{1}{2}\lambda^8\right] = \frac{\sqrt{3}}{4}\begin{pmatrix} 0 & 0 & -1 \\ 0 & 0 & 0 \\ 1 & 0 & 0 \end{pmatrix} = -\frac{i\sqrt{3}}{4}\lambda^5 \qquad ; f^{485} = -\frac{\sqrt{3}}{2}$$

For $a = 5$ there are 3 independent commutators

$$\left[\frac{1}{2}\lambda^5, \frac{1}{2}\lambda^6\right] = -\frac{i}{4}\begin{pmatrix} 0 & 1 & 0 \\ 1 & 0 & 0 \\ 0 & 0 & 0 \end{pmatrix} = -\frac{i}{4}\lambda^1 \qquad ; f^{561} = -\frac{1}{2}$$

$$\left[\frac{1}{2}\lambda^5, \frac{1}{2}\lambda^7\right] = \frac{1}{4}\begin{pmatrix} 0 & 1 & 0 \\ -1 & 0 & 0 \\ 0 & 0 & 0 \end{pmatrix} = \frac{i}{4}\lambda^2 \qquad ; f^{572} = \frac{1}{2}$$

$$\left[\frac{1}{2}\lambda^5, \frac{1}{2}\lambda^8\right] = \frac{i\sqrt{3}}{4}\begin{pmatrix} 0 & 0 & 1 \\ 0 & 0 & 0 \\ 1 & 0 & 0 \end{pmatrix} = \frac{i\sqrt{3}}{4}\lambda^4 \qquad ; f^{584} = \frac{\sqrt{3}}{2}$$

For $a = 6$ there are 2 independent commutators

$$\left[\frac{1}{2}\lambda^6, \frac{1}{2}\lambda^7\right] = \frac{i}{2}\begin{pmatrix} 0 & 0 & 0 \\ 0 & 1 & 0 \\ 0 & 0 & -1 \end{pmatrix} = -\frac{i}{4}\lambda^3 + \frac{i\sqrt{3}}{4}\lambda^8 \quad ; f^{673} = -\frac{1}{2}$$

$$; f^{678} = \frac{\sqrt{3}}{2}$$

$$\left[\frac{1}{2}\lambda^6, \frac{1}{2}\lambda^8\right] = \frac{\sqrt{3}}{4}\begin{pmatrix} 0 & 0 & 0 \\ 0 & 0 & -1 \\ 0 & 1 & 0 \end{pmatrix} = -\frac{i\sqrt{3}}{4}\lambda^7 \qquad ; f^{687} = -\frac{\sqrt{3}}{2}$$

Finally, for $a = 7$ there is just one commutator

$$\left[\frac{1}{2}\lambda^7, \frac{1}{2}\lambda^8\right] = \frac{i\sqrt{3}}{4}\begin{pmatrix} 0 & 0 & 0 \\ 0 & 0 & 1 \\ 0 & 1 & 0 \end{pmatrix} = \frac{i\sqrt{3}}{4}\lambda^6 \qquad ; f^{786} = \frac{\sqrt{3}}{2}$$

By direct inspection of these results we see that the structure constants are antisymmetric in the three indices. They are also invariant with respect to cyclic permutations of the indices.

As a result of these calculations, all the non-vanishing structure constants can be obtained from the 9 structure constants

$$f^{123} = 1$$

$$f^{147} = f^{165} = f^{246} = f^{572} = f^{345} = f^{376} = \frac{1}{2}$$

$$f^{458} = f^{678} = \frac{\sqrt{3}}{2}$$

Problem B.6 (a) Show that it follows from Eq. (9.17) that in matrix notation $\underline{\Psi}^\dagger$ is the row vector $\underline{\Psi}^\dagger = (\psi_1^\star, \psi_2^\star, \cdots, \psi_n^\star)$ and Eq. (9.9) is then just $\rho = \underline{\Psi}^\dagger \underline{\Psi}$;

(b) Show that the new adjoint in Eq. (9.44) is $\overline{\underline{\Psi}} = (\psi_1^\star, \psi_2^\star, -\psi_3^\star, -\psi_4^\star)$.

Solution to Problem B.6

(a) The transpose of a column vector is a row vector, with the same elements; moreover, the adjoint of a vector is the transpose conjugate of that vector. Thus

$$\underline{\Psi}^\dagger = (\psi_1^\star, \psi_2^\star, \cdots, \psi_n^\star)$$

The product of a row vector with a column vector provides a scalar. In particular, if we use the vectors $\underline{\Psi}^\dagger$ and $\underline{\Psi}$ we obtain

$$\underline{\Psi}^\dagger \underline{\Psi} = \sum_{\sigma=1}^n \psi_\sigma^\star \, \psi_\sigma = \rho$$

which coincides with Eq. (9.9).

(b) The proof here just involves calculating the product of the row vector of part (a) with the matrix $\gamma_4 = \beta$, given explicitly in Eq. (9.27)

$$\overline{\underline{\Psi}} = (\psi_1^\star, \psi_2^\star, \psi_3^\star, \psi_4^\star) \begin{pmatrix} 1 & 0 & 0 & 0 \\ 0 & 1 & 0 & 0 \\ 0 & 0 & -1 & 0 \\ 0 & 0 & 0 & -1 \end{pmatrix} = (\psi_1^\star, \psi_2^\star, -\psi_3^\star, -\psi_4^\star)$$

Appendix C

Fourier Series and Fourier Integrals

Problem C.1 Verify Eqs. (C.12).

Solution to Problem C.1

The Fourier series for the odd function $g(-x) = -g(x)$ is

$$g(x) = \sum_{n=1}^{\infty} a_n \left(\frac{2}{L}\right)^{1/2} \sin\left(\frac{n\pi x}{L}\right)$$

where

$$a_n = \left(\frac{2}{L}\right)^{1/2} \int_0^L dx\, g(x) \sin\left(\frac{n\pi x}{L}\right)$$

Notice that here $a_{-n} = -a_n$.

Introduce $k_n \equiv n\pi/L$, and take the limit $L \to \infty$. Then the sum over n can be replaced by an integral over k as follows

$$\sum_{n=1}^{\infty} F(k_n) = \frac{L}{\pi} \sum_{n=1}^{\infty} \left(\frac{\pi \Delta n}{L}\right) F(k_n)$$

$$\to \frac{L}{\pi} \int_0^{\infty} dk\, F(k) \qquad ; L \to \infty$$

In this limit, with $a_n \equiv a(k)$, we obtain the result that

$$g(x) = \frac{L}{\pi} \int_0^{\infty} dk\, a(k) \left(\frac{2}{L}\right)^{1/2} \sin(kx)$$

$$= \frac{L}{2\pi} \int_{-\infty}^{\infty} dk\, a(k) \left(\frac{2}{L}\right)^{1/2} \sin(kx)$$

The second line follows from the symmetry of the Fourier coefficient $a(-k) = -a(k)$. In a similar manner

$$a(k) = \left(\frac{2}{L}\right)^{1/2} \int_0^\infty dx\, g(x) \sin(kx)$$

$$= \left(\frac{2}{L}\right)^{1/2} \frac{1}{2} \int_{-\infty}^\infty dx\, g(x) \sin(kx)$$

With the re-definition

$$\left(\frac{L}{\pi}\right)^{1/2} a(k) \equiv \tilde{g}(k)$$

these expressions are re-written as a Fourier sine transform pair

$$g(x) = \frac{1}{\sqrt{2\pi}} \int_{-\infty}^\infty dk\, \tilde{g}(k) \sin(kx)$$

$$\tilde{g}(k) = \frac{1}{\sqrt{2\pi}} \int_{-\infty}^\infty dx\, g(x) \sin(kx)$$

Problem C.2 Consider the Fourier cosine series in Eqs. (C.6):

(a) Show that it defines a function in the interval $-L \leq x \leq L$ that is even about the origin with $f(-x) = f(x)$;

(b) Show that the *slope* of this function vanishes on the boundaries;

(c) Show that this series defines an extension of this function to all x that is *periodic in x* with period $2L$.

(d) What are the corresponding features of the Fourier sine series in Eqs. (2.45) and (2.48)?

Solution to Problem C.2

(a) Since the functions $\cos(n\pi x/L)$ are even in x, the function $f(x)$ defined by

$$f(x) = \sum_{n=1}^\infty a_n \left(\frac{2}{L}\right)^{1/2} \cos\left(\frac{n\pi x}{L}\right) + a_0 \left(\frac{1}{L}\right)^{1/2}$$

is also even in x, with $f(-x) = f(x)$.

(b) We have

$$f'(x) = -\sum_{n=1}^\infty a_n \frac{n\pi}{L} \left(\frac{2}{L}\right)^{1/2} \sin\left(\frac{n\pi x}{L}\right)$$

and therefore $f'(L) = f'(-L) = 0$.

(c) We have

$$f(x + 2L) = \sum_{n=1}^{\infty} a_n \left(\frac{2}{L}\right)^{1/2} \cos\left[\frac{n\pi(x + 2L)}{L}\right] + a_0 \left(\frac{1}{L}\right)^{1/2}$$

$$= \sum_{n=1}^{\infty} a_n \left(\frac{2}{L}\right)^{1/2} \cos\left(\frac{n\pi x}{L}\right) + a_0 \left(\frac{1}{L}\right)^{1/2}$$

$$= f(x)$$

Therefore $f(x)$ is periodic in x with period $2L$.

(d) We now turn to the function defined in Eq. (2.45)

$$f(x) = \sum_{n=1}^{\infty} a_n \left(\frac{2}{L}\right)^{1/2} \sin\left(\frac{n\pi x}{L}\right)$$

We easily see that the function is now odd, with $f(-x) = -f(x)$. Its derivative is

$$f'(x) = \sum_{n=1}^{\infty} a_n \frac{n\pi}{L} \left(\frac{2}{L}\right)^{1/2} \cos\left(\frac{n\pi x}{L}\right)$$

This does not vanish on the boundaries

$$f'(L) = \sum_{n=1}^{\infty} a_n (-1)^n \frac{n\pi}{L} \left(\frac{2}{L}\right)^{1/2} = f'(-L)$$

Finally, $f(x)$ is again periodic in x with period $2L$

$$f(x + 2L) = \sum_{n=1}^{\infty} a_n \left(\frac{2}{L}\right)^{1/2} \sin\left[\frac{n\pi(x + 2L)}{L}\right] = f(x)$$

Problem C.3x Compute the Fourier transform $\tilde{F}(k)$ in Eqs. (C.20) for the following functions:

(a) $F(x) = e^{-\gamma x}$ where γ is a constant;

(b) $F(x) = \sin \lambda x$ where λ is a constant. [*Hint*: Use Eq. (A.12) to write $\sin \phi = (e^{i\phi} - e^{-i\phi})/2i$.]

Solution to Problem C.3x

The Fourier transform $\tilde{F}(k)$ is given by

$$\tilde{F}(k) = \frac{1}{\sqrt{2\pi}} \int_{-\infty}^{\infty} e^{-ikx} F(x)dx$$

(a) For $F(x) = e^{-\gamma|x|}$, we have[1]

$$\tilde{F}(k) = \frac{1}{\sqrt{2\pi}} \left\{ \int_0^\infty e^{-ikx} e^{-\gamma x} dx + \int_{-\infty}^0 e^{-ikx} e^{\gamma x} dx \right\}$$

$$= \frac{1}{\sqrt{2\pi}} \left\{ \frac{1}{ik+\gamma} + \frac{1}{-ik+\gamma} \right\} = \left(\frac{2}{\pi}\right)^{1/2} \frac{\gamma}{k^2+\gamma^2}$$

(b) For $F(x) = \sin \lambda x$, We have

$$\tilde{F}(k) = \frac{1}{\sqrt{2\pi}} \int_{-\infty}^\infty e^{-ikx} \sin \lambda x \, dx$$

$$= \frac{1}{2i} \frac{1}{\sqrt{2\pi}} \int_{-\infty}^\infty e^{-ikx} \left(e^{i\lambda x} - e^{-i\lambda x} \right) dx$$

$$= \frac{1}{i} \left(\frac{\pi}{2}\right)^{1/2} [\delta(k-\lambda) - \delta(k+\lambda)]$$

Problem C.3 (a) Use Eq. (A.12) and the assumption that $f(x)$ is real and symmetric with $f(-x) = f(x)$ to rewrite the Fourier cosine series in Eqs. (C.6) in the following form

$$f(x) = \frac{1}{\sqrt{l}} \sum_{n=-\infty}^\infty a_e(k_n) e^{ik_n x} \qquad\qquad ; \; k_n = \frac{2\pi n}{l}$$

$$a_e(k_n) = \frac{1}{\sqrt{l}} \int_{-l/2}^{l/2} dx \, e^{-ik_n x} f(x) \qquad\qquad n = 0, \pm 1, \pm 2, \cdots, \pm\infty$$

where $l \equiv 2L$. This forms a *complex Fourier series*;
  (b) Show that in this case $a_e(k_n)^\star = a_e(-k_n) = a_e(k_n)$.

Solution to Problem C.3

(a) We start from Eqs. (C.6)

$$f(x) = \sum_{n=1}^\infty a_n \left(\frac{2}{L}\right)^{1/2} \cos\left(\frac{n\pi x}{L}\right) + a_0 \left(\frac{1}{L}\right)^{1/2}$$

$$a_n = \left(\frac{2}{L}\right)^{1/2} \int_0^L dx \, f(x) \cos\left(\frac{n\pi x}{L}\right) \qquad ; \; n = 1, 2, \cdots, \infty$$

$$a_0 = \left(\frac{1}{L}\right)^{1/2} \int_0^L dx \, f(x)$$

---

[1]The integrals are only well-defined for negative x if $F(x) = e^{-\gamma|x|}$; we assume this to be the case. The problem should have so-stated.

We write

$$f(x) = \sum_{n=1}^{\infty} a_n \left(\frac{2}{L}\right)^{1/2} \cos\left(\frac{n\pi x}{L}\right) + a_0 \left(\frac{1}{L}\right)^{1/2}$$

$$= \sum_{n=1}^{\infty} a_n \left(\frac{1}{2L}\right)^{1/2} \left[e^{i(n\pi x/L)} + e^{-i(n\pi x/L)}\right] + a_0 \left(\frac{1}{L}\right)^{1/2}$$

where we have used Eq. (A.12) to write $\cos a = (e^{ia} + e^{-ia})/2$. Notice that since $f(x)$ is real, the Fourier coefficients a_n are real, with $a_n^* = a_n$.

Consider the series containing $e^{-i(n\pi x/L)}$, and re-write it after making the substitution $n \to -n$

$$\sum_{n=1}^{\infty} a_n \left(\frac{1}{2L}\right)^{1/2} e^{-i(n\pi x/L)} = \sum_{n=-\infty}^{-1} a_n \left(\frac{1}{2L}\right)^{1/2} e^{i(n\pi x/L)}$$

where we have used the property $a_{-n} = a_n$, which follows from the definition of the a_n. Therefore, with $l \equiv 2L$,

$$f(x) = \frac{1}{\sqrt{l}} \sum_{n=-\infty}^{\infty} a_e(k_n) \, e^{i(2n\pi x/l)}$$

Here $a_e(k_n) \equiv a_n$ for $n \neq 0$ and $a_e(0) \equiv \sqrt{2}a_0$.

Let us now consider the Fourier coefficients a_n given above

$$a_n = \left(\frac{2}{L}\right)^{1/2} \int_0^L dx \, f(x) \cos\left(\frac{n\pi x}{L}\right)$$

$$= \left(\frac{1}{l}\right)^{1/2} \int_0^L dx \, f(x) \left(e^{ik_n x} + e^{-ik_n x}\right)$$

We now perform the change of variable $x \to -x$ in the part of integral that contains $e^{ik_n x}$, and use the fact that $f(x)$ is symmetric to obtain

$$a_e(k_n) = \frac{1}{\sqrt{l}} \int_{-l/2}^{l/2} dx \, f(x) \, e^{-ik_n x}$$

where we have made explicit the dependence on $k_n = 2\pi n/l$.

(b) If we take the complex conjugate of $a_e(k_n)$ we obtain

$$a_e(k_n)^* = \frac{1}{\sqrt{l}} \int_{-l/2}^{l/2} dx \, f(x) e^{ik_n x} = a_e(-k_n)$$

On the other hand we may also rewrite $a_e(k_n)$ performing the change of variable $x \to -x$ in the integral, and using the symmetry of $f(x)$ to obtain

$$a_e(k_n) = \frac{1}{\sqrt{l}} \int_{-l/2}^{l/2} dx\, f(x) e^{ik_n x} = a_e(-k_n)$$

The relations

$$f(x) = \frac{1}{\sqrt{l}} \sum_{n=-\infty}^{\infty} a_e(k_n) e^{ik_n x} \qquad ; \; k_n = \frac{2\pi n}{l}$$

$$a_e(k_n) = \frac{1}{\sqrt{l}} \int_{-l/2}^{l/2} dx\, e^{-ik_n x} f(x) \qquad n = 0, \pm 1, \pm 2, \cdots, \pm \infty$$

form a *complex Fourier series*, here established for an even function $f(-x) = f(x)$.

Problem C.4 Extend the results in Prob. C.3 to a real *antisymmetric* function $g(-x) = -g(x)$.

(a) Show the modification of the result in Prob. C.3(a) is to replace $f(x)$ by $g(x)$, and $a_e(k_n)$ by $-ia_o(k_n)$, with $a_o(0) = 0$;

(b) Show that in this case $a_o(k_n)^* = -a_o(-k_n) = a_o(k_n)$.

Solution to Problem C.4

(a) In this case, the odd function $g(-x) = -g(x)$ may be expressed in terms of the series

$$g(x) = \sum_{n=1}^{\infty} a_n \left(\frac{2}{L}\right)^{1/2} \sin\left(\frac{n\pi x}{L}\right)$$

$$= \sum_{n=1}^{\infty} \frac{a_n}{i} \left(\frac{1}{2L}\right)^{1/2} \left[e^{i(n\pi x/L)} - e^{-i(n\pi x/L)} \right]$$

where we have used Eq. (A.12) to write $\sin a = (e^{ia} - e^{-ia})/2i$. Here

$$a_n = \left(\frac{2}{L}\right)^{1/2} \int_0^L dx\, g(x) \sin\left(\frac{n\pi x}{L}\right)$$

and therefore $a_{-n} = -a_n$.

We now consider the second part of the series above, containing $e^{-i(n\pi x/L)}$, and we write it as

$$-\sum_{n=1}^{\infty} \frac{a_n}{i} \left(\frac{1}{2L}\right)^{1/2} e^{-i(n\pi x/L)} = \sum_{n=-\infty}^{-1} \frac{a_n}{i} \left(\frac{1}{2L}\right)^{1/2} e^{i(n\pi x/L)}$$

Here we have performed the change $n \to -n$ and used the property of the Fourier coefficients $a_{-n} = -a_n$. Notice that this property implies $a_0 = 0$.

Finally, with the definition $a_o(k_n) \equiv a_n$, we may write

$$g(x) = -\frac{i}{\sqrt{l}} \sum_{n=-\infty}^{\infty} a_o(k_n)\, e^{ik_n x} \qquad ; \; k_n = \frac{2\pi n}{l}$$

where, as in Prob. C.4,

$$a_o(k_n) = \frac{i}{\sqrt{l}} \int_{-l/2}^{l/2} dx\, g(x) e^{-ik_n x}$$

(b) We have from the above

$$a_o(k_n)^\star = -\frac{i}{\sqrt{l}} \int_{-l/2}^{l/2} dx\, g(x) e^{ik_n x} = -a_o(-k_n)$$

and from a change of variable $x \to -x$

$$a_o(k_n)^\star = -\frac{i}{\sqrt{l}} \int_{-l/2}^{l/2} dx\, g(-x) e^{-ik_n x} = a_o(k_n)$$

Hence

$$a_o(k_n)^\star = -a_o(-k_n) = a_o(k_n)$$

In summary, the relations

$$g(x) = \frac{-i}{\sqrt{l}} \sum_{n=-\infty}^{\infty} a_o(k_n) e^{ik_n x} \qquad ; \; k_n = \frac{2\pi n}{l}$$

$$-ia_o(k_n) = \frac{1}{\sqrt{l}} \int_{-l/2}^{l/2} dx\, e^{-ik_n x} g(x) \qquad n = 0, \pm 1, \pm 2, \cdots, \pm\infty$$

again form a *complex Fourier series*, here established for an odd function $g(-x) = -g(x)$.

Problem C.5 (a) Use the results in the previous two problems to extend the complex Fourier series in Prob. C.3(a) to an *arbitrary* real function $f(x)$ with Fourier coefficients $a(k_n)$;

(b) Show $a(k_n)^\star = a(-k_n)$ in this case;

(c) Show the extension of $f(x)$ to all x is then periodic with period l;[2]

---

[2] As in the physical string problem, we are content here to assume that $f(x)$ is everywhere continuous with a continuous first derivative; the Fourier series is capable of representing a broader class of functions.

(d) Take the limit $l \to \infty$, and re-derive the Fourier integral representation in Eqs. (C.20).

Solution to Problem C.5

(a) For an arbitrary real function $f(x)$ we may write

$$f(x) = \frac{f(x) + f(-x)}{2} + \frac{f(x) - f(-x)}{2} \equiv f_e(x) + f_o(x)$$

where $f_e(x)$ and $f_o(x)$ are even and odd real functions respectively, to which the results of the previous two problems apply. Therefore

$$f(x) = \frac{1}{\sqrt{l}} \sum_{n=-\infty}^{\infty} [a_e(k_n) - i a_o(k_n)]\, e^{i k_n x}$$

$$= \frac{1}{\sqrt{l}} \sum_{n=-\infty}^{\infty} a(k_n) e^{i k_n x}$$

where the Fourier coefficients are given by

$$a(k_n) \equiv a_e(k_n) - i a_o(k_n) = \frac{1}{\sqrt{l}} \int_{-l/2}^{l/2} dx\, e^{-i k_n x} f(x)$$

This establishes the complex Fourier series for an arbitrary real function $f(x)$ on the interval $[-l/2,\, l/2]$.

(b) Use the expression for $a(k_n)$, and remember that $f(x)$ is real. Then

$$a(k_n)^\star = \frac{1}{\sqrt{l}} \int_{-l/2}^{l/2} dx\, e^{i k_n x} f(x) = a(-k_n)$$

(c) Given $k_n = 2\pi n/l$, we have

$$f(x + l) = \frac{1}{\sqrt{l}} \sum_{n=-\infty}^{\infty} a(k_n) e^{i(k_n x + 2\pi n)} = f(x)$$

and therefore $f(x)$ is periodic with period l.

(d) In the limit $l \to \infty$, we can use $k_n = 2\pi n/l$ and convert the sum on n to an integral on k through

$$\sum_n F(k_n) = \frac{l}{2\pi} \sum_n \left(\frac{2\pi \Delta n}{l} \right) F(k_n)$$

$$\to \frac{l}{2\pi} \int dk\, F(k) \qquad\qquad ;\, l \to \infty$$

This gives

$$
\begin{aligned}
f(x) &= \lim_{l \to \infty} \frac{\sqrt{l}}{2\pi} \int_{-\infty}^{\infty} dk \left[\frac{1}{\sqrt{l}} \int_{-l/2}^{l/2} dy \, e^{-iky} f(y) \right] e^{ikx} \\
&= \frac{1}{\sqrt{2\pi}} \int_{-\infty}^{\infty} dk \left[\frac{1}{\sqrt{2\pi}} \int_{-\infty}^{\infty} dy \, e^{-iky} f(y) \right] e^{ikx}
\end{aligned}
$$

from which Eqs. (C.20) follow

$$
f(x) = \frac{1}{\sqrt{2\pi}} \int_{-\infty}^{\infty} e^{ikx} \tilde{f}(k) \, dk
$$

$$
\tilde{f}(k) = \frac{1}{\sqrt{2\pi}} \int_{-\infty}^{\infty} e^{-ikx} f(x) \, dx
$$

This re-derives the Fourier transform relations for an arbitrary real function $f(x)$.

Problem C.6 Use the results in Prob. 4.3 to show that the Fourier integral representation in Eqs. (C.20) can be extended to hold for an *arbitrary, well-behaved, complex function* $F(x)$.

Solution to Problem C.6

An arbitrary complex function, $F(x)$, may be expressed as

$$
F(x) = \frac{F(x) + F(x)^\star}{2} + \frac{F(x) - F(x)^\star}{2} \equiv F_R(x) + iF_I(x)
$$

where $F_R(x) \equiv [F(x) + F(x)^\star]/2$ and $F_I(x) \equiv -i[F(x) - F(x)^\star]/2$ are both real functions, each having a Fourier integral representation as in Eqs. (C.20). We thus conclude that the Fourier integral representation in Eqs. (C.20) holds for an *arbitrary, well-behaved, complex function* $F(x)$.

If we substitute the second of Eqs.(C.20) inside the first we obtain

$$
\begin{aligned}
F(x) &= \frac{1}{\sqrt{2\pi}} \int_{-\infty}^{\infty} dk \, e^{ikx} \left\{ \frac{1}{\sqrt{2\pi}} \int_{-\infty}^{\infty} dy \, e^{-iky} [F_R(y) + iF_I(y)] \right\} \\
&= \int_{-\infty}^{\infty} dy [F_R(y) + iF_I(y)] \left[\frac{1}{2\pi} \int_{-\infty}^{\infty} dk \, e^{ik(x-y)} \right] \\
&= \int_{-\infty}^{\infty} dy [F_R(y) + iF_I(y)] \delta(x-y) \\
&= F(x)
\end{aligned}
$$

where the last relation follows from the properties of the Dirac delta function seen in Prob. 4.3.

Appendix D

Some Thermodynamics

There are no problems included with this Appendix.

Appendix E

Some Statistical Mechanics

Problem E.1 The discussion in appendix E centers on independent *localized* particles. If the particles are *non-localized*, then a configuration that simply corresponds to an interchange of particle *labels* does not give rise to a new, distinct complexion. The number of such interchanges is $N!$, and thus one should replace $\Omega \to \Omega/N!$ in these arguments. Show the only effect of this change is to replace the Helmholtz free energy in Eq. (E.31) by

$$A(N, V, T) = -k_B T \ln \frac{(\text{p.f.})^N}{N!} \qquad ; \text{ non-localized particles}$$

Solution to Problem E.1

For non-localized particles, we replace

$$S = k_B \ln \Omega \to k_B \ln \frac{\Omega}{N!}$$

in the calculation of $k_B \ln \Omega$ in appendix E. Since $\ln N!$ is a constant, it does not affect the argument leading to the Boltzmann distribution numbers in Eqs. (E.16)–(E.25). Equation (E.27) then takes the form

$$S(N, V, E) = k_B \left[\ln \frac{\left(\sum_{l=1}^{\infty} e^{\beta \varepsilon_l} \right)^N}{N!} - \beta E \right]$$

Again, since $\ln N!$ is a constant, it does not affect the argument leading to the identification of the temperature $\beta = -1/k_B T$ in Eq. (E.29). A rearrangement of the above then gives the Helmholtz free energy $A = E - TS$ as

$$A(N, V, T) = -k_B T \ln \frac{(\text{p.f.})^N}{N!} \qquad ; \text{ non-localized particles}$$

283

Problem E.2 The classical partition function for a particle in a box of volume V in three dimensions is given by the first of Eqs. (E.33) as

$$(\text{p.f.}) = \frac{1}{h^3} \int \cdots \int e^{-(p_1^2 + p_2^2 + p_3^2)/2mk_\mathrm{B}T} \, dp_1 \cdots dp_3 \, dq_1 \cdots dq_3$$

Show

$$(\text{p.f.}) = V \left(\frac{2\pi m k_\mathrm{B} T}{h^2} \right)^{3/2} \qquad \text{; classical limit}$$

Solution to Problem E.2

The multiple integral can be performed in a straightforward fashion. The integral over coordinates $(q_1, q_2, q_3) = (x, y, z)$ gives the volume $\int \int \int dq_1 \, dq_2 \, dq_3 = V$. The integrals over momenta factor into three identical gaussian integrals, which are evaluated in Eq (2.65)

$$\int_{-\infty}^{\infty} e^{-p^2/2mk_\mathrm{B}T} dp = \sqrt{2\pi m k_\mathrm{B} T}$$

Thus

$$(\text{p.f.}) = V \left(\frac{2\pi m k_\mathrm{B} T}{h^2} \right)^{3/2}$$

Problem E.3 Use the results in Probs. E.1 and E.2, and Eqs. (D.13), to show that the classical limit of the chemical potential for an ideal gas of density $n = N/V$ is given by

$$\frac{\mu}{k_\mathrm{B}T} = -\ln \left\{ \frac{1}{n} \left(\frac{2\pi m k_\mathrm{B} T}{h^2} \right)^{3/2} \right\} \qquad \text{; classical limit}$$

Solution to Problem E.3

With the use of the results in Probs. E.1 and E.2, and Eqs. (D.13), we have

$$\mu = \left(\frac{\partial A}{\partial N} \right)_{V,T} = -k_B T \frac{\partial}{\partial N} \ln \frac{(\text{p.f.})^N}{N!}$$

$$= -k_B T \frac{\partial}{\partial N} \left\{ N \ln \left[V \left(\frac{2\pi m k_\mathrm{B} T}{h^2} \right)^{3/2} \right] - \ln N! \right\}$$

We may use Stirling's formula $\ln N! = N \ln N - N$ in Eq (E.9) in the above expression

$$\mu = -k_B T \frac{\partial}{\partial N} \left\{ N \ln \left[V \left(\frac{2\pi m k_B T}{h^2} \right)^{3/2} \right] - N \ln N + N \right\}$$

The derivative then gives

$$\mu = -k_B T \left\{ \ln \left[V \left(\frac{2\pi m k_B T}{h^2} \right)^{3/2} \right] - \ln N \right\}$$

Therefore, with $n \equiv N/V$, we have for the chemical potential

$$\frac{\mu}{k_B T} = -\ln \left\{ \frac{1}{n} \left(\frac{2\pi m k_B T}{h^2} \right)^{3/2} \right\} \qquad ; \; n \equiv \frac{N}{V}$$

as anticipated.

Problem E.4 The quantum partition function for a particle in a box is given by

$$(\text{p.f.}) = \sum_{n_1=1}^{\infty} \sum_{n_2=1}^{\infty} \sum_{n_3=1}^{\infty} e^{-E_{n_1 n_2 n_3}/k_B T}$$

where $E_{n_1 n_2 n_3}$ is the eigenvalue in Eq. (4.116). Introduce the quantities $u_i \equiv \hbar \pi n_i / \sqrt{2 m k_B T}$ with $i = (1, 2, 3)$. In the high-temperature limit where $T \to \infty$, the sum over n_i can be replaced by an integral over u_i, using the technique employed many times in this text.

(a) Show that the high-temperature limit of this quantum partition function is the classical result in Prob. E.2.

(b) Verify the identification of the constant h, here put in by hand in the classical partition function, with *Planck's constant*.

Solution to Problem E.4

(a) The eigenvalues of Eq. (4.116) are

$$E_{n_1, n_2, n_3} = \frac{\hbar^2 \pi^2}{2m} \left(\frac{n_1^2}{a^2} + \frac{n_2^2}{b^2} + \frac{n_3^2}{c^2} \right) \qquad ; \; n_i = 1, 2, 3, \cdots$$

These take a simple form when expressed in terms of the quantities u_i

$$E_{u_1, u_2, u_3} = k_B T \left(\frac{u_1^2}{a^2} + \frac{u_2^2}{b^2} + \frac{u_3^2}{c^2} \right) \qquad ; \; u_i = \frac{\hbar \pi n_i}{\sqrt{2 m k_B T}}$$

In the limit of high-temperature, the sum over n_i in the quantum partition function may be replaced by an integral over the u_i

$$\sum_{n_i=1}^{\infty} F(u_i) = \sum_{n_i=1}^{\infty} \Delta n_i F(u_i) = \frac{\sqrt{2mk_BT}}{\hbar\pi} \sum_{n_i=1}^{\infty} \Delta u_i F(u_i)$$

$$\rightarrow \frac{\sqrt{2mk_BT}}{\hbar\pi} \int_0^{\infty} du\, F(u) \qquad ; T \rightarrow \infty$$

The high-temperature partition function factors into three such integrals

$$(\text{p.f.}) = \left(\frac{\sqrt{2mk_BT}}{\hbar\pi}\right)^3 \int_0^{\infty} du_1\, e^{-u_1^2/a^2} \int_0^{\infty} du_2\, e^{-u_2^2/b^2} \int_0^{\infty} du_3\, e^{-u_3^2/c^2}$$

The gaussian integrals in this expression are easily obtained using Eq. (2.65). Since our region of integration only goes from 0 to ∞, each gaussian integral is only half the integral over the whole real axis. Therefore[1]

$$(\text{p.f.}) = \left(\frac{\sqrt{2mk_BT}}{\hbar\pi}\right)^3 \left(\frac{\sqrt{\pi}}{2}\right)^3 abc = V \left(\frac{2\pi mk_BT}{h^2}\right)^{3/2}$$

(b) The eigenvalues E_{n_1,n_2,n_3} for a particle in a box depend on Planck's constant h. In the high-temperature limit, the quantum partition function obtained using these eigenvalues reduces to the same form as the classical partition function. Thus we see that the constant h appearing in the classical partition function in Prob. E.2, which had to be put in by hand, is indeed Planck's constant.

Problem E.5 (a) Consider the periodic orbit in phase space for the simple harmonic oscillator of energy E defined by Eqs. (2.12) and (2.15). The area of an ellipse is given by $A = \pi ab$ where (a, b) are the semi-major and semi-minor axes respectively. Show the area of the orbit in phase space is

$$A = \frac{2\pi E}{\omega} \qquad ; \kappa \equiv m\omega^2$$

(b) When quantized, the energy spectrum of the oscillator is that in Eq. (4.102). Show that the area in phase space between two adjacent levels is given by

$$A_{n+1} - A_n = h$$

---

[1] Recall $2\pi\hbar = h$.

where h is Planck's constant. This is a special case of the early Bohr-Sommerfeld quantization condition for periodic motion

$$\oint p \, dq = nh \qquad ; \text{ Bohr-Sommerfeld}$$

While not an exact result, it does hold quite generally for large quantum numbers, or in the classical limit.

(c) At low temperatures, the partition function in Eqs. (E.31) receives a contribution from only a few states, while at high T, many states contribute. In this case, one can group the contributions by using a mean value $e^{-\varepsilon(p,q)/k_B T}$ times the number of states $dn = dA/h$ in the small area of phase space $dA = dp \, dq$ from part (b). Thus, justify the first of Eqs. (E.32), and derive the first of Eqs. (E.33) for the partition function with $s = 1$ in the high T, or classical limit.

Solution to Problem E.5

(a) The constant-energy orbit in phase space of a particle in the simple harmonic oscillator orbit is described by an ellipse

$$\frac{p^2}{2m} + \frac{1}{2} m \omega^2 q^2 = E$$

This ellipse in standard form $(q/a)^2 + (p/b)^2 = 1$ has semi-axes of length

$$a = \frac{1}{\omega} \sqrt{\frac{2E}{m}} \qquad ; \ b = \sqrt{2mE}$$

The area of the ellipse is then

$$A = \pi a b = \frac{2\pi E}{\omega}$$

(b) The area in phase space corresponding to the energy $E_n = \hbar\omega(n + 1/2)$ is

$$A_n = \frac{2\pi E_n}{\omega} = h(n + 1/2)$$

The states in the simple harmonic oscillator can therefore be put into one-to-one correspondence with an element of phase-space area of

$$A_{n+1} - A_n = h$$

(c) In the high-temperature, or classical, limit many states contribute to the partition function. In this limit, the sum in the quantum partition function may be approximated with an integral. In particular

$$\sum_{i=1}^{\infty} \rightarrow \int dn$$

where dn is the number of states in a small area of phase space $dA = dpdq$, that is, $dn = dpdq/h$. Thus

$$\sum_{i=1}^{\infty} \rightarrow \frac{1}{h} \int \int dpdq$$

If we use this result in the quantum partition function of Eq. (E.31) for one degree of freedom ($s = 1$), we obtain

$$(\text{p.f.}) \rightarrow \frac{1}{h} \int \int dpdq \; e^{-\varepsilon(p,q)/k_{\mathrm{B}}T} \qquad ; T \rightarrow \infty$$

which is just the classical partition function in the first of Eqs. (E.33).

Problem E.6 Use the equipartition theorem to show that the internal energy of an ideal gas of N structureless constituents at a temperature T is given by $E(T) = 3Nk_{\mathrm{B}}T/2$. Note that this only depends on the temperature.

Solution to Problem E.6

Since we have an ideal gas of N structureless constituents at temperature T, the internal energy of the system is simply the total kinetic energy of the system. Therefore there are three quadratic degrees of freedom associated with each constituent, and according to the equipartition theorem we get

$$E(T) = \frac{3}{2}Nk_{B}T$$

Appendix F

Some Vector Calculus

Problem F.1 (a) Complete the demonstration of Gauss' theorem in two dimensions in Eq. (F.6) by extending the analysis to include the other three quadrants;

(b) Complete the demonstration of Stokes' theorem in two dimensions in Eq. (F.9) by extending the analysis to include the other three quadrants.

Solution to Problem F.1

(a) We consider the convex region of area A surrounded by a curve C represented in Fig. F.1 in the text. Let us consider the second quadrant (II). In this case we have

$$
\int_{\mathrm{II}} d^2x\, \boldsymbol{\nabla}\cdot\mathbf{v} = \int_{\mathrm{II}} dxdy\left(\frac{\partial v_x}{\partial x} + \frac{\partial v_y}{\partial y}\right)
$$

$$
= \int_0^{y_{\max}} dy \int_{x_{\min}(y)}^0 dx\,\frac{\partial v_x}{\partial x} + \int_{x_{\min}}^0 dx \int_0^{y_{\max}(x)} dy\,\frac{\partial v_y}{\partial y}
$$

$$
\rightarrow -\int_0^{y_{\max}} dy\, v_x[x_{\min}(y)] + \int_{x_{\min}}^0 dx\, v_y[y_{\max}(x)]
$$

where we have omitted the contributions from the definite integral corresponding to the horizontal and vertical axes, since these cancel between contiguous quadrants. Let us call θ the angle that the tangent unit vector $\mathbf{n}_\parallel$ forms with $\mathbf{e}_y$ and the normal unit vector $\mathbf{n}_\perp$ forms with $\mathbf{e}_x$. Therefore[1]

$$
\mathbf{n}_\perp = \mathbf{e}_x\cos\theta + \mathbf{e}_y\sin\theta
$$

Here θ is measured in the positive direction from the positive x-axis. It

---

[1] We will later consistently use a caret over a symbol to indicate an operator in abstract Hilbert space. In anticipation, we here delete all hats over the unit vectors.

is easy to see that in this case $dx = dl \sin(\pi - \theta) = dl \sin \theta$ and $dy = dl \cos(\pi - \theta) = -dl \cos \theta$. With the use of these relations, we may write the above as

$$\int_{II} d^2x\, \boldsymbol{\nabla} \cdot \mathbf{v} \to \int_{II} dl\, [\cos \theta\, v_x + \sin \theta\, v_y] = \int_{II} dl\, \mathbf{n}_\perp \cdot \mathbf{v}$$

where the notation $\int_{II} dl$ is used to indicate the portion of integral over the curve C falling in the second quadrant.

The cases of the third and fourth quadrant are obtained in a similar fashion. For the third quadrant we have

$$\int_{III} d^2x\, \boldsymbol{\nabla} \cdot \mathbf{v} = \int_{III} dxdy \left(\frac{\partial v_x}{\partial x} + \frac{\partial v_y}{\partial y} \right)$$

$$= \int_{y_{min}}^{0} dy \int_{x_{min}(y)}^{0} dx \frac{\partial v_x}{\partial x} + \int_{x_{min}}^{0} dx \int_{y_{min}(x)}^{0} dy \frac{\partial v_y}{\partial y}$$

$$\to - \int_{y_{min}}^{0} dy\, v_x[x_{min}(y)] - \int_{x_{min}}^{0} dx\, v_y[y_{min}(x)]$$

In the third quadrant we have $dx = dl\ \sin(\theta - \pi) = -dl \sin \theta$ and $dy = dl \cos(\theta - \pi) = -dl \cos \theta$. Thus we have

$$\int_{III} d^2x\, \boldsymbol{\nabla} \cdot \mathbf{v} \to \int_{III} dl\, [\cos \theta\, v_x + \sin \theta\, v_y] = \int_{III} dl\, \mathbf{n}_\perp \cdot \mathbf{v}$$

Finally, for the fourth quadrant we have

$$\int_{IV} d^2x\, \boldsymbol{\nabla} \cdot \mathbf{v} = \int_{IV} dxdy \left(\frac{\partial v_x}{\partial x} + \frac{\partial v_y}{\partial y} \right)$$

$$= \int_{y_{min}}^{0} dy \int_{0}^{x_{max}(y)} dx \frac{\partial v_x}{\partial x} + \int_{0}^{x_{max}} dx \int_{y_{min}(x)}^{0} dy \frac{\partial v_y}{\partial y}$$

$$\to \int_{y_{min}}^{0} dy\, v_x[x_{max}(y)] - \int_{0}^{x_{max}} dx\, v_y[y_{min}(x)]$$

In the fourth quadrant we have $dx = dl \sin(2\pi - \theta) = -dl \sin \theta$ and $dy = dl \cos(2\pi - \theta) = dl \cos \theta$. Therefore

$$\int_{IV} d^2x\, \boldsymbol{\nabla} \cdot \mathbf{v} \to \int_{IV} dl\, [\cos \theta\, v_x + \sin \theta\, v_y] = \int_{IV} dl\, \mathbf{n}_\perp \cdot \mathbf{v}$$

A combination of the results from the four quadrants then gives Gauss' theorem in two dimensions in Eq. (F.7)

$$\int_{A} d^2x\, \boldsymbol{\nabla} \cdot \mathbf{v} = \oint_{C} d\mathbf{S} \cdot \mathbf{v}$$

(b) The proof of Stokes' theorem in two dimensions can be carried out in analogy to what we have done in part (a). We need to notice that in this case $\mathbf{n}_\| = \mathbf{e}_x \cos(\theta + \pi/2) + \mathbf{e}_y \sin(\theta + \pi/2) = -\mathbf{e}_x \sin\theta + \mathbf{e}_y \cos\theta$, where θ is the angle that $\mathbf{n}_\perp$ forms with $\mathbf{e}_x$.

We notice that

$$\int_A d\mathbf{A} \cdot \nabla \times \mathbf{v} = \int_A dx dy \left(\frac{\partial v_y}{\partial x} - \frac{\partial v_x}{\partial y} \right)$$

With the definition $\mathbf{w} = [w_x(x,y),\, w_y(x,y)] \equiv [v_y(x,y),\, -v_x(x,y)]$ we may write

$$\int_A d\mathbf{A} \cdot \nabla \times \mathbf{v} = \int_A d^2x\, \nabla \cdot \mathbf{w}$$

An application of Gauss' theorem from part (a) then yields

$$\int_A d\mathbf{A} \cdot \nabla \times \mathbf{v} = \oint_C dl\, \mathbf{n}_\perp \cdot \mathbf{w}$$

Since θ is also the angle that $\mathbf{n}_\|$ forms with $\mathbf{e}_y$, we may write

$$\mathbf{n}_\perp = \mathbf{e}_x \cos\theta + \mathbf{e}_y \sin\theta$$
$$\mathbf{n}_\| = -\mathbf{e}_x \sin\theta + \mathbf{e}_y \cos\theta$$

and thus

$$\int_A d\mathbf{A} \cdot \nabla \times \mathbf{v} = \oint_C dl\, \mathbf{n}_\| \cdot \mathbf{v}$$

This is Stokes' theorem in two dimensions in Eq. (F.10)

$$\int_A d\mathbf{A} \cdot \nabla \times \mathbf{v} = \oint_C d\mathbf{l} \cdot \mathbf{v}$$

Problem F.2 Extend the proof of Gauss' theorem from two to three dimensions and derive Eq. (F.7). It is sufficient here to confine the analysis to the first octant.

Solution to Problem F.2

Let $\mathbf{n}$ be the outward pointing normal unit vector on the bounding surface S. Then

$$\mathbf{n} = \cos\phi \sin\theta\, \mathbf{e}_x + \sin\phi \sin\theta\, \mathbf{e}_y + \cos\theta\, \mathbf{e}_z$$

We have

$$\int_V d^3x\, \boldsymbol{\nabla}\cdot\mathbf{v} = \int_V d^3x\left(\frac{\partial v_x}{\partial x} + \frac{\partial v_y}{\partial y} + \frac{\partial v_z}{\partial z}\right)$$

As in two dimensions, we may carry out an integration on each separate term in the above expression to obtain

$$\int_V d^3x\left(\frac{\partial v_x}{\partial x} + \frac{\partial v_y}{\partial y} + \frac{\partial v_z}{\partial z}\right) \to \int_0^{y_{\max}} dy \int_0^{z_{\max}} dz\, v_x[x_{\max}(y,z),y,z]$$
$$+ \int_0^{x_{\max}} dx \int_0^{z_{\max}} dz\, v_y[x,y_{\max}(x,z),z]$$
$$+ \int_0^{x_{\max}} dx \int_0^{y_{\max}} dy\, v_z[x,y,z_{\max}(x,y)]$$

Now consider the surface element $d\mathbf{S} = dS\mathbf{n}$. The projection of $d\mathbf{S}$ on the x-y plane is given by $d\mathbf{S}\cdot\mathbf{e}_z = dS\cos\theta = dxdy$; similarly $d\mathbf{S}\cdot\mathbf{e}_x = dS\cos\phi\sin\theta = dydz$ and $d\mathbf{S}\cdot\mathbf{e}_y = dS\sin\phi\sin\theta = dxdz$ are the projections of $d\mathbf{S}$ on the y-z and x-z planes, respectively. Hence we can write

$$d\mathbf{S} = dS\mathbf{n} = dS\,(\cos\phi\sin\theta\,\mathbf{e}_x + \sin\phi\sin\theta\,\mathbf{e}_y + \cos\theta\,\mathbf{e}_z)$$
$$= \mathbf{e}_x dydz + \mathbf{e}_y dxdz + \mathbf{e}_z dxdy$$

The integrand on the above r.h.s. is thus simply the projection of $\mathbf{v}$ on $d\mathbf{S}$

$$d\mathbf{S}\cdot\mathbf{v} = v_x dydz + v_y dxdz + v_z dxdy$$

Gauss' law in three dimensions can thus be written as an integral over the surface S bounding the volume V with surface element $d\mathbf{S} = dS\mathbf{n}$

$$\int_V d^3x\, \boldsymbol{\nabla}\cdot\mathbf{v} = \int_S (v_x dydz + v_y dxdz + v_z dxdy)$$
$$= \int_S d\mathbf{S}\cdot\mathbf{v}$$

Problem F.3 Assume $\phi(x,y,z)$ is some differentiable scalar function of position, and use the definition of the gradient $\boldsymbol{\nabla}$ in Eq. (4.112).

(a) Show

$$\boldsymbol{\nabla}\times\boldsymbol{\nabla}\phi = 0$$

(b) Let $d\mathbf{l} = \mathbf{e}_x\, dx + \mathbf{e}_y\, dy + \mathbf{e}_z\, dz$ be an infinitesimal displacement. Show the total differential of ϕ is given by

$$d\phi = \boldsymbol{\nabla}\phi\cdot d\mathbf{l}$$

(c) Show that in cylindrical coordinates (z, r, θ) the gradient $\boldsymbol{\nabla}$ takes the form

$$\boldsymbol{\nabla} = \mathbf{e}_z \frac{\partial}{\partial z} + \mathbf{e}_r \frac{\partial}{\partial r} + \mathbf{e}_\theta \frac{1}{r} \frac{\partial}{\partial \theta}$$

Solution to Problem F.3

(a) Let us work on the *ith* component of the vector $\boldsymbol{\nabla} \times \boldsymbol{\nabla} \phi$

$$(\boldsymbol{\nabla} \times \boldsymbol{\nabla} \phi)_i = \sum_{jk} \epsilon_{ijk} \frac{\partial}{\partial x_j} \frac{\partial}{\partial x_k} \phi$$

where ϵ_{ijk} is the antisymmetric Levi-Civita tensor. On the other hand, we notice that $\partial^2 \phi / \partial x_j \partial x_k$ is a symmetric tensor. It is a general result that the contraction of an antisymmetric tensor with a symmetric tensor vanishes. As a matter of fact, we are free to re-name the indices that are summed over, and therefore we may re-cast the above expression as

$$(\boldsymbol{\nabla} \times \boldsymbol{\nabla} \phi)_i = \sum_{jk} \epsilon_{ikj} \frac{\partial}{\partial x_k} \frac{\partial}{\partial x_j} \phi = -\sum_{jk} \epsilon_{ijk} \frac{\partial}{\partial x_j} \frac{\partial}{\partial x_k} \phi$$

where the last line follows from the antisymmetry of ϵ_{ijk}, and the fact that partial derivatives commute. A comparison with the previous result clearly implies that

$$(\boldsymbol{\nabla} \times \boldsymbol{\nabla} \phi)_i = 0$$

(b) We have

$$\boldsymbol{\nabla} \phi = \mathbf{e}_x \frac{\partial \phi}{\partial x} + \mathbf{e}_y \frac{\partial \phi}{\partial y} + \mathbf{e}_z \frac{\partial \phi}{\partial z}$$

Therefore

$$\boldsymbol{\nabla} \phi \cdot d\mathbf{l} = \frac{\partial \phi}{\partial x} dx + \frac{\partial \phi}{\partial y} dy + \frac{\partial \phi}{\partial z} dz = d\phi(x, y, z)$$

(c) We use cylindrical coordinates

$$x = r \cos \theta \qquad ; \ y = r \sin \theta \qquad ; \ z = z$$

After switching to cylindrical coordinates $(x, y, z) \to (r, \theta, z)$, we obtain

$$\boldsymbol{\nabla} = \mathbf{e}_x \left(\frac{\partial r}{\partial x} \frac{\partial}{\partial r} + \frac{\partial \theta}{\partial x} \frac{\partial}{\partial \theta} \right) + \mathbf{e}_y \left(\frac{\partial r}{\partial y} \frac{\partial}{\partial r} + \frac{\partial \theta}{\partial y} \frac{\partial}{\partial \theta} \right) + \mathbf{e}_z \frac{\partial}{\partial z}$$

We now observe that

$$\frac{\partial r}{\partial x} = \frac{x}{r} \qquad ; \qquad \frac{\partial r}{\partial y} = \frac{y}{r}$$

$$\frac{\partial \theta}{\partial x} = -\frac{\sin \theta}{r} \qquad ; \qquad \frac{\partial \theta}{\partial y} = \frac{\cos \theta}{r}$$

Therefore

$$\boldsymbol{\nabla} = (\mathbf{e}_x \cos \theta + \mathbf{e}_y \sin \theta) \frac{\partial}{\partial r} + (-\mathbf{e}_x \sin \theta + \mathbf{e}_y \cos \theta) \frac{1}{r} \frac{\partial}{\partial \theta} + \mathbf{e}_z \frac{\partial}{\partial z}$$

$$= \mathbf{e}_z \frac{\partial}{\partial z} + \mathbf{e}_r \frac{\partial}{\partial r} + \mathbf{e}_\theta \frac{1}{r} \frac{\partial}{\partial \theta}$$

where

$$\mathbf{e}_r = \mathbf{e}_x \cos \theta + \mathbf{e}_y \sin \theta$$

$$\mathbf{e}_\theta = -\mathbf{e}_x \sin \theta + \mathbf{e}_y \cos \theta$$

Problem F.4 Let $\mathbf{v}(\mathbf{x})$ be an arbitrary vector field. Work in cartesian coordinates, and prove the following vector identity

$$\boldsymbol{\nabla} \times (\boldsymbol{\nabla} \times \mathbf{v}) = \boldsymbol{\nabla}(\boldsymbol{\nabla} \cdot \mathbf{v}) - \nabla^2 \mathbf{v}$$

Solution to Problem F.4

We express the *ith* component of $\boldsymbol{\nabla} \times (\boldsymbol{\nabla} \times \mathbf{v})$ using the Levi-Civita symbol as

$$[\boldsymbol{\nabla} \times (\boldsymbol{\nabla} \times \mathbf{v})]_i = \sum_{jk} \epsilon_{ijk} \nabla_j (\boldsymbol{\nabla} \times \mathbf{v})_k = \sum_{jklm} \epsilon_{ijk} \epsilon_{klm} \nabla_j \nabla_l v_m$$

Now use the identity[2]

$$\sum_k \epsilon_{ijk} \epsilon_{klm} = \delta_{il} \delta_{jm} - \delta_{im} \delta_{jl}$$

to write

$$[\boldsymbol{\nabla} \times (\boldsymbol{\nabla} \times \mathbf{v})]_i = \sum_j [\nabla_j \nabla_i v_j - \nabla_j \nabla_j v_i] = \left[\boldsymbol{\nabla}(\boldsymbol{\nabla} \cdot \mathbf{v}) - \nabla^2 \mathbf{v} \right]_i$$

---

[2]The simplest proof is to just write out both sides for various choices of components.

Appendix G

Black Body Flux

Problem G.1 Start from Eqs. (12.39)–(12.42) and derive Eqs. (3.24) and (3.23) for the free radiation field in a big box with periodic boundary conditions.

Solution to Problem G.1

Consider the modes in a big box with periodic boundary conditions where $\mathbf{k} = 2\pi(n_x, n_y, n_z)/L$, with $n_i = 0, \pm 1, \pm 2, \cdots$. We have

$$\mathbf{k}^2 = \left(\frac{2\pi}{L}\right)^2 \left(n_x^2 + n_y^2 + n_z^2\right) \qquad ; \; n_i = 0, \pm 1, \pm 2, \cdots$$

and

$$\mathbf{n}^2 = \left(n_x^2 + n_y^2 + n_z^2\right) = \left(\frac{L}{2\pi}\right)^2 \mathbf{k}^2$$

Now calculate the total number of modes with frequency less than ν, where

$$\nu \equiv \frac{c}{2\pi}|\mathbf{k}|_{\max} = \frac{c}{L}|\mathbf{n}|_{\max}$$

Since there is a one-to-one correspondence between neighboring modes (n_x, n_y, n_z) and a unit volume in n-space, one simply needs to evaluate the total volume of the sphere of radius $|\mathbf{n}|_{\max} = L\nu/c$ in n-space, multiplying it by the degeneracy factor g

$$\mathcal{N} = g\frac{4\pi}{3}\frac{L^3\nu^3}{c^3}$$

The number of normal modes with frequency between ν and $\nu + d\nu$ per unit volume is simply obtained by differentiating this result

$$\frac{d\mathcal{N}}{V} = g\frac{4\pi\nu^2 d\nu}{c^3}$$

where we use $L^3 = V$. The energy density in the cavity is then given by the classical equipartition theorem as

$$\frac{dE}{V} = k_B T \left[g\frac{4\pi\nu^2 d\nu}{c^3} \right]$$

These reproduce the expressions in Eqs. (3.23)–(3.25).

Appendix H

Wave Functions for Identical Particles

Problem H.1 The symmetrizing operator $\sum_{(\mathcal{P})} \mathcal{P}$ produces a wave function that is symmetric under the interchange of any pair of identical bosons. Assume three spin-zero bosons and an appropriate set of single-particle wave functions $\phi_\alpha(\mathbf{x})$.

    (a) Write the wave function $\Phi_{\alpha_1 \alpha_2 \alpha_3}(\mathbf{x}_1, \mathbf{x}_2, \mathbf{x}_3)$ when the α_i are distinct;

    (b) Repeat part (a) for $\alpha_1 = \alpha_2$;

    (c) Determine the normalization of the wave function in both cases;

    (d) Write the expectation value of the kinetic energy in both cases.

Solution to Problem H.1

    (a) We consider three spin-zero bosons, each described by a single-particle wave function $\phi_\alpha(\mathbf{x})$, with distinct quantum numbers, $\alpha_1 \neq \alpha_2 \neq \alpha_3$. In the absence of any interaction between the bosons, the (unnormalized) total wave function of the system $\Phi_{\alpha_1 \alpha_2 \alpha_3}(\mathbf{x}_1, \mathbf{x}_2, \mathbf{x}_3)$ is obtained by applying the symmetrizing operator $\sum_{(\mathcal{P})} \mathcal{P}$ to the product $\phi_{\alpha_1}(\mathbf{x}_1)\phi_{\alpha_2}(\mathbf{x}_2)\phi_{\alpha_3}(\mathbf{x}_3)$

$$
\begin{aligned}
\Phi_{\alpha_1 \alpha_2 \alpha_3}(\mathbf{x}_1, \mathbf{x}_2, \mathbf{x}_3) &= \sum_{(\mathcal{P})} \mathcal{P}\, \phi_{\alpha_1}(\mathbf{x}_1)\phi_{\alpha_2}(\mathbf{x}_2)\phi_{\alpha_3}(\mathbf{x}_3) \\
&= \phi_{\alpha_1}(\mathbf{x}_1)\phi_{\alpha_2}(\mathbf{x}_2)\phi_{\alpha_3}(\mathbf{x}_3) + \phi_{\alpha_1}(\mathbf{x}_1)\phi_{\alpha_3}(\mathbf{x}_2)\phi_{\alpha_2}(\mathbf{x}_3) \\
&\quad + \phi_{\alpha_2}(\mathbf{x}_1)\phi_{\alpha_1}(\mathbf{x}_2)\phi_{\alpha_3}(\mathbf{x}_3) + \phi_{\alpha_2}(\mathbf{x}_1)\phi_{\alpha_3}(\mathbf{x}_2)\phi_{\alpha_1}(\mathbf{x}_3) \\
&\quad + \phi_{\alpha_3}(\mathbf{x}_1)\phi_{\alpha_2}(\mathbf{x}_2)\phi_{\alpha_1}(\mathbf{x}_3) + \phi_{\alpha_3}(\mathbf{x}_1)\phi_{\alpha_1}(\mathbf{x}_2)\phi_{\alpha_2}(\mathbf{x}_3)
\end{aligned}
$$

    (b) Now assume that $\alpha_1 = \alpha_2$. Here the unnormalized wave function is

$$
\begin{aligned}
\Phi_{\alpha_1 \alpha_1 \alpha_3}(\mathbf{x}_1, \mathbf{x}_2, \mathbf{x}_3) &= \phi_{\alpha_1}(\mathbf{x}_1)\phi_{\alpha_1}(\mathbf{x}_2)\phi_{\alpha_3}(\mathbf{x}_3) + \phi_{\alpha_1}(\mathbf{x}_1)\phi_{\alpha_3}(\mathbf{x}_2)\phi_{\alpha_1}(\mathbf{x}_3) \\
&\quad + \phi_{\alpha_3}(\mathbf{x}_1)\phi_{\alpha_1}(\mathbf{x}_2)\phi_{\alpha_1}(\mathbf{x}_3)
\end{aligned}
$$

(c) We assume that the single-particle wave functions are normalized

$$\int d^3x \, |\phi_\alpha(\mathbf{x})|^2 = 1$$

In the case of part (a), we then determine the normalization from

$$\int d^3x_1 \int d^3x_2 \int d^3x_3 \, |\Phi_{\alpha_1\alpha_2\alpha_3}(\mathbf{x}_1, \mathbf{x}_2, \mathbf{x}_3)|^2 = 6$$

In the case of part (b) we have

$$\int d^3x_1 \int d^3x_2 \int d^3x_3 \, |\Phi_{\alpha_1\alpha_1\alpha_3}(\mathbf{x}_1, \mathbf{x}_2, \mathbf{x}_3)|^2 = 3$$

(d) The total kinetic-energy operator is

$$T = -\frac{\hbar^2}{2m} \left(\nabla_1^2 + \nabla_2^2 + \nabla_3^2\right) \equiv T_1 + T_2 + T_3$$

We now calculate the expectation value of T in the case of part (a). Notice that for the operator $-\hbar^2\nabla_1^2/2m$

$$\langle T_1 \rangle = \frac{1}{6} \left[\int d^3x_1 \phi^*_{\alpha_1}(\mathbf{x}_1) T_1 \phi_{\alpha_1}(\mathbf{x}_1) \int d^3x_2 \, |\phi_{\alpha_2}(\mathbf{x}_2)|^2 \int d^3x_3 \, |\phi_{\alpha_3}(\mathbf{x}_3)|^2 \right.$$
$$\left. + \text{permutations of } (\alpha_1, \alpha_2, \alpha_3) \right]$$
$$= \frac{1}{3} \left[\langle \alpha_1|T_1|\alpha_1 \rangle + \langle \alpha_2|T_1|\alpha_2 \rangle + \langle \alpha_3|T_1|\alpha_3 \rangle \right]$$

Thus the expectation value of the total kinetic-energy operator in the state of part (a) is just the sum of the individual contributions

$$\langle T \rangle = -\frac{\hbar^2}{2m} \left[\langle \alpha_1|\nabla^2|\alpha_1 \rangle + \langle \alpha_2|\nabla^2|\alpha_2 \rangle + \langle \alpha_3|\nabla^2|\alpha_3 \rangle \right]$$

The expectation value of T in the state of part (b) is similarly calculated to be

$$\langle T \rangle = -2\frac{\hbar^2}{2m} \langle \alpha_1|\nabla^2|\alpha_1 \rangle - \frac{\hbar^2}{2m} \langle \alpha_3|\nabla^2|\alpha_3 \rangle$$

Problem H.2 Derive Eq. (H.11).

Solution to Problem H.2

We consider the many-body hamiltonian of Eq. (H.10)

$$H = \sum_{i=1}^{N} T(i) + \frac{1}{2} \sum_{i \neq j=1}^{N} V(|\mathbf{x}_i - \mathbf{x}_j|)$$

and evaluate its expectation value in the many-particle ground-state wave function

$$\Phi_{\alpha_1 \alpha_2 \cdots \alpha_N}(1, 2, \cdots, N) = \phi_0(1)\phi_0(2) \cdots \phi_0(N)$$

where $\phi_0(i)$ is the single-particle wave function for a spin-zero boson in the state with $\mathbf{k} = 0$ in a big box of volume $\Omega = L^3$ with periodic boundary conditions

$$\phi_0(i) = \frac{1}{\sqrt{\Omega}}$$

The expectation value of the kinetic-energy operator is clearly zero, since all bosons are in the zero-momentum state. On the other hand, the potential part of the many-body hamiltonian only involves two-body interactions, and therefore we need to calculate the possible ways of picking any two particles out of a total of N particles, which is $N(N-1)/2$. Therefore

$$\langle H \rangle = \frac{N(N-1)}{2\Omega^2} \int_\Omega d^3 x_1 \int_\Omega d^3 x_2 \ V(|\mathbf{x}_1 - \mathbf{x}_2|)$$

If we assume that $V(x)$ is of finite range, and Ω is large enough, we may write

$$\langle H \rangle \approx \frac{N(N-1)}{2\Omega} \left[\int_\Omega d^3 x \ V(x) \right]$$

which proves Eq. (H.11).

Problem H.3 The Hartree-Fock (H-F) approximation determines the "best" single-particle wave functions to use in a Slater determinant to describe an arbitrary system of many identical fermions. The resulting H-F eqns can be derived from a variational calculation that minimizes the energy functional in Eq. (H.31) and (H.34). We forgo that derivation here, and simply state the H-F equations. In the notation of appendix H they read

$$t\underline{\phi}_\alpha(\mathbf{x}) + \sum_\beta \left[\int \underline{\phi}^\dagger_\beta(\mathbf{y}) V(|\mathbf{x} - \mathbf{y}|) \underline{\phi}_\beta(\mathbf{y}) d^3 y \right] \underline{\phi}_\alpha(\mathbf{x})$$

$$- \sum_\beta \left[\int \underline{\phi}^\dagger_\beta(\mathbf{y}) V(|\mathbf{x} - \mathbf{y}|) \underline{\phi}_\alpha(\mathbf{y}) d^3 y \right] \underline{\phi}_\beta(\mathbf{x}) = \varepsilon_\alpha \underline{\phi}_\alpha(\mathbf{x}) \quad ; \text{H-F eqns}$$

Here the indices (α, β) run over the occupied states, and ε_α is an eigen-value.[1] As they stand, these form a set of N coupled, non-linear, integro-differential equations with appropriate boundary conditions. The first po-tential term represents the *direct* interaction, and the second the *exchange* interaction.

Show that if the exchange term is neglected, the Hartree-Fock equations reduce to the Hartree equations of chapter 5.

Solution to Problem H.3

We write the Hartree-Fock equations in terms of the density of Eq. (5.74)

$$n(\mathbf{x}) = \sum_\alpha^F |\underline{\phi}_\alpha(\mathbf{x})|^2$$

With neglect of the exchange term, they become

$$(t + U)\underline{\phi}_\alpha(\mathbf{x}) + \left[\int n(\mathbf{y}) V(|\mathbf{x} - \mathbf{y}|) d^3 y \right] \underline{\phi}_\alpha(\mathbf{x}) = \varepsilon_\alpha \underline{\phi}_\alpha(\mathbf{x})$$

where $t = -\hbar^2 \nabla^2 / 2m$ is the single-particle kinetic-energy operator, and U is a possible additional one-body potential. This equation corresponds to the Hartree Eq. (5.76) for the case where the one-body potential is the attractive Coulomb potential of the nucleus $U(r) = -Ze^2 / 4\pi\varepsilon_0 r$, and $V(|\mathbf{x} - \mathbf{y}|)$ is the repulsive Coulomb potential acting between the electrons.

Problem H.4 (a) Multiply the H-F eqns of Prob. H.3 by $\underline{\phi}^\dagger_\gamma(\mathbf{x})$, and then carry out $\int d^3 x$, to derive the following result (in the notation of appendix H)

$$\langle \gamma | t | \alpha \rangle + \sum_\beta [\langle \gamma\beta | V | \alpha\beta \rangle - \langle \gamma\beta | V | \beta\alpha \rangle] = \varepsilon_\alpha \langle \gamma | \alpha \rangle$$

(b) Assume that t and V are hermitian (Prob. 4.5). Derive the following

---

[1] An additional one-body potential U can always be included by replacing $t \rightarrow t + U$.

relations on the one- and two-body matrix elements of Eqs. (H.30) and (H.33)

$$\langle\gamma|t|\alpha\rangle^* = \langle\alpha|t|\gamma\rangle$$

$$\langle\alpha\beta|V|\gamma\delta\rangle^* = \langle\gamma\delta|V|\alpha\beta\rangle$$

$$\langle\alpha\beta|V|\gamma\delta\rangle = \langle\beta\alpha|V|\delta\gamma\rangle$$

(c) Set $\gamma = \alpha$ in part (a), take the complex conjugate, assume normalized single-particle wave functions with $\langle\alpha|\alpha\rangle = 1$, and prove that the eigenvalues in the H-F eqns are *real*

$$\varepsilon_\alpha^* = \varepsilon_\alpha \qquad ; \text{ eigenvalues real}$$

(d) Now interchange $\gamma \leftrightarrow \alpha$ in part (a), take the complex conjugate, and subtract the resulting expression. Show the solutions to the H-F eqns corresponding to different eigenvalues are *orthogonal*. Since the solutions are to be normalized (and one can always orthogonalize degenerate solutions), show that

$$\langle\alpha|\beta\rangle = \delta_{\alpha\beta} \qquad ; \text{ solutions orthonormal}$$

Solution to Problem H.4

(a) In line with the suggestion in the problem, we multiply the H-F equations on the left by $\underline{\phi}_\gamma^\dagger(\mathbf{x})$, and then carry out $\int d^3x$ to obtain[2]

$$\int d^3x\, \underline{\phi}_\gamma^\dagger(\mathbf{x})t\underline{\phi}_\alpha(\mathbf{x}) + \sum_\beta \int d^3x \int d^3y\, \underline{\phi}_\gamma^\dagger(\mathbf{x})\underline{\phi}_\beta^\dagger(\mathbf{y})\, V(|\mathbf{x}-\mathbf{y}|)\underline{\phi}_\alpha(\mathbf{x})\underline{\phi}_\beta(\mathbf{y})$$

$$- \sum_\beta \int d^3x \int d^3y\, \underline{\phi}_\gamma^\dagger(\mathbf{x})\underline{\phi}_\beta^\dagger(\mathbf{y})V(|\mathbf{x}-\mathbf{y}|)\underline{\phi}_\beta(\mathbf{x})\underline{\phi}_\alpha(\mathbf{y})$$

$$= \varepsilon_\alpha \int d^3x\, \underline{\phi}_\gamma^\dagger(\mathbf{x})\underline{\phi}_\alpha(\mathbf{x})$$

If we use the notation of appendix H, these equations can be written in the desired form as

$$\langle\gamma|t|\alpha\rangle + \sum_\beta [\langle\gamma\beta|V|\alpha\beta\rangle - \langle\gamma\beta|V|\beta\alpha\rangle] = \varepsilon_\alpha\langle\gamma|\alpha\rangle$$

---

[2] Here the spinors with the same spatial label are paired off.

(b) To prove the relations on the matrix elements, we have for the kinetic energy[3]

$$\langle \gamma | t | \alpha \rangle^\star = -\frac{\hbar^2}{2m} \int d^3x \left[\underline{\phi}^\dagger_\gamma(\mathbf{x}) \nabla^2 \underline{\phi}_\alpha(\mathbf{x}) \right]^\star$$

$$= \frac{\hbar^2}{2m} \int d^3x \left[\nabla\underline{\phi}^\dagger_\gamma(\mathbf{x}) \cdot \nabla\underline{\phi}_\alpha(\mathbf{x}) \right]^\star$$

$$= -\frac{\hbar^2}{2m} \int d^3x \left\{ \left[\nabla^2\underline{\phi}^\dagger_\gamma(\mathbf{x}) \right] \underline{\phi}_\alpha(\mathbf{x}) \right\}^\star = \langle \alpha | t | \gamma \rangle$$

Then for the potential energy

$$\langle \alpha\beta | V | \gamma\delta \rangle^\star = \int d^3x \int d^3y \left[\underline{\phi}^\dagger_\alpha(\mathbf{x})\underline{\phi}^\dagger_\beta(\mathbf{y}) V(|\mathbf{x} - \mathbf{y}|) \, \underline{\phi}_\gamma(\mathbf{x})\underline{\phi}_\delta(\mathbf{y}) \right]^\star$$

$$= \langle \gamma\delta | V | \alpha\beta \rangle$$

Also, since V is symmetric,

$$\langle \alpha\beta | V | \gamma\delta \rangle = \int d^3x \int d^3y \, \underline{\phi}^\dagger_\alpha(\mathbf{x})\underline{\phi}^\dagger_\beta(\mathbf{y}) V(|\mathbf{x} - \mathbf{y}|)\underline{\phi}_\gamma(\mathbf{x})\underline{\phi}_\delta(\mathbf{y})$$

$$= \int d^3x \int d^3y \, \underline{\phi}^\dagger_\beta(\mathbf{y})\underline{\phi}^\dagger_\alpha(\mathbf{x}) V(|\mathbf{y} - \mathbf{x}|) \, \underline{\phi}_\delta(\mathbf{y})\underline{\phi}_\gamma(\mathbf{x})$$

$$= \langle \beta\alpha | V | \delta\gamma \rangle$$

(c) We set $\gamma = \alpha$ in the expression of part (a), with $\langle \alpha | \alpha \rangle = 1$, to obtain

$$\langle \alpha | t | \alpha \rangle + \sum_\beta \left[\langle \alpha\beta | V | \alpha\beta \rangle - \langle \alpha\beta | V | \beta\alpha \rangle \right] = \varepsilon_\alpha$$

Now take the complex conjugate of these equations, and use the properties proven in part (b)

$$\varepsilon^\star_\alpha = \langle \alpha | t | \alpha \rangle^\star + \sum_\beta \left[\langle \alpha\beta | V | \alpha\beta \rangle^\star - \langle \alpha\beta | V | \beta\alpha \rangle^\star \right]$$

$$= \langle \alpha | t | \alpha \rangle + \sum_\beta \left[\langle \alpha\beta | V | \alpha\beta \rangle - \langle \alpha\beta | V | \beta\alpha \rangle \right] = \varepsilon_\alpha$$

Therefore the eigenvalues of the H-F equations are real.

(d) Now interchange $\gamma \rightleftharpoons \alpha$ in the equations of part (a), and take the complex conjugate to obtain

$$\langle \alpha | t | \gamma \rangle^\star + \sum_\beta \left[\langle \alpha\beta | V | \gamma\beta \rangle^\star - \langle \alpha\beta | V | \beta\gamma \rangle^\star \right] = \varepsilon_\gamma \langle \alpha | \gamma \rangle^\star$$

---

[3]This is just an explicit demonstration that t is hermitian.

With the use of the properties in part (b), this equation can be re-cast in the form

$$\langle \gamma | t | \alpha \rangle + \sum_{\beta} [\langle \gamma \beta | V | \alpha \beta \rangle - \langle \gamma \beta | V | \beta \alpha \rangle] = \varepsilon_\gamma \langle \gamma | \alpha \rangle$$

Now subtract this equation from the equation in part (a) to obtain the condition

$$(\varepsilon_\alpha - \varepsilon_\gamma) \langle \gamma | \alpha \rangle = 0$$

This implies that $\langle \gamma | \alpha \rangle = 0$ for $\varepsilon_\gamma \neq \varepsilon_\alpha$, that is, solutions corresponding to different eigenvalues are orthogonal. If the single-particle wave functions are normalized so that $\langle \alpha | \alpha \rangle = 1$, and degenerate wave functions are orthogonalized, then we have

$$\langle \gamma | \alpha \rangle = \delta_{\gamma \alpha}$$

Problem H.5 Set $\alpha = \gamma$ in Prob. H.4(a), and use the result in Eq. (H.34) to show that the H-F single-particle energy and total energy of the system can be written as

$$\varepsilon_\alpha = t_\alpha + v_\alpha \qquad ; \text{ H-F energy}$$

$$E = \sum_\alpha t_\alpha + \frac{1}{2} \sum_\alpha v_\alpha$$

where

$$t_\alpha \equiv \langle \alpha | t | \alpha \rangle$$

$$v_\alpha \equiv \sum_\beta [\langle \alpha \beta | V | \alpha \beta \rangle - \langle \alpha \beta | V | \beta \alpha \rangle]$$

Solution to Problem H.5

If we set $\gamma = \alpha$ in Prob. H.4(a), and use $\langle \alpha | \alpha \rangle = 1$, we obtain the H-F single-particle energy as

$$\varepsilon_\alpha = \langle \alpha | t | \alpha \rangle + \sum_\beta [\langle \alpha \beta | V | \alpha \beta \rangle - \langle \alpha \beta | V | \beta \alpha \rangle]$$

$$\equiv t_\alpha + v_\alpha$$

We may now use Eq. (H.31) and Eq. (H.34) to express the expectation values of one- and two-body operators (in this case the kinetic-energy operator and the inter-particle potential respectively) as

$$\langle T \rangle = \sum_\alpha \langle \alpha | t | \alpha \rangle$$

$$\langle V \rangle = \frac{1}{2} \sum_{\alpha\beta} [\langle \alpha\beta | V | \alpha\beta \rangle - \langle \alpha\beta | V | \beta\alpha \rangle]$$

The total energy is therefore[4]

$$E = \sum_\alpha t_\alpha + \frac{1}{2} \sum_\alpha v_\alpha$$

Problem H.6 Assume a uniform system of identical spin-1/2 fermions in a big box of volume Ω with p.b.c., and a spin-independent two-particle potential of finite range. Show that the single-particle wave functions of Eqs. (4.159) provide the *solutions to the H-F eqns* in the form[5]

$$[t + V_D - V_E(\mathbf{k})] \underline{\phi}_{\mathbf{k}\lambda}(\mathbf{x}) = \varepsilon_{\mathbf{k}} \underline{\phi}_{\mathbf{k}\lambda}(\mathbf{x})$$

In this expression the direct and exchange potentials are given by

$$V_D = \frac{1}{\Omega} \sum_{\mathbf{k}'\lambda'}^{k_F} \int V(z)\, d^3z = n \int V(z)\, d^3z$$

$$V_E(\mathbf{k}) = \frac{1}{\Omega} \sum_{\mathbf{k}'\lambda'}^{k_F} \delta_{\lambda\lambda'} \int e^{i(\mathbf{k}-\mathbf{k}')\cdot\mathbf{z}}\, V(z)\, d^3z$$

where $n = N/V$ is the particle density. The eigenvalues are then

$$\varepsilon_{\mathbf{k}} = \frac{\hbar^2 \mathbf{k}^2}{2m} + \frac{1}{\Omega} \sum_{\mathbf{k}'\lambda'}^{k_F} \int V(z)\, d^3z \left[1 - \delta_{\lambda\lambda'}\, e^{i(\mathbf{k}-\mathbf{k}')\cdot\mathbf{z}} \right]$$

Solution to Problem H.6

We consider a uniform system of identical spin-1/2 fermions in a big box with periodic boundary conditions. The single-particle wave-functions in this case are

$$\phi_{\mathbf{k},\lambda}(\mathbf{x}) = \frac{1}{\sqrt{\Omega}}\, e^{i\mathbf{k}\cdot\mathbf{x}}\, \eta_\lambda$$

---

[4]Note the factor of 1/2 in the second term; it arises from the fact that while each two-body interaction gets counted in the H-F energy, it only gets counted once in the total energy.

[5]This result follows from the *translational invariance* of this system.

where η_λ are the spinors[6]

$$\eta_\uparrow = \begin{pmatrix} 1 \\ 0 \end{pmatrix} \qquad ; \eta_\downarrow = \begin{pmatrix} 0 \\ 1 \end{pmatrix}$$

We now write the H-F equations using these wave functions, assuming a spin-independent two-particle potential of finite range

$$t\phi_{\mathbf{k}\lambda}(\mathbf{x}) + \sum_{\mathbf{k}'\lambda'}^{k_F} \left[\int \phi^\dagger_{\mathbf{k}'\lambda'}(\mathbf{y})V(|\mathbf{x}-\mathbf{y}|)\phi_{\mathbf{k}'\lambda'}(\mathbf{y})d^3y \right] \phi_{\mathbf{k}\lambda}(\mathbf{x})$$

$$- \sum_{\mathbf{k}'\lambda'}^{k_F} \left[\int \phi^\dagger_{\mathbf{k}'\lambda'}(\mathbf{y})V(|\mathbf{x}-\mathbf{y}|)\phi_{\mathbf{k},\lambda}(\mathbf{y})d^3y \right] \phi_{\mathbf{k}'\lambda'}(\mathbf{x}) = \varepsilon_\mathbf{k}\phi_{\mathbf{k}\lambda}(\mathbf{x})$$

Let us first examine the direct term in this equation

$$V_D\phi_{\mathbf{k}\lambda}(\mathbf{x}) \equiv \sum_{\mathbf{k}'\lambda'}^{k_F} \left[\int \phi^\dagger_{\mathbf{k}'\lambda'}(\mathbf{y})V(|\mathbf{x}-\mathbf{y}|)\phi_{\mathbf{k}'\lambda'}(\mathbf{y})d^3y \right] \phi_{\mathbf{k}\lambda}(\mathbf{x})$$

$$= \sum_{\mathbf{k}'\lambda'}^{k_F} \left[\frac{1}{\Omega} \int V(z)d^3z \right] \phi_{\mathbf{k}\lambda}(\mathbf{x}) = n \int V(z)d^3z \, \phi_{\mathbf{k}\lambda}(\mathbf{x})$$

We next consider the exchange term

$$V_E(\mathbf{k})\phi_{\mathbf{k}\lambda}(\mathbf{x}) \equiv \sum_{\mathbf{k}'\lambda'}^{k_F} \left[\int \phi^\dagger_{\mathbf{k}'\lambda'}(\mathbf{y})V(|\mathbf{x}-\mathbf{y}|)\phi_{\mathbf{k}\lambda}(\mathbf{y})d^3y \right] \phi_{\mathbf{k}'\lambda'}(\mathbf{x})$$

$$= \frac{1}{\Omega} \sum_{\mathbf{k}'\lambda'}^{k_F} \delta_{\lambda\lambda'} \int V(|\mathbf{x}-\mathbf{y}|)e^{i(\mathbf{k}-\mathbf{k}')\cdot\mathbf{y}}d^3y \, \frac{1}{\sqrt{\Omega}}e^{i\mathbf{k}'\cdot\mathbf{x}}\eta_{\lambda'}$$

Now introduce the relative coordinate $\mathbf{z} \equiv \mathbf{y} - \mathbf{x}$, and use the spin-independence of the interaction, to recover $\phi_{\mathbf{k}\lambda}(\mathbf{x})$

$$V_E(\mathbf{k})\phi_{\mathbf{k}\lambda}(\mathbf{x}) = \frac{1}{\Omega} \sum_{\mathbf{k}'\lambda'}^{k_F} \delta_{\lambda\lambda'} \int V(z)e^{i(\mathbf{k}-\mathbf{k}')\cdot\mathbf{z}}d^3z \, \frac{1}{\sqrt{\Omega}}e^{i\mathbf{k}\cdot\mathbf{x}}\eta_\lambda$$

Therefore the H-F equations take the anticipated form

$$[t + V_D - V_E(\mathbf{k})] \underline{\phi}_{\mathbf{k}\lambda}(\mathbf{x}) = \varepsilon_\mathbf{k} \underline{\phi}_{\mathbf{k}\lambda}(\mathbf{x})$$

where the quantities appearing in this relation are as stated in the problem.

---

[6]As in the text, we suppress the underlining of these spinors.

Problem H.7 Start from the result for $\varepsilon_{\mathbf{k}}$ in Prob. H.6:

(a) Show that for "short-range" potentials, the exchange contribution to the H-F energy will be comparable to the direct contribution. What does "short-range" mean?

(b) Give an argument that for "long-range" potentials, the exchange term can be expected to be much smaller than the direct term. Use this argument to justify the Hartree approximation used for atoms in chap. 5. [*Hint*: Consider the role of $\exp\{i(\mathbf{k} - \mathbf{k}') \cdot \mathbf{z}\}$ in the integrand.]

Solution to Problem H.7

(a) In Prob. H.6 the eigenvalues of the H-F equation in the plane-wave basis have been obtained as

$$\varepsilon_{\mathbf{k}} = \frac{\hbar^2 \mathbf{k}^2}{2m} + \frac{1}{\Omega} \sum_{\mathbf{k}'\lambda'}^{k_F} \int V(z)\, d^3 z \left[1 - \delta_{\lambda\lambda'}\, e^{i(\mathbf{k}-\mathbf{k}')\cdot \mathbf{z}}\right]$$

In the case of a short-range potential, two fermions interact only when they are very close to each other: in this case the integral inside the expression for $\varepsilon_{\mathbf{k}}$ is dominated by the contributions around $z = 0$, and the exponential term is nearly maximal, $e^{i(\mathbf{k}-\mathbf{k}')\cdot \mathbf{z}} \approx 1$. The exchange contribution is then comparable to the direct contribution.

(b) In the case of a long-range potential, two fermions interact even when they are well separated, as the potential falls off slowly as the relative distance increases. For large $|\mathbf{z}|$, however, the argument of the exponential now takes large values, and the exponential is then highly oscillatory, thus leading to large cancellations. In this case, the direct contribution, which does not contain oscillatory terms, is much larger than the exchange contribution.

In the case of atoms and molecules, the interparticle potential is the Coulomb potential, which is a long-range potential, and the argument above may then be used to justify the Hartree approximation, which corresponds to retaining only the direct contribution.

Appendix I

Transition Rate

Problem I.1 This problem presents one application of Eqs. (I.35) and (I.36). Consider the scattering of a spin-zero particle of mass m, charge ze_p, and coordinate $\mathbf{x}_p$ through the Coulomb interaction with the nuclear charge distribution $Z\rho_N(\mathbf{x})$. Here H_1 is given by

$$H_1 = \frac{zZe^2}{4\pi\varepsilon_0} \int d^3x \, \frac{\rho_N(\mathbf{x})}{|\mathbf{x}_p - \mathbf{x}|}$$

The initial and final wave functions for the particle are

$$\phi_i(\mathbf{x}_p) = \frac{1}{\sqrt{\Omega}} e^{i\mathbf{k}_i \cdot \mathbf{x}_p} \qquad ; \; \phi_f(\mathbf{x}_p) = \frac{1}{\sqrt{\Omega}} e^{i\mathbf{k}_f \cdot \mathbf{x}_p}$$

(a) Show the matrix element of the perturbation is[1]

$$\langle \phi_f | H_1 | \phi_i \rangle = \frac{zZe^2}{\varepsilon_0\Omega} \frac{F(\mathbf{q})}{\mathbf{q}^2} \qquad ; \; \mathbf{q} = \mathbf{k}_i - \mathbf{k}_f$$

$$F(\mathbf{q}) = \int d^3x \, e^{i\mathbf{q}\cdot\mathbf{x}} \rho_N(\mathbf{x}) \qquad ; \; \text{form factor}$$

(b) Assume non-relativistic kinematics for the scattering particle, and show

$$\int dk_f \, \delta(E_f - E_i) = \frac{m}{\hbar^2 k} \qquad ; \; k \equiv k_i = k_f$$

(c) Show the incident flux in the laboratory frame is $I_{\text{inc}} = \hbar k/m\Omega$;

(d) Hence show the differential cross section is given by

$$\frac{d\sigma}{d\Omega_k} = Z^2 z^2 \frac{4\alpha^2}{\mathbf{q}^4} \left(\frac{mc}{\hbar}\right)^2 |F(\mathbf{q})|^2 \qquad ; \; \mathbf{q} = \mathbf{k}_i - \mathbf{k}_f$$

---

[1] Use $\int d^3x \, e^{i\mathbf{q}\cdot\mathbf{x}}/x = \text{Lim}_{\mu\to 0} \int d^3x \, e^{i\mathbf{q}\cdot\mathbf{x}} e^{-\mu x}/x = \text{Lim}_{\mu\to 0} 4\pi/(\mathbf{q}^2 + \mu^2) = 4\pi/\mathbf{q}^2$.

Here α is the fine structure constant, $d\Omega_k$ is the differential solid angle for the scattered particle, and $\mathbf{q}$ is the momentum transfer. Note that the quantization volume Ω cancels from this result.

(e) Show the result in part (d) can be rewritten as

$$\frac{d\sigma}{d\Omega_k} = z^2 Z^2 \sigma_{\text{Ruth}} |F(\mathbf{q})|^2$$

$$\sigma_{\text{Ruth}} = \left| \frac{\hbar c \, \alpha}{4E \sin^2 \theta/2} \right|^2 \qquad ; \text{Rutherford}$$

Here $E = \hbar^2 k^2 / 2m$ is the incident energy, and σ_{Ruth} is the Rutherford cross section (see [Fetter and Walecka (2003)]).

Solution to Problem I.1

(a) We write the matrix element of the perturbation H_1 taken between the initial and final states as

$$\langle \phi_f | H_1 | \phi_i \rangle = \frac{1}{\Omega} \int d^3 x_p \, e^{i(\mathbf{k}_i - \mathbf{k}_f) \cdot \mathbf{x}_p} H_1$$

$$= \frac{1}{\Omega} \frac{zZe^2}{4\pi\varepsilon_0} \int d^3 x_p \int d^3 x \, e^{i\mathbf{q} \cdot \mathbf{x}_p} \frac{\rho_N(\mathbf{x})}{|\mathbf{x}_p - \mathbf{x}|}$$

Since the integrals over $\mathbf{x}_p$ and $\mathbf{x}$ are done over all space, we may introduce a simple change of variable $\mathbf{y} \equiv \mathbf{x}_p - \mathbf{x}$ and cast the matrix element in the form

$$\langle \phi_f | H_1 | \phi_i \rangle = \frac{1}{\Omega} \frac{zZe^2}{4\pi\varepsilon_0} \int d^3 y \, \frac{e^{i\mathbf{q} \cdot \mathbf{y}}}{|\mathbf{y}|} \int d^3 x \, e^{i\mathbf{q} \cdot \mathbf{x}} \rho_N(\mathbf{x})$$

The integral over $\mathbf{x}$ provides the form factor $F(\mathbf{q})$ of the nuclear charge distribution, while the integral over $\mathbf{y}$ may be evaluated as explained in the footnote. We therefore obtain

$$\langle \phi_f | H_1 | \phi_i \rangle = \frac{zZe^2}{\varepsilon_0 \Omega} \frac{F(\mathbf{q})}{\mathbf{q}^2} \qquad ; \mathbf{q} = \mathbf{k}_i - \mathbf{k}_f$$

(b) The assumption of non-relativistic kinematics gives

$$\int dk_f \delta(E_f - E_i) = \int dk_f \, \delta \left[\frac{\hbar^2}{2m} (k_f^2 - k_i^2) \right]$$

With a change of variable in the integral, $k_f = \sqrt{2mE_f}/\hbar$, the integral reduces to

$$\int dk_f \delta(E_f - E_i) = \int dE_f \frac{\sqrt{2m}}{2\hbar\sqrt{E_f}} \delta(E_f - E_i) = \frac{m}{\hbar^2 k}$$

where now $k_f = k_i \equiv k$.

(c) The incident particle is described by the wave function $\psi_{inc}(\mathbf{x}) = e^{i\mathbf{k}_i \cdot \mathbf{x}}/\sqrt{\Omega}$. The incident flux in the laboratory is then obtained as

$$I_{inc} = \frac{\hbar}{2mi} \left\{ \psi_{inc}^*(\mathbf{x}) \boldsymbol{\nabla} \psi_{inc}(\mathbf{x}) - [\boldsymbol{\nabla}\psi_{inc}(\mathbf{x})]^* \psi_{inc}(\mathbf{x}) \right\} \cdot \mathbf{e}_z$$

$$= \frac{\hbar k_i}{m} \frac{1}{\Omega} = \frac{v_i}{\Omega}$$

where $v_i = \hbar k_i/m$ is the velocity of the incident particle, assumed to be in the z-direction.

(d) The differential cross section is given by Eq. (I.36)

$$I_{inc} d\sigma_{fi} = R_{fi} \frac{\Omega d^3 k_f}{(2\pi)^3}$$

where the transition rate into one state is

$$R_{fi} = \frac{2\pi}{\hbar} |\langle \phi_f | H_1 | \phi_i \rangle|^2 \delta(E_f - E_i)$$

The differential element $d^3 k_f$ may be expressed as $d^3 k_f = d\Omega_k k_f^2 dk_f$, and carrying out an integration over k_f as described above, one obtains

$$\frac{d\sigma}{d\Omega_k} = \left| \frac{zZe^2}{\varepsilon_0} \frac{F(\mathbf{q})}{q^2} \right|^2 \left(\frac{m}{2\pi\hbar^2} \right)^2 = Z^2 z^2 \frac{4\alpha^2}{q^4} \left(\frac{mc}{\hbar} \right)^2 |F(\mathbf{q})|^2$$

where the fine-structure constant is $\alpha = e^2/4\pi\varepsilon_0\hbar c$.

(e) We may express q^2 in terms of the scattering angle θ between $\mathbf{k}_i$ and $\mathbf{k}_f$ as

$$q^2 = (\mathbf{k}_i - \mathbf{k}_f)^2 = 2k^2(1 - \cos\theta) = 4k^2 \sin^2 \frac{\theta}{2}$$

Substitution into the expression for $d\sigma/d\Omega_k$ obtained in part (d) then provides the stated result

$$\frac{d\sigma}{d\Omega_k} = Z^2 z^2 |F(\mathbf{q})|^2 \left(\frac{\hbar c\alpha}{4E \sin^2 \theta/2} \right)^2 = z^2 Z^2 \sigma_{\text{Ruth}} |F(\mathbf{q})|^2$$

Problem I.2 Compare Eqs. (I.32) and (I.35) with Eq. (7.46), and conclude that to within an overall phase, the T-matrix for a two-body scattering process is given to first order in the interaction H_1 by[2]

$$\frac{\Omega \, \delta_{\mathbf{K}_f, \mathbf{K}_i}}{\Omega^2} T_{fi}^{(1)} = \langle \phi_f | H_1 | \phi_i \rangle$$

Solution to Problem I.2

Equation (I.32) states the following

$$R_{fi} = \frac{P_{fi}(T)}{T} = \frac{2\pi}{\hbar} |\langle \phi_f | H_1 | \phi_i \rangle|^2 \delta(E_f - E_i) \quad ; \text{ transition rate}$$

The transition rate into the group of states dn_f that get into the detector is then given by Eqs. (I.34)–(I.35)

$$d\omega_{fi} \equiv R_{fi} dn_f$$

The cross section is then given by Eq. (I.36)

$$I_{\text{inc}} \, d\sigma_{fi} = d\omega_{fi} \quad\quad\quad ; \text{ cross section}$$

On the other hand, Eq. (7.46) says that the transition rate for a two-body scattering process is

$$d\omega_{fi} = \frac{2\pi}{\hbar} \delta(W_f - W_i) \frac{1}{\Omega^2} |T_{fi}|^2 \frac{\Omega d^3 k_1'}{(2\pi)^3} \quad\quad ; \text{ transition rate}$$

while the cross section then follows in the same fashion.

In the matrix element $\langle \phi_f | H_1 | \phi_i \rangle$, the initial and final wave functions for the two-body system contain an overall normalization factor of $1/\sqrt{\Omega^4}$ for the plane waves, and factors of $e^{i\mathbf{K}_i \cdot \mathbf{X}}$ and $e^{i\mathbf{K}_f \cdot \mathbf{X}}$ for the C-M motion. With p.b.c., the integral over the C-M coordinate gives

$$\int_\Omega d^3 X \, e^{i(\mathbf{K}_i - \mathbf{K}_f) \cdot \mathbf{X}} = \Omega \delta_{\mathbf{K}_f, \mathbf{K}_i}$$

Thus, with the volume factors exhibited and momentum conservation extracted, one has a first-order T-matrix given by

$$\langle \phi_f | H_1 | \phi_i \rangle \equiv \frac{\Omega \, \delta_{\mathbf{K}_f, \mathbf{K}_i}}{\Omega^2} T_{fi}^{(1)}$$

---

[2]The proper treatment of the volume factors Ω in this case proceeds as in chapter 7.

When squared, and summed over the final particle momenta, this gives[3]

$$\sum_{\mathbf{k}_1'} \sum_{\mathbf{k}_2'} \frac{1}{\Omega^2} \left| T_{fi}^{(1)} \right|^2 \delta_{\mathbf{k}_1+\mathbf{k}_2,\, \mathbf{k}_1'+\mathbf{k}_2'} = \sum_{\mathbf{k}_1'} \frac{1}{\Omega^2} \left| T_{fi}^{(1)} \right|^2$$

$$\rightarrow \frac{1}{\Omega^2} \left| T_{fi}^{(1)} \right|^2 \frac{\Omega d^3 k_1'}{(2\pi)^3}$$

which reproduces the result in Eq. (7.46).

---

[3]Note $[\delta_{\mathbf{K}_f, \mathbf{K}_i}]^2 = \delta_{\mathbf{K}_f, \mathbf{K}_i}$. Also, here the total energy is $E \equiv W$.

Appendix J

Neutrino Mixing

There are no problems included with this Appendix.

Appendix K

Units

Problem K.1 (a) Show that $e_{cgs} = 4.803 \times 10^{-10}$ esu

(b) How many esu are there in 1 C?

(c) What are the dimensions of e and e_{cgs}?

Solution to Problem K.1

It is easiest to do part (b) first.

(b) A charge of 1 esu at a distance of 1 cm gives, through Coulomb's law in cgs units, a force of 1 dyne

$$\frac{1 \, (\text{esu})^2}{1 \, \text{cm}^2} = 1 \, \frac{\text{g-cm}}{\text{s}^2}$$

In SI units, the force in Coulomb's law is

$$\frac{e^2}{4\pi\varepsilon_0 r^2} = F$$

where e is in Coulombs, and $1 \, \text{N} = 1\text{kg-m/s}^2$. From appendix L

$$\frac{1}{4\pi\varepsilon_0} = 8.988 \times 10^9 \, \frac{\text{Nm}^2}{\text{C}^2}$$

The same configuration as above is then described in SI units as

$$e^2 \times 8.988 \times 10^9 \, \frac{\text{Nm}^2}{\text{C}^2} \times \frac{1}{10^{-4}\text{m}^2} = 10^{-5}\frac{\text{kg-m}}{\text{s}^2} = 10^{-5} \, \text{N}$$

Hence

$$e^2 = 1.113 \times 10^{-19} \, \text{C}^2$$

Therefore

$$1\,(\text{esu})^2 = 1.113 \times 10^{-19}\,\text{C}^2$$
$$1\,\text{C} = 2.998 \times 10^9\,\text{esu}$$

The Coulomb and esu simply measure *different amounts of charge*.

(a) From appendix L, the measured charge on the electron is

$$|e| = 1.602 \times 10^{-19}\,\text{C}$$

Hence

$$e_{\text{cgs}} = 4.803 \times 10^{-10}\,\text{esu}$$

(c) From part (b), the dimensions of both $e^2/4\pi\varepsilon_0$ and e_{cgs}^2 are $[\text{ML}^3\text{T}^{-2}]$, making the fine-structure constant $\alpha = e^2/4\pi\varepsilon_0\hbar c = e_{\text{cgs}}^2/\hbar c$ dimensionless in both systems of units. It is evident from the results in parts (a,b) that the proportionality factor $1/4\pi\varepsilon_0$ is also dimensionless.

Problem K.2 Use the Lorentz force equation for an electron in a magnetic field in both SI and cgs units to show that $1\,\text{T} = 10^4\,\text{Gauss}$.

Solution to Problem K.2

The magnetic force in SI units is

$$\mathbf{F} = e\mathbf{v} \times \mathbf{B} \qquad ;\ \text{SI units}$$

In cgs units, the force is

$$\mathbf{F} = e_{\text{cgs}}\frac{\mathbf{v}}{c} \times \mathbf{B} \qquad ;\ \text{cgs units}$$

A field of 1 Tesla on 1 Coulomb of charge moving at 1 m/s perpendicular to the field produces a force of 1 Newton. In cgs units the same configuration is described by

$$|\mathbf{F}| = 2.998 \times 10^9\,\text{esu} \times \frac{10^2\,\text{cm/s}}{2.998 \times 10^{10}\,\text{cm/s}} \times 1\,\text{T}$$
$$= 10^5\,\frac{\text{g-cm}}{\text{s}^2}$$

Thus

$$1\,\text{esu} \times 10^{-4} \times 1\,\text{T} \equiv 1\,\text{esu} \times 1\,\text{Gauss}$$
$$= 1\,\frac{\text{g-cm}}{\text{s}^2}$$

where the first line defines the unit of magnetic field in cgs units, the *Gauss*. Hence

$$1\,\text{T} = 10^4\,\text{Gauss}$$

Bibliography

Arfken, G. B., Weber, H. J., and Harris, F. E., (2012). *Mathematical Methods for Physicists: A Comprehensive Guide*, Academic Press, New York, NY

Bjorken, J. D., and Paschos, E. A., (1969). *Phys. Rev.* **185**, 1975

Fetter, A. L., and Walecka, J. D., (2003). *Theoretical Mechanics of Particles and Continua*, Dover Publications, Mineola, NY; originally published by McGraw-Hill, New York, NY (1980)

Fetter, A. L., and Walecka, J. D., (2003a). *Quantum Theory of Many-Particle Systems*, Dover Publications, Mineola, NY; originally published by McGraw-Hill, New York, NY (1971)

Feynman, R. P., Metropolis, N., and Teller, E., (1949). *Phys. Rev.* **75**, 1561

Feynman, R. P., and Gell-Mann, M., (1958). *Phys. Rev.* **109**, 193

Friedman, J. I., and Kendall, H. W., (1972). *Ann. Rev. Nucl. Sci.* **22**, 203

Hamermesh, M., (1989). *Group Theory and Its Applications to Physical Problems*, Dover Publications, Mineola, NY

Horowitz, C. J., and Serot, B. D., (1981). *Nucl. Phys.* **A368**, 503

Lawrence Berkeley Laboratory, (2003). *Atomic Masses*, http://ie.lbl.gov/toi2003/MassSearch.asp

Lawrence Berkeley Laboratory, (2005). *Nuclear Moments*, http://ie.lbl.gov/toipdf/mometbl.pdf

National Nuclear Data Center, (2007). *Nuclear Data*, http://www.nndc.bnl.gov/

Ohanian, H. C., (1995). *Modern Physics, 2nd ed.*, Prentice-Hall, Upper Saddle River, NJ

Pauli, W., (1948). *Meson Theory of Nuclear Forces, 2nd ed.*, Interscience, New York, NY

Schiff, L. I., (1968). *Quantum Mechanics, 3rd ed.*, McGraw-Hill, New York, NY

Walecka, J. D., (2001). *Electron Scattering for Nuclear and Nucleon Structure*, Cambridge University Press, Cambridge, UK

Walecka, J. D., (2004). *Theoretical Nuclear and Subnuclear Physics, 2nd ed.*, World Scientific Publishing Company, Singapore

Walecka, J. D., (2007). *Introduction to General Relativity*, World Scientific Publishing Company, Singapore

Walecka, J. D., (2008). *Introduction to Modern Physics: Theoretical Foundations*, World Scientific Publishing Company, Singapore

Walecka, J. D., (2010). *Advanced Modern Physics: Theoretical Foundations*, World Scientific Publishing Company, Singapore

Walecka, J. D., (2013). *Topics in Modern Physics: Theoretical Foundations*, World Scientific Publishing Company, Singapore

Index